ALSO BY ADRIAN RAINE

The Psychopathology of Crime

Violence and Psychopathy

Crime and Schizophrenia

The Anatomy of Violence

THE ANATOMY OF VIOLENCE

The Biological Roots of Crime

Adrian Raine

PANTHEON BOOKS   NEW YORK

Pantheon Books and colophon are registered trademarks of
Random House, Inc.

Library of Congress Cataloging-in-Publication Data
Raine, Adrian.
The anatomy of violence : the biological roots of crime / Adrian Raine.
Pages cm
Includes bibliographical references and index.
ISBN 978-0-307-37884-2 (hardback)
1. Violence—Physiological aspects. 2. Violence—Psychological aspects.
I. Title.
RC569.5.V55R35 2013 616.85′82—dc23 2012036952

www.pantheonbooks.com
Jacket design and illustration by Kelly & Cardon Webb
Book design by Soonyoung Kwon
Printed in the United States of America
First Edition
9 8 7 6 5 4 3 2 1

To my sons, Andrew and Philip, in the hope that you will never fall by the wayside as so many in this book have, but will instead move along into happy and fulfilled lives. Don't worry too much about where the train is going—just decide to get on board for wherever it will take you on life's adventures. Believe in the spirit of giving at Christmas, remember Tintin, and never forget Sammy Jankis!

"Oh, Agent Starling, you think you can dissect me with this blunt little tool?"

Hannibal Lecter admonishing Clarice Starling for using a self-report instrument to assess him in Jonathan Demme's movie *Silence of the Lambs*

Contents

Preface

It's July 19, 2012, and it's as hot as the hobs of hell here in Philadelphia. The air-conditioning in my work office conked out, so I came home to an airy upstairs library room to write this preface. I should have been filming a crime documentary this afternoon with a crew from Chicago, but they had their equipment stolen this morning. That's not a surprise, though, as crime strikes all the time here in Philadelphia. Yesterday, I was dealing with two police detectives—Lydon and Boyle—here at my house, which had been burgled yesterday. Just what you want when you come back after midnight from Hong Kong. But I live close to my data, which is one reason I reside here in West Philadelphia.

Looking around this upstairs library, I'm surrounded by hundreds of rare-edition books on crime and violence that the burglar didn't take. I suppose he's not as interested as we are in what causes crime. They're not my books, mind you. They belong to the people who lived here during the seventy-year period before I moved in. Most belong to Marvin Wolfgang, a world-renowned criminologist who, beginning in 1969, sat and wrote in this very library room. For the thirty years before that, Thorsten Sellin, another world-leading criminologist and Wolfgang's PhD supervisor, lived here, having bought the house just seven weeks before the outbreak of World War II. I am at his desk. For three-quarters of a century between the two of them—professor and mentor—these intellectual giants in sociology redefined the field of criminology at the University of Pennsylvania, where I myself now work.

Given that remarkable criminological legacy, my mind inevitably turns to a historical perspective on the fundamental question addressed by this book. Is there a significant biological contribution to the causes and cures of crime? It turns out that that idea was all the rage 150 years ago, when an Italian doctor named Cesare Lombroso broke with intellectual tradition and, taking a novel empirical approach to studying crime, tried to persuade the world of a basis to crime residing in the brain. But as the twentieth century progressed, what was once an innovative viewpoint quickly fizzled out and sociological perspectives took center stage. During that time no criminologist worth his or her salt would have anything to do with an anatomy of violence or the biology of bad behavior.

Except, that is, the sociologist whose ghost lingers close to me beside the fireplace in this upstairs library overlooking Locust Street. Marvin Wolfgang documented in a far-reaching historical analysis of Cesare Lombroso that never in the history of criminology has a person been simultaneously more eulogized and more condemned.[1] He noted how Lombroso continues to be held up as a straw man for attack by those hostile to a biological theory of crime causation. He recognized the clear limitations in Lombroso's research, yet simultaneously saw the enormous contributions that this Italian made.

Toward the end of his own career, Wolfgang himself became convinced that there was—in part—a biological, cerebral basis to crime. His mentor Thorsten Sellin similarly believed that Lombroso's biological perspective, focusing as it did on the criminal rather than the crime, was unprecedented in its vitality and influence.[2] Sharing their home and library as I do at this moment, I can hardly disagree with them.

Yet most in the field of criminology would disagree. Biological research on violence was vilified in the 1970s and 1980s, during my formative years as a scientist. Amid interdisciplinary rivalries the perception was that researchers like me were at best biological determinists who ignored social processes—and at worst racist eugenicists.

Perhaps because of a rebellious and stubborn streak running through me, that negative perspective has never deterred me throughout my thirty-five years of researching the biology of crime. Nevertheless, working as I have within the confines of top-security prisons and ivory-tower universities, I have been shut off from a wider audience who might be just as excited as I am about what new insights a biological perspective can offer. It is that desire to share this research with a wider audience that inspired me to write this book.

In that context I owe an enormous debt of thanks to Jonathan Kellerman for encouraging me to write a popular book about my work. Jonathan, as one of the world's foremost writers of crime fiction, has himself written a provocative nonfiction science book, *Savage Spawn*, on the causes of crime in the wake of a horrific schoolyard shooting.[3]

About fifteen years ago we had lunch together. Jonathan has a PhD in clinical psychology, had read and absorbed my academic work, and believed I had something important to share with others. He put me in touch with his own agent, and I wrote a proposal. It came to nothing. At that time, no matter how I tried, I could not get any publisher interested.

But times changed in those fifteen years. On the tails of the genome project, societies across the world have begun to realize the importance of genetic and biological factors in a whole host of processes—and not just medical conditions. Serendipity struck. Eric Lupfer, an alumnus of the University of Pennsylvania and a literary agent at William Morris Endeavor, read a question-and-answer article about my work in our university's magazine. Eric too recognized the potential public interest in a book on the anatomy of violence, and thanks to his outreach and vision, here I am completing the book in this historic room. I could not have had a more supportive, helpful agent. Sincere thanks are also due to Jeff Alexander at Pantheon for his splendid edits, vision, and guidance in the final throes of my writing—the time spent with him has been magical. Josie Kals and Jocelyn Miller at Pantheon provided invaluable support and help, and I am particularly indebted to my copy editor, Kate Norris, for her meticulous and careful fine-tuning of the manuscript. Thanks also to Helen Conford at Penguin for her strong enthusiasm and encouragement throughout this long march. Eric, Jeff, and Helen have together provided me with a wonderful opportunity for which I am truly grateful.

That sea change in opinion I mentioned is also filtering through into academia. Leading criminologists across the world are now beginning to follow in Wolfgang and Sellin's footsteps. They are recognizing the cross-disciplinary potential of a biological approach not as a competitive challenge, but as a cross-fertilizing joint enterprise that combines social with biological perspectives. Even the world's premier sociology journal, *American Sociological Review*, is beginning to publish molecular genetic research on crime and violence. Nobody would have dreamed that just fifteen years ago. Now the new subdiscipline of neurocriminology is quickly sweeping us back to the future.

Friedrich Lösel, the director of the Institute of Criminology at the University of Cambridge, was a kind host to me there while I completed this book. In Cambridge I benefited enormously from discussions with Sir Anthony Bottoms, Manuel Eisner, David Farrington, and Per-

Olof Wikström, as well as Friedrich himself. At the University of Pennsylvania, Bill Laufer worked with me to bridge my imaging research with his expertise on white-collar crime. Martha Farah was pivotal in introducing me to neuroethics, while Stephen Morse has tutored me patiently in neurolaw. It has been an honor to work with such extraordinary colleagues. I should also thank Richard Perry, who endowed my chair, as well as Amy Gutmann, who had faith in my controversial work and hired me into her Penn Integrates Knowledge initiative.

Interest in the biology of violence goes well beyond academia and into the media. Erin Conroy at William Morris Endeavor had masterly intuition in showing *Anatomy of Violence* to Howard Gordon and Alex Gansa, who then obtained a pilot production commitment for it from CBS. My thanks to you, Erin, and also to you, Howard, for finding something in this book to spark your interest for a new TV series; it has truly meant a lot to me.

So very many research collaborators, colleagues, and academic friends have helped and inspired me over the years. Among these I am especially indebted in different ways to Freda Adler, Rebecca Ang, Josef Aoun, Laura Baker, Irv Biederman, John Brekke, Patty Brennan, Monte Buchsbaum, Ty Cannon, Avshalom Caspi, Antonio and Hannah Damasio, Mike Dawson, Barbra Dickerman, Ken Dodge, Annis Fung, Daniel Fung, Lisa Gatzke-Kopp, Chenbo Han, Robert Hare, Lori LaCasse, Jerry Lee, Tatia Lee, Rolf and Magda Loeber, Zhong-lin Lu, Don Lynam, John MacDonald, Tashneem Mahoomed, Sarnoff Mednick, Terrie Moffitt, Joe Newman, Chris Patrick, Angela Scarpa, Richard Tremblay, and Stephanie van Goozen. Their friendship, support, and inspiration have meant a lot to me over the years. My students at the University of Pennsylvania have been a true joy to instruct and supervise. Among many I must particularly acknowledge the "Gang of Four"—Yu Gao, Andrea Glenn, Robert Schug, and Yaling Yang—for the privilege of learning from such a talented, gifted, and productive research team.

We gain inspiration from many sources in different ways. I am especially indebted to my PhD supervisor, Peter Venables, at York University, for his support and encouragement over the past thirty-five years, particularly during the four years I spent working in prison, where for seven months I simply gave up on completing my PhD. He has been a very special person in my life. Dick Passingham did more than anyone in tutoring me to think clearly and simply when I was an undergraduate at Oxford University. In a different vein, Larry Sherman was pivotal in bringing me to criminology at the University of Pennsylvania five years ago. To him I owe an enormous debt of thanks. His vision in believing that neurocriminology is a field of the future has been truly inspirational. Marty Seligman gave me thoughtful advice on writing this book and sparked in my mind one of the futuristic scenarios in the final chapter.

I learned a great deal from discussions with Julia Lisle, Ed Lock, and John, Marcus, and Sally Sims on social and legal issues in the last chapters. But most of all, I'm extremely grateful to my family—Philip, Andrew, and Jianghong—for being so very patient with me and understanding why I have had so little free time with them of late. They have given me the joy, support, and love that have kept me moving throughout the course of this writing.

The Anatomy of Violence

INTRODUCTION

It was the summer of 1989 in Bodrum, a beautiful seaside resort on the southwestern coast of Turkey, soaked in sun, history, and nightlife. I was on vacation and it had been a long day. I had taken the bus from Iráklion, where I had caught the second-worst case of food poisoning I had ever had in my life, including two days in bed throwing up with backbreaking pain.

It was very hot that July night, and I could not sleep. I had kept the windows open to try to stay cool. I tossed and turned, still somewhat sick and sleepless—in and out of consciousness, as my girlfriend slept in the room's other single bed. It was just after three a.m. when I became aware of a stranger standing above me. At that time I was teaching a class on criminal behavior, and I would tell my students that when they became aware of an intruder in their apartment, they should feign being asleep. Ninety percent of the time thieves just wanted to grab the goods and then get out. Let them go—then call 911. You run no risk and have a fighting chance of getting your possessions back without a violent confrontation.

So what did I do when I saw the intruder at my bedside? I fought. In the milliseconds that it took my visual cortex to interpret the shadowy figure and signal this to the amygdala, which jump-starts the fight-flight response, I leaped out of my bed. In little more than a second, I had instinctively grabbed the intruder. I was on automatic pilot.

Information from the senses reaches the amygdala twice as fast as it gets to the frontal lobe. So before my frontal cortex could rein back the amygdala's aggressive response, I'd already made a threatening move toward the burglar. This in turn immediately activated the intruder's fight-flight system. Unfortunately for me, his instinct to fight also kicked in.

The next thing I knew I was being hit so quickly that it felt like the man had four fists. He hit me so hard on the head that I saw a streak of white light flash before my eyes. He also hit me in the throat. He seemed to hit me all over.

I was violently thrown against the door. I felt the doorknob and I must confess the thought of escape sprang into my mind. But at that instant I heard piercing screams from my girlfriend, struggling with the man. She eventually ended up with bruises on her arms, but I think these were defense wounds and that the intruder only wanted to keep her quiet. Seeing them struggle, the instinctive reaction that had originally come upon me when I was in bed returned. I leaped at him again and somehow managed to push him out of the open window.

In that instant I felt an immediate sense of safety and relief. But the euphoria evaporated after I turned on the light switch and saw the blood running down my chest. I tried to shout out, but what came out of my mouth was a hoarse whimper.

Completely unknown to me in the midst of that mismatched contest was that the assailant had been holding a knife. Quite a long one, with a red handle and a six-inch blade, it turned out.[1] But I was lucky. As I warded off his blows with my arms, the blade of the cheap knife had snapped off, leaving only a few millimeters of metal left on the handle. So when he attempted to cut my throat, the damage was far less than it might have been.

The police arrived surprisingly quickly. The hotel was right beside an army barracks. There had been a sentry on duty who had heard the shouts and screams and he raised the alarm. The hotel had been quickly surrounded, so that when the police arrived they believed that the perpetrator was still inside the hotel.

Meanwhile I was taken to the hospital. It was rudimentary and bare. I was laid on my back on what felt like a hard concrete slab, while the doctor put a few stitches in my throat. The window of the hospital room was open, and I could hear in the distance that a party was still going on. The strains of the music wafted through the window, the Beatles' "Hard Day's Night," of all songs.

Afterward, the police wanted me back at the hotel to go over what had happened. All the residents were now standing in the lobby, even though it must have been about five a.m. by then.

The police had thoroughly gone through the rooms of all the residents in search of my assailant. I learned later that one man had looked a bit flushed when the police pulled him from his bed, and he had a red mark on his torso that looked fresh. He was in the upstairs room right next to me. So he was one of the two suspects waiting for me when I entered the lobby.

Both were young Turkish men. Both were naked from the waist up—just as the attacker had been. One was quite a good-looking man, but otherwise there was nothing out of the ordinary about him. The second suspect had a rougher look. He was also stocky and muscular, and what flashed through my mind at that moment was that he had the classic mesomorphic physique that early criminologists believed typified criminals.[2] He also had a striking scar on his upper arm. His nose looked as if it had been broken. His looks persuaded me. He had to be the man who'd tried to cut my throat.

The police pulled him aside and had a quiet word with him. But not so quiet that the manager of the hotel couldn't overhear and translate the conversation back to me. The police told him they simply wanted to clear up the case, and if he'd admit that he was the perpetrator, they would let him go. So the gullible guy made his admission, and was promptly arrested.

At that point, I'd had enough of Bodrum and Turkey, and I told the police I was off to the neighboring island of Kos in Greece in the next two days. Remarkably, they decided to expedite the trial. It was something of a ceremony at the outset. It started off at the police station. I was placed next to my assailant, and we were marched through the center of the town, side by side, to the courthouse. Quite a number of people came out to watch, as I had been featured in Bodrum's local newspaper the previous day, pictured with a prominent white bandage on my throat. Many of them pointed at us and yelled at the defendant. Although whatever they said was incomprehensible to me, it was clear that the defendant was not a popular man.

The trial itself was novel, to say the least. The courtroom looked like a scene out of the Nuremberg trials, but in a distorted dream. There was no jury at all. Instead, there were three judges in scarlet robes seated loftily above us. The defendant did not have an attorney.

Neither did I, for that matter. Adding to the strangeness, none of the judges could speak or understand any English, and I certainly could not speak Turkish. So they procured a cook who could speak some English and serve as my interpreter. It was all very surreal.

I gave my testimony. The judges asked me how I could identify the assailant given that the incident had occurred just after three a.m. and it had been dark. I described to them how the moonlight was streaming through the window by my bed, illuminating one side of the assailant's face as we struggled. That I had frantically wrestled with him and that that gave me a sense of his stature and build. I said that I could not be completely sure—but frankly, whether that part ever got translated, I'll never know.

After I gave my testimony through the cook, the defendant gave his testimony. Whatever he said in Turkish, the judges were not persuaded. They found him guilty as charged. It was as simple as that.

After the verdict one of the judges ushered me and my translator over to the bench. He told us that the defendant would be brought back later for sentencing, and that it would be a prison sentence of several years' duration. Justice is swift and efficient in Turkey, I thought. I had seen on that trip more than one elderly man with a hand missing, a vestige of the days when theft was punished by detaching the offending part of the perpetrator's anatomy. That had seemed harsh when I had seen it earlier on my trip. But at that moment in the courtroom, in spite of the seeming lack of due process, hearing that my attacker would see significant prison time was music to my ears. Justice, as they say, is sweet.

Until that experience in Bodrum, violence had been primarily an academic concern for me. I'd tolerated my fair share of small-scale crime up to that point—two burglaries, theft, and an assault—but having one's throat cut can change the way one looks at the world, or at least at one's self. My girlfriend and I left the next day for Greece, but as I simmered under the hot sun on the beach in Kos, I remember suddenly feeling a surge of anger about the whole ordeal. The thief, who easily could have killed me, had gotten off easy. *He* should have been beaten up. *His* throat should be cut. *He* should spend the rest of his life a fitful sleeper, hypersensitive to the slightest sound in the night. A few years inside did not seem like justice. It perhaps should have been enough, but to me, especially at that moment, it wasn't.

This experience had a powerful effect on me. It broke through my outer façade of liberal humanitarian values and put me in touch with a deep, primitive sense of retributive justice. From an assured English-bred opponent of the death penalty, I became a person who could no longer be ruled out of a jury pool for a capital crime in the United States. An evolutionary instinct for vengeance was triggered inside me, and it has stayed with me for years.

Consequently, I have something of a Jekyll-and-Hyde attitude about my work investigating the biological basis to crime. One conclusion I've drawn from the research presented in this book is that biological factors early in life can propel some kids toward adult violence. Risk factors like poor nutrition, brain trauma from childhood abuse, and genetics are beyond an individual's control, and when those factors are combined with social disadvantages and our society's anemic ability to spot and treat potential offenders, the odds are that people with these disadvantages will turn to crime. That means I likely should cut my assailant some slack. And if the standards of that hospital I was in are anything to go by, I'm sure a grim Turkish prison is very unlikely to change his criminal behavior. Are we doing justice to the offender? That's the Dr. Jekyll in me speaking, and it's the spirit in which my scientific work is conducted.

But another man inside me doesn't give a damn about what caused my attacker to develop into a violent offender. Mr. Hyde retorts that the man nearly killed me and he should be nearly killed too. To hell with forgiveness and pseudoscientific drivel about early biological risk factors that constrain free will. Out of professional interest, I should have investigated further, but at the time, in his specific case, I did not care. I do know that during the summer months before attacking me he had already committed nineteen thefts—he owned up to the police after his capture so he would not later be prosecuted for them. None of these victims had been injured—so I put down my bad luck to Mr. Hyde's instinct of leaping up at him and grabbing him by the throat. In any event, Hyde rants that a recidivistic criminal like him should be locked up and the key thrown away forever—we need to protect ourselves from these dangerous villains.

In the intervening years I've had more time to reflect on my reactions to that attack. Is defensive aggression genetically built into us? Can my brain be wired to aggressively respond even though my rational mind, trained by years of experience, tells me that's just not the right response? And what do I make of the fact that my physical perception of that suspect in the identity parade biased me to conclude he was the culprit? During that instant there in the hotel lobby, as I gazed on his torso and face, there was literally a "body of evidence" standing in front of me, a man with the anatomy of violence written all over him—a body I'd had tangible experience of during my struggle.

That body of evidence, and the sliver of moonlight streaking into the dark bedroom allowing me to see my attacker's face, symbolizes to me in a metaphorical sense the dawning of a new beacon of research light helping us to identify the violent offender—and what makes him tick. A radical change has been taking place in recent years regarding our understanding of how and why people become violent criminals. That change is what *The Anatomy of Violence* is all about.

The dominant model for understanding criminal behavior has been, for most of the twentieth century, one built almost exclusively on social and sociological models. My main argument is that sole reliance on these social perspectives is fundamentally flawed. Biology is also critically important in understanding violence, and probing through its anatomical underpinnings will be vital for treating the epidemic of violence and crime afflicting our societies.

Today this perspective is slowly but surely seeping into public consciousness, largely because of two recent scientific developments. First, molecular and behavioral genetics is increasingly demonstrating that many behaviors have in part a genetic basis. Genes shape physiological functioning, which in turn affects our thinking, personality, and behavior—including the propensity to break the laws of the land, whatever those laws may be. Second, revolutionary advances in brain imaging are opening a new window into the biological basis of crime. Together these two advances are prodding us to redefine our sense of self. They have jointly placed us on the threshold of the new discipline that I call neurocriminology—the neural basis to crime—which involves the application of the principles and techniques of neuroscience to understand the origins of antisocial behavior. By better understanding these origins, we will improve our ability to prevent the misery and harm crime causes. The anatomy of violence encapsulates this exciting and vibrant new approach to the discipline of criminology that Lombroso himself spawned but that had been all but abandoned throughout the twentieth century.

There is a third development that is not so much scientific as an undeniable historical fact. The heavy emphasis on an exclusively social approach to crime and violence throughout the last century did nothing to turn the rising tide of this perennial problem. It is widely acknowledged in criminology that as crime went up throughout the 1970s and 1980s our society largely gave up on the rehabilitation of inmates. Prisons became holding bays for the unrepentant—not retreats for the rehabilitation of lost souls, as the Pennsylvania Prison Society espoused in the early nineteenth century. That single-minded approach has just not worked.

Thinking of human behavior from a biological perspective is no longer controversial—you can hardly open a newspaper or magazine today without reading about a new breakthrough in how genes and the brain shape our personality and influence the moral and financial decisions we make, or what we buy, or whether we turn out to vote or not. So why would they not also influence whether we commit a crime or not? The pendulum is slowly but surely swinging us back to Lombroso's dramatic nineteenth-century intuition, and forcing us to revisit the tangled ethical quandaries and legitimate social fears inherent in applying a neurocriminological approach. But when one considers the myriad ways in which violence plagues us, the stakes are too high, and the potential good is too great, to ignore the compelling scientific evidence we are discovering about the biological roots of crime.

I have three central objectives in writing this book: First, to inform readers of the intriguing new scientific research that I and other scientists have conducted in recent years, focusing on the biological basis for crime and violence. Second, I want to stress that social factors are critical both in interacting with biological forces in causing crime, and in directly producing the biological changes that predispose a person to violence. Third, I want to explore with you the practical implications of this emerging neurocriminological knowledge, ranging from treatment to the legal system to social policy—both today and in the future.

I have written this book for the general reader who has at least a passing interest in crime, as well as for undergraduate and graduate students who want an accessible introduction to a new and exciting perspective on crime and violence. Anyone with an inquisitive mind, who is curious about what makes the criminal offender tick will, I hope, find something of interest in these pages. In *The Anatomy of Violence* I'm going to reveal the internal mechanisms of violent crime as well as the way external forces interact with them to produce criminals. I will lay out what biological research is revealing on the root causes of crime. These deep roots are now being dug up using neuroscience tools, exposing the biological culprits giving rise to violence. Throughout I have included case studies of a rogues' gallery of killers to illustrate my points.

More than anything I hope that this book will open your mind not just to how biological research can contribute to our understanding of violence, but also how it may lead to benign and acceptable ways of reducing the suffering violence causes to societies throughout the world. Biology is not destiny. We can unlock the causes of crime with a set of biosocial keys forged from a new generation of integrative inter-disciplinary research combined with a public-health perspective.

But we need to exchange views in an open and honest dialogue in order to ensure sensible use of this new knowledge for the good of everyone, to develop a framework for further research, and to firmly grasp the neuroethical issues surrounding neurocriminology to more effectively apply this new knowledge. We'll begin our discussion with that pivotal moment when a scientist other than myself stared at the anatomy of a different violent offender, and began the long and precarious journey along the causeway of neurocriminology.

I

BASIC INSTINCTS

How Violence Evolved

The scientific study of biological criminology started on a cold, gray November morning in 1871 on the east coast of Italy. Cesare Lombroso, a former Italian army medic, was working as a psychiatrist and prison doctor at an asylum for the criminally insane in the town of Pesaro.[1] During a routine autopsy he peered into the skull of an infamous Calabrian brigand named Giuseppe Villella. At that moment he experienced an epiphany that was to radically alter both his life and the course of criminology. He described this pivotal experience in the following way:

> I seemed to see all at once, standing out clearly illuminated as in a vast plain under a flaming sky, the problem of the nature of the criminal, who reproduces in civilized times characteristics, not only of primitive savages, but of still lower types as far back as the carnivores.[2]

What did Lombroso see as he gazed deep into Villella's skull? He detected an unusual indentation at its base, which he interpreted as reflecting a smaller cerebellum—or "little brain"—seated under the two larger hemispheres of the brain. From this singular and almost ghoulish observation, Lombroso went on to become the founding father of criminology, producing an extraordinarily controversial theory that was to quickly have significant cross-continental influence.

Lombroso's theory had two pivotal points: that there was a basis to crime originating in the brain, and that criminals were an evolutionary throwback to more primitive species. Criminals, Lombroso believed, could be identified on the basis of "atavistic stigmata"—physical characteristics from more primitive stages of human evolution, such as a large jaw, a sloping forehead, and a single palmar crease. Based on his measurements of such traits, Lombroso created an evolutionary hierarchy that placed Jews and Northern Italians at the top and Southern Italians (including Villella), along with Bolivians and Peruvians, at the bottom. Perhaps not coincidentally, at the time there was much higher crime in the poorer, more agricultural south of Italy, one of the many symptoms of the "southern problem" besetting the recently unified nation.

These beliefs, which were based partly on Franz Gall's phrenological theories, flourished throughout Europe in the late nineteenth and early twentieth centuries. They were discussed in parliaments and throughout public administrations as well as in universities. Contrary to appearances, Lombroso was a famous, well-meaning intellectual, as well as a staunch supporter of the Italian Socialist Party. He wished to employ his research to serve the public good. He abhorred retribution and instead placed the emphasis of punishment on the protection of society.[3] He strongly advocated rehabilitation of offenders. Yet at the same time he felt that the "born criminal" was, to paraphrase Shakespeare's Prospero, "a devil, a born devil, upon whose nature nurture can never stick,"[4] and consequently favored the death penalty for such offenders.

Perhaps because of these views, Lombroso has become infamous in the annals of criminological history. The theory he spawned turned out to be socially disastrous, feeding the eugenics movement in the early twentieth century and directly influencing the persecution of the Jewish people. The thinking and vocabulary of Mussolini's racial laws of 1938, which excluded Jews from public schools and ownership, owes a rhetorical debt to Lombroso's writings and theories, as well as those of the students who followed him into the early twentieth century.[5] The major difference in Mussolini's laws was that Aryans replaced Jews at the top of the racial hierarchy, and Jews were relegated to the bottom alongside Africans and below Southern Italians. The dreadful irony in this—a fact carefully avoided in almost all references to Lombroso in contemporary criminological texts—is that Lombroso himself was Jewish.

Understandably, Lombrosian thinking fell into disrepute in the twentieth century and was replaced by a sociological perspective on human behavior—including crime—which still holds sway today. It is not too difficult to see how this biological-to-social pendulum swing came about. Crime, after all, is a social construction. It is defined by the law, and socio-legal processes hold sway over conviction and punishment. Laws change across time and space, and acts such as prostitution that are illegal in one country are both legal and condoned in others. So how can there possibly be a biological and genetic contribution to a social construction? Surely social causation *must* be central to crime? This simple argument has made a compelling case for an almost exclusive sociological and social-psychological perspective on crime, a seemingly sound bedrock on which to build workable principles for social control and treatment.

What do I make of Lombroso's claims? Of course I reject Lombroso's evolutionary scale that placed Northern Italians at the top and Southern Italians at the bottom. Not least because I am half Italian, through my mother, who was from Arpino in the southern half of Italy—I'm not an evolutionary throwback to a more primitive species. And yet, unlike other criminologists, I do believe that Lombroso, stumbling as he did amid his offensive racial stereotyping and fumbling with the hundreds of macabre prisoner skulls he had collected, was on the path toward a sublime truth.

We'll now see how modern-day sociobiologists have made a far more coherent and compelling argument than Lombroso ever could have that there is, in part, an evolutionary basis to crime that provides the foundations for a genetic and brain basis to crime—the anatomy of violence. We'll explore violence in its many shapes and forms, from homicide to infanticide to rape, and suggest from an anthropological perspective how different ecological niches may have given rise to the ultimate in selfish, cheating behavior—psychopathy.

LOOKING AFTER NUMBER ONE—THE CHEATING GAME

So why are people more than a hundred times more likely to be murdered on the day they are born than to be murdered on an average day in their life? Why are they fifty times more likely to be murdered by their stepfather than by their natural father? Why do some men, not content to rape only strangers, also want to rape their wives? And why on earth do some parents kill their kids?

These are among a host of questions that baffle society and that seem impenetrable from a social perspective. But there is an answer: the dark forces of our evolutionary past. Despite what we may think of our good-naturedness, we are, it could be argued, little more than selfish gene machines that will, when the time and place is ripe, readily use violence and rape to ensure that our genes will be reproduced in the next generation.

In evolutionary terms, the human capacity for antisocial and violent behavior wasn't a random occurrence. Even as early hominids developed the ability to reason, communicate, and cooperate, brute violence remained a successful "cheating" strategy. Most criminal acts can be seen, directly or indirectly, as a way to take resources away from others. The more resources or status a man has, the better able he is to attract young, fertile females. These women in turn are on the lookout for men who can give them the protection and the resources they need to raise their future children.

Many violent crimes may sound mindless, but they are informed by a primitive evolutionary logic. The mugger who kills for $1.79 is not getting much for his efforts, yet the general strategy of theft can pay off in the long run in terms of acquiring goods. Drive-by shootings may seem senseless, but they help establish dominance and status in the neighborhood. And while a barroom brawl over who's next at the pool table may sound to you like fighting over nothing, the real game being played has nothing to do with pool.

From rape to robbery and even to theft, evolution has made violence and antisocial behavior a profitable way of life for a small minority of the population. The ultimate capacity for our antisocial misdeeds can be understood with reference to evolutionary biology. And it is from fundamental evolutionary mechanisms that genetic differences among us have come into play and shaped the anatomy of violence.

We think of aggression today as *maladaptive* and aberrant. We give heavy legal sentences to violent offenders to deter them and others from committing such crimes, so surely it cannot be viewed as adaptive. But evolutionary psychologists think differently. Aggression is used to grab resources from others, and resources are the name of the evolutionary game. Resources are needed to live, reproduce, and care for offspring. There is an evolutionary root to actions that run the gamut from bullies threatening other kids for candy to men robbing banks for money. And aggression—more specifically defensive aggression—is also important in warding off others who may wish to steal our precious resources. Bar fights help establish a pecking order of dominance and power, helping to put down rivals in the eyes of desirable women and other potential competitors. The mating game for males is about developing desirable status in society. Gaining a reputation for aggression not only increases status in one's social group and allows more access to resources but also deters aggression from others. And that is true whether we are talking about a child in a playground or an inmate in a prison.

From a chubby-faced baby to a crooked-faced criminal, there is a development and unfolding of antisocial behavior predicated on biology and a cheating strategy to living out life. As a tiny kid, you took what you wanted without a care. All that mattered in the world was you and your selfish desires. You may have forgotten those days, but in that untamed, uncivilized period of your life, you were standing on the threshold of a life of crime.

Of course culture quickly took care of that. You were taught by parents, and maybe your older siblings, the rules of social behavior—"Don't hit your sister," "Don't take your brother's toys"—and your evolving brain began to slowly learn not just that there were others in the world, but that selfishness was not always a wise guiding principle on life's long, arduous journey. You never exactly gave up on looking out for yourself and what was good for you, but at least you began to take into account others' feelings and to express appropriate concern for others at appropriate moments—at times genuinely, and perhaps at other times disingenuously. But is there more to explaining antisocial behavior than the presence or absence of familial socializing forces?

There is. The thesis that really challenges our perspective on ourselves and our evolutionary history first appeared in 1976, in a radical book called *The Selfish Gene,* by Richard Dawkins.[6] I'll not forget this book, or Richard Dawkins, for that matter. As an undergraduate I had one-on-one tutorials with him on evolutionary theory. They were thrilling lessons on the all-embracing influence of evolution on behavior, and they led me to start thinking of violence and crime in evolutionary terms.

The central thesis in his landmark book was that "successful" genes are ruthlessly selfish in their struggle for survival, giving rise to selfish individual behavior. In this context, human and animal bodies are little more than containers, or "survival machines," for armies of ruthless renegade genes. These machines plot a merciless campaign of success in the world, where success is defined solely in terms of survival and achieving greater representation in the next gene pool. However, the gene is the basic unit of "selfishness" rather than the individual. The individual eventually dies, but selfish genes are passed on from body to body, from generation to generation, and potentially from millennium to millennium.

It all boils down to how "fit" you are. Not so much whether you can run a marathon or how much you can lift, but how many children you can produce that are yours. The more kids you have that are genetically yours, the more copies of your genes there will be in the following gene pool. That, and only that, is success in the gene's-eye view of the world. If more lofty perspectives come to mind when you contemplate the meaning of "success"—like doing well in school, having a great job, or writing a book—then consider this: your gene machine has been built to generate these fanciful ideas to maliciously motivate you into gaining status and resources that will translate into reproductive success. It's a genetic con.

As a male you can maximize your genetic fitness in one of two ways. One, you can invest a lot of parental effort and resources into just a few offspring. You put all your eggs into a small basket, nurturing and protecting a couple of kids, ensuring their survival into full maturity, and even helping them look after their own children. Alternatively, you can put all your eggs, or rather sperm, into a lot of baskets. Here you maximize the number of your offspring without really doing very much to support them, spreading your parental effort more thinly.

A male can much more easily adopt this latter reproductive strategy of high offspring–low effort if he "cheats" on his many female partners by misrepresenting his ability to acquire resources and his long-term parenting intentions. Mate support and resources are critical for women. Once fertilized, females are largely lumbered with their progeny. They make the bigger investment in raising the child, so they are on the lookout for men who can come up with the goods, and will commit to long-term support.

So fitness—an organism's ability to pass on its genetic material—is central to the evolution of all behavior and the driving force behind selfishness. Certainly in the animal world, it is easy to see how antisocial and aggressive behaviors have evolved. Animals fight for food and they fight for mates. And whether we like it or not, it's not too much of a stretch from the animal kingdom to us humans. The temptation to "cheat"—whether it is not sharing resources after having accepted them from others or manipulation of others to selfishly acquire resources—is always there.

But surely we humans are different from animals. We have a strong capacity for social cooperation, altruism, and selflessness. Reciprocal altruism has indeed evolved because in the long run it benefits the performer. It ultimately pays you to help save a stranger if that stranger will reciprocate your help in the future, and save your life.[7] Today, by and large, we live in a world populated by reciprocal altruists. And yet, at the same time, reciprocal altruism can itself give rise to "cheating." If you accept acts of altruism from others, but fail to reciprocate in the future, you're cheating. There is room for a bit of cheating—truth be told, we all do it from time to time. But a small number of us cheat a lot—and in this group we find the psychopath. The trouble for psychopaths, however, is that sooner or later they get a bad reputation. People stop helping them out, and potential mates pass them over. In this scenario the psychopathic cheat is on a downward spiral.

Fortunately for the psychopath there is a slippery way out. After he's been spotted by reciprocal altruists he leaves this social network and migrates to a new population, where he can begin to fleece a different set of unsuspecting victims. It's easy to see in this analysis, therefore, how a small minority of antisocial cheats could survive in a world largely populated by reciprocal altruists. The proportion of cheats within any population would have to stay relatively small—cheats lose out when they meet one another—but otherwise cheats can survive, as long as they are prepared to tough it out and take a few hits before moving on.

Such a scenario would lead to the prediction that these hard-core antisocials drift from population to population. Consistent with this prediction, the modern-day psychopath has been characterized as an impulsive, sensation-seeking individual who fails to follow any life plan, aimlessly drifting from person to person, job to job, and town to town.[8] Probably the best assessment tool for psychopathy—the Psychopathy Checklist—makes reference to the psychopath's short-term plans and goals, nomadic existence, frequent breaking off of relationships, poor parenting, moving from one place to another, frequent changes of jobs and addresses, and parasitic lifestyle.[9] The "pure" cheat strategy is therefore entirely consistent with present-day psychopaths who manifest a nomadic lifestyle.

In any game there is more than one winning strategy, and that holds true in the game of reproductive fitness. Reciprocal altruism can pay for most, and for a few the psychopathic cheating strategy wins out. We'll now turn to how certain environmental conditions could nudge some whole societies to become altruistic or selfish, and how psychopathic behaviors could have evolved. Given certain environmental circumstances, whole populations of cheats could evolve, and studies of primitive societies provide some interesting clues on the evolution of psychopathic behavior.

PSYCHOPATHS ACROSS CULTURES

Environmental conditions vary greatly across the world, and throughout prehistory behaviors have evolved in an adaptive response to changing environmental circumstances. Building on this notion, some anthropological studies lend support to the idea that whole populations can develop an antisocial trait. The main method of these studies has been to compare cultures differing in antisocial conduct on ecological and environmental factors that give rise to different reproductive strategies and social behaviors. If certain ecological niches are associated with certain types of behavior, this could support the notion that what we call antisocial traits could be advantageous in cultures found in certain environments. Such cultures could have jump-started the evolution of antisocial, psychopathic-like lifestyles.

When comparing, for instance, the cultures of the !Kung Bushmen of the Kalahari Desert in Southern Africa and the Mundurucú villagers in the Amazon Basin, anthropologists have found that the strikingly different environments they inhabit correlate with altruistic and antisocial behavior, respectively.[10] The !Kung Bushmen live in a relatively inhospitable desert environment. Due to the extremely difficult living conditions, cooperation is prized. Men need to hunt together in search of food, and game is shared in the camp.[11] There is also a high degree of parental investment in children, who are highly supervised and weaned gradually. Because of that high parental investment, fertility is relatively low. A disruption of a pair bond by either partner could have fatal consequences for the offspring, who are highly dependent on parental care. The personal characteristics adapted to the !Kung's environment are good hunting skills, reliable reciprocation of altruistic acts, the careful choosing of mates, and high parental investment in offspring. This personality profile is clearly more aligned to altruism than to cheating, a trait that is argued to be in part an adaptation to an inhospitable environment.

In contrast, the Mundurucú are low-intensity tropical gardeners living in a relatively rich ecological niche along the Tapajós and Trombetas Rivers in the Amazon basin. Everything grows there, and life is relatively easy. In an interesting role-reversal, women carry out most of the food production.[12] This environment makes for a very different way of life and a different male personality profile. The relatively greater availability of food frees males to engage in male-male competitive interactions centered around politics, planning raids and warfare, gossiping, fighting, and elaborate ritual ceremonies. Occasionally they engage in hunting game that they trade for sex with the village women. Men sleep together in a house separate from the women, whom they hold in disdain. Indeed, females are viewed as sources of pollution and danger. Males in the Gainj tribe, low-intensity gardeners in the highlands of New Guinea, also view sexual contact with women as dangerous, especially during menstruation.

In contrast to the !Kung, Mundurucú mothers provide little care to their infants once they are weaned, and these children must quickly learn to fend for themselves. Mundurucú men play a minimal role in caring for their offspring. Personal characteristics of the successful Mundurucú male in this competitive society consist of good verbal skills for political oratory, fearlessness, skill at fighting and carrying out raids, bluff and bravado to avoid the risk of battle, and the ability to manipulate and deceive prospective mates on what resources he can offer to maximize offspring. Furthermore, he should not be gullible, since belief in the folklore regarding the dangers of sex and women as a source of pollution would not foster the passing on of one's genes.[13]

Similarly, for females living in a social context of low parental investment, those who can manipulate their menfolk by deception over an offspring's paternity, exaggeration of requirements, and resistance to the development of monogamous bonds are the most successful. The Mundurucú's way of life is then more associated with a cheating, antisocial strategy than with reciprocal altruism. Figure 1.1 summarizes

	!Kung Bushmen	Mundurucú
Location	Kalahari Desert	Amazon basin
Ecological niche	Harsh	Rich
Social climate	Cooperative	Competitive
Parental investment	High	Low
Fertility	Low	High
Male activities	Group hunting	Competition, raids
Favored traits	Reciprocal altruism, careful mate selection, good parenting	Manipulation, fearlessness, fighting

Figure 1.1 Contrasting environmental features of two societies that shape different personality traits

the key features of these two societies and how they stand in sharp contrast.

The nature of the Mundurucú's social environment clearly favors the expression of aggressive, psychopathic-like behavior. Certainly when one considers the fact that the Mundurucú were in the past fiercely aggressive headhunters, this parallel to psychopathy becomes clearer. Intriguingly, many of the features of the Mundurucú have parallels with features of psychopathic behavior in modern industrialized societies.[14] For example, psychopaths show lack of conscience, superficial charm, high verbal skills, promiscuity, and lack of long-term interpersonal bonds.[15] While these traits are advantageous in the Mundurucú environment, they are clearly disadvantageous in the milieu of the !Kung Bushmen, which demands high male parental effort, reciprocal altruism, and monogamous relationships.

The Yanomamo Indians in the tropical rain forests of northern Brazil and southern Venezuela provide another parallel culture to the Mundurucú. With a total population of about 20,000, they live in villages that can range in size from 90 people to about 300. As with the Mundurucú, they subsist on plants and vegetables and only need to do about three hours of work a day. They too live in a rich ecological niche.

Napoleon Chagnon, in his intensive anthropological studies on the Yanomamo, has documented a number of striking features of this culture.[16] They'll break rules when it's in their interest. They participate in the forcible appropriation of women. They call themselves *waiteri*—meaning "fierce." And they are indeed both fearless and highly aggressive. Boys are socialized into acts of aggression from a surprisingly young age, with their "play" consisting of throwing spears and shooting arrows at other boys. Initially they are scared by this initiation into violence, but soon they come to revel in the adrenaline rush that the mock battles provide.

To give you a perspective on their level of aggression, 30 percent of all male deaths among the Yanomamo are due to violence, an astonishing level. If you think the United States is a violent society, consider that 44 percent of all Yanomamo men over the age of twenty-five have killed someone, thus achieving the status of being a *unokai*. Some kill more than once, and one *unokai* had killed sixteen times. The source of the killing in the majority of cases is sexual jealousy—exactly what you'd expect from an evolutionary perspective and a species whose females make the greater parental investment. They also conduct raids on other villages for revenge killings that can take up to four days to execute, involving from ten to twenty men in the raiding party.

From our perspective on the evolution of violence, however, the most interesting element of the Yanomamo is what happens to *unokais*, the men who kill. They have an average of 1.63 wives compared with 0.63 wives of men who do not kill. The *unokais* have an average of 4.91 children compared with an average of 1.59 children for non-killers. In terms of reproductive fitness, serious violence pays handsomely in two critical resources. First, lots of kids. Second, lots of wives to look after them. We can see how planned violence and the lack of remorse over killing others have been rewarded in the *unokais'* society. These are precisely the features of Western psychopaths,[17] who also commit more aggressive acts than non-psychopaths, and are more likely to commit homicide for gain.[18]

Inevitably, Western society does not condone such violence. We hardly applaud and reward people who kill others. Or do we? With significant pomp and ceremony we decorate and reward soldiers who have taken significant risks to kill others in warfare. Crowds cheer wildly as boxers punch each other senseless in a sport that we know results in brain damage. We certainly revel in kung fu movies or other film genres when the good guy beats the living daylights out of the bad guy.

Whatever our cultivated minds may publicly say about the senselessness of warfare, do not our primitive hearts still thrill to the drums of combat? Is this why we enjoy sports competitions, to watch the dominant winner end up on top? Is that what gives us the vicarious thrill and excitement of seeing someone win a gold medal at the Olympics? Or when a violent tackle occurs in a football game? Our present-day cultured minds weave an alternative story to explain the feeling—we just love sports, that's all. But why? Isn't it because selection pressures have built into us a mechanism to carefully observe who ranks where, empathic skills to imagine ourselves as a winner, basking in that reflected glory, giving us that "feel-good" mood and a desire to emulate such achievements?

Mundurucú women are clearly attracted to men around them who kill. Have you ever wondered why seemingly sensible, peaceful women want to marry serial killers in prison? Their primitive heartstrings are being plucked by the siren's call of the serial-killer status. They yearn to be with a strong male, even when their modern minds might logically object. At a milder level we have a morbid fascination with true crime. Something attracts us to violence. That evolutionary pull may even have explained why you bought this book.

Part of the attraction we have to violence is that when executed in the right place and the right time, it's adaptive—even today. The vestiges of our evolutionary backgrounds persist, far more than we care to imagine. Let's take this a step further into the here and the now to examine in what specific situations aggression is adaptive, and what aspects of crime can be explained from an evolutionary perspective.

KILLING YOUR KIDS

I mentioned earlier that people in general are a hundred times more likely to be killed on the day they are born than on any other day.[19] Murders of children and adolescents are most likely to occur in the first year of life.[20] And within that year, eighteen times more children are murdered on the day they were born than on any other day.[21] In 95 percent of these cases, the babies were not born in a hospital. They are mostly the product of undesired, unplanned pregnancies. They are battered to death (32.9 percent), physically assaulted (28.1 percent), drowned (4.3 percent), burned (2.3 percent), stabbed (2.1 percent), or shot (3.0 percent).[22] It all flies in the face of the exhilaration that most couples experience on the day of their child's birth. But an explanation for this seeming contradiction can be found within the layers of evolutionary psychology theory.

Indeed, once we step across the threshold of the home, there are facts that seem to fly in the face of an evolutionary perspective on violence. For example, people are more likely to be killed in their home by a family member than by a stranger. How can that make sense from an evolutionary standpoint? Don't we expect solid protection of everyone at home to ensure that the family's genes are passed on to future generations? Martin Daly and Margo Wilson are two Canadian evolutionary psychologists who have done more than anyone else to resolve enigmas like this and to further demonstrate the power of an evolutionary psychological perspective on violence.

What they demonstrated was an inverse relationship between the degree of genetic relatedness and being a victim of homicide. So the less genetically related two individuals are, the more likely it is that a homicide will take place. For example, in Miami, 10 percent of all homicides were the killings of a spouse—a family killing—but of course, spouses are almost always genetically unrelated. In fact, Daly and Wilson found that the offender and the victim are genetically related in only 1.8 percent of all homicides of all forms.[23] So 98 percent of all homicides are killings of people who do not share their killer's genes.

Selfish genes in their strivings for immortality wish to increase—not decrease—their representation in the next gene pool. Hence this inverse relationship between genetic relatedness and homicide. On the other hand, if you are living with someone not genetically related to you, you are eleven times more likely to be killed by that unrelated person than by someone genetically related to you.

Stepparents are a particularly pernicious case in hand, a fact captured in countless myths and fairy tales. Remember the grim story of Hansel and Gretel, whose wicked stepmother badgered their natural father into leaving his children deep in the woods to die of starvation? Or Sleeping Beauty's evil and vain stepmother, who ordered a hunter to take her into the woods and slaughter her? Recall Cinderella's cruel stepmother? Actually, the reality is so potent that our childhood lives are full of images of mean stepmothers—real or imaginary—almost as an eerie warning call for us to be on our guard.

Did you grow up as a child with a stepparent? If you did and you survived unscathed, you've done pretty well. In England, only 1 percent of babies live with a stepparent,[24] and yet 53 percent of all baby killings are perpetrated by a stepparent.[25] Data from the United States show a similar pattern—a child is a hundred times more likely to be killed as a result of abuse by a stepparent than by a genetically related parent. If we look at child abuse, we see the same thing. Stepparents are six times more likely to abuse their genetically unrelated child under the age of two than genetic parents.

It's a finding that makes you wonder if in cases of death from abuse by someone thought to be the biological parent, that person may not be the genetic parent after all. In cases where the children and the father believe that they are genetically related, it is estimated that in about 10 percent of cases the father is not the genetic father. Could at some subconscious, evolutionary level the father sense genetic unrelatedness and pick on the unrelated child? Such abuse would be a paternal strategy to push that child out, to minimize the resources given to him, and instead maximize resources for other, genetically related children. We know that stepparents sometimes selectively abuse their stepchildren, sparing the children in the family who are genetically related to them.[26]

Such actions of some stepparents can thus be comprehensible from an evolutionary perspective. But more perplexing are parents who kill children they *are* genetically related to. How can evolutionary theory come to grips with these killings?

The basic concept to remember here, if you think back to your own parents when you were growing up, is that they likely worked hard to raise you—and don't they just let you know it sometimes! They worked their fingers to the bone and sacrificed much for your future betterment. Okay, so that's par for the course when it comes to looking after your own genes. But also bear in mind that the longer a child lives, the more her parents invest in her. But suppose someone's genetic parents change their minds about their investment? If they do, they ought to do it early on before they waste more energy. And that's exactly what we see.

Take a look at the top graph in Figure 1.2, showing the age at which a child will be killed by its mother *if* she is indeed going to kill it. It shows homicides per million children per year averaged over a period from 1974 to 1983 in Canada. You'll see that the peak age for killing is in

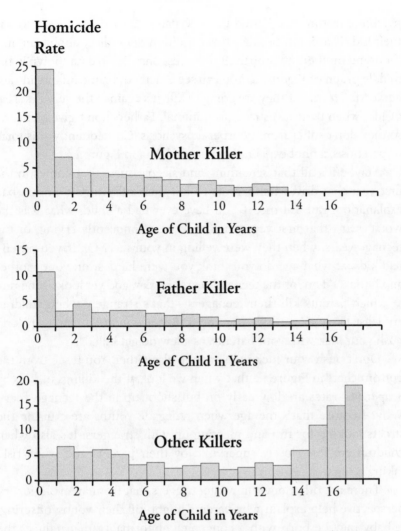

Figure 1.2 Age at which Canadian children are murdered by their mother, father, and others

the very first few months of that little baby's life.[27] After that time, the homicide rate drops dramatically and keeps on declining right throughout adolescence. Soon after birth the mother bails out on her own baby. Maybe she wants to move on. Maybe her mate has moved out and she knows she's better off without this baggage, better able to attract a new mate. Whatever the reason, there is a strong age effect to be explained.

I think I know what you're thinking. Some mothers just after birth have puerperal psychosis. They sink into a very deep depression with psychotic features, and amid their despair and madness they may kill their kid. Fair point, because this condition does affect about one in a thousand mothers after birth. But the response lies in data shown in the middle graph of Figure 1.2. You can see exactly the same infanticide age curve for *fathers*.[28] If they are going to kill, it's again in the very first year of life, when their investment is minimal. Fathers don't give birth and so they don't suffer from puerperal psychosis. Consequently, this form of psychosis cannot explain the maternal data in Figure 1.2.

Maybe it's all that screaming and sleeplessness that comes in the first year that drives the parents to kill their offspring. It's not a bad explanation. But tell me, if you have ever had a child, what was the worst year—that first year when they were innocently crying, or the teenage years, when they were yelling in your face? Or, if you haven't had kids, at what age do you think you were hardest on your mother and father? I'd go for the teenage years any day, and yet look at the rate at which parents kill their teenagers—that's strangely when children are least likely to get killed by them. But if you are a teenager don't push your luck with your parents, as a few do get killed.

Don't push your luck with anyone else either. You'll see from the bottom chart in Figure 1.2 that when we look at the killings of kids by *nonparents*, rates are low early on but shoot up in the teenage years. Why? Because that's the age when renegade youths are cruising the streets looking for fun and meeting up with strangers. It's also when children are less closely supervised by their parents and when risk-taking is highest.

There are other environmental triggers that from an evolutionary perspective help explain why parents might kill their young offspring. A baby may be born with a congenital abnormality that reduces the odds of survival or reproduction, or it may have a chronic illness that saps parental resources. Even with normal offspring, if food is short it may pay the parents in terms of genetic investment to spend scarce resources on the survival of an older sibling closer to the age of maturity and independence, rather than spreading the butter too thinly, trying to support both the newborn and the older sib.

Even if there is no older sibling, killing the baby could make evolutionary sense. In some bird species where both parents forage for their offspring, the death of one parent can result in the other parent abandoning the offspring. The load is just too hard to bear, and it's better for the remaining parent to look after number one and try again

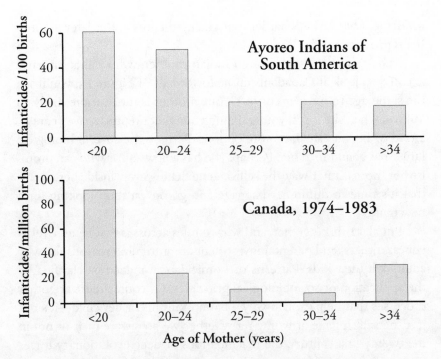

Figure 1.3 Age of mother when she kills her own child

in the reproductive success game. Don't we sometimes get a sense of that in stories of young mothers abandoning their babies? We tend to interpret their actions as due to social processes like immaturity, shame, or teenage impulsivity. Shame may be the superficial explanation, but at a deeper level the underlying cause may be cold-blooded maximization of reproductive success. The negative emotions and behaviors that we attribute to the mother in trying to explain the homicide may not be the whole story. The selfish genes inside the teenage killer mom may be the ultimate source of such callous, cold-blooded behavior.

There's one more point to make about parents killing their children: how old the mother is when she kills her own child. The upper graph of Figure 1.3 shows the rate of child homicides as a function of the mother's age among the Ayoreo Indians of South America. It's highest when the mother is under the age of twenty, and it goes down after that. Why would that be? The mother is more fertile when she's younger—and more attractive in drawing a desirable mate to her. The older she is, the more it makes sense to hold on to her long-term genetic investment because it's harder to make up the loss at this later point in her reproductive life.

And it's not just the Ayoreo Indian mothers who kill at an early age. If you look at Canadians in the lower half of Figure 1.3, you'll see the same age-to-murder curve.[29] Your mother is much more likely to kill you when she is still young. Being young, her reproductive years lie ahead of her and she has more options. Perhaps the current biological father has abandoned her. Perhaps she has a new suitor who can promise her more. Either way, the selfish gene ticking away inside her signals that it's time to dump her baggage and go on vacation looking for a new mate.

Put all of this together, and what comes across is that genetic relatedness, fitness, and parental investment are intriguing reasons for why adults kill their kids. Patterns of homicide can indeed be clarified by the application of sociobiological principles. Of course there are other processes that help explain why a parent kills his or her child—it's not just the selfish gene at work. Yet whether we are aware of it or not in the twenty-first century, the machinations of deep evolutionary forces are laboring away down in the depths of our humanity, forging devious tools to maximize our genetic potential. And behind those closed doors in the family home, those forces don't end with killing your kids.

RAPING YOUR WIFE

Is rape an act of hate? A malicious and derisory act against women condoned by a patriarchal society where men attempt to control and regulate their womenfolk? Or can this act of violence be partly explained by evolutionary psychology?

We can view the rape of a nonrelative as the ultimate genetic cheating strategy. Rather than striving to accrue resources to attract a female and investing years in the upbringing of their offspring, a male can cut through this tedious process in the twinkling of an eye. He just needs to rape a woman. Men have hundreds of millions of sperm that are always at the ready to inseminate a woman. The sex act is quick. And the male can immediately walk away, never to see that woman again. He knows that if pregnancy does occur, there is a decent chance that the female will care for their joint progeny. His selfish genes have reproduced.

How often will a rape result in a pregnancy? This was estimated in one study of 405 women aged twelve to forty-five who had suffered penile-vaginal rape. The total base rate was 6.42 percent, which was twice as high as the 3.1 percent base rate for unprotected penile-vaginal intercourse in consensual couples. After correction for the use of contraceptives, the pregnancy base rate from rapes was estimated at 7.98 percent.[30] The rates of pregnancies from rape can only be estimates because paternity is not investigated with definitive DNA evidence. Some women could "invent" a rape as a cover-up for an unwanted pregnancy. However, other studies have also reported higher rape-pregnancy rates than consensual-sex-pregnancy rates. It is nevertheless surprising. If we accept the findings, why would rape be more likely to result in a pregnancy?

One conceivable hypothesis is that rapists are more likely to insemi-nate fertile women. Rapists select their victims, and we certainly know that they are far more likely to select women at their peak reproductive age than other women.[31] Furthermore, putting age aside, the possibility that a rapist may be more visibly drawn to women who are the most fertile is not impossible. Females with a smaller waist relative to their hips are viewed as more attractive in many cultures throughout the world. This smaller waist-to-hip ratio is also associated with increased fertility as well as better health.[32] Consequently, male rapists could in theory select a more fertile female, consciously or subconsciously, based on how she looks.

Not all rapists choose victims they find attractive. It can even be the other way around. When I worked with prisoners in England, one rap-ist told me that he specifically picked out *unattractive* women to rape. Why would he do this? His argument was that an unattractive woman does not get enough sex, so it's okay to give her the sex that she really wants. This is just one example of a number of cognitive distortions that some rapists have.[33] Their perverted belief is that women actually enjoy the act of rape and interpret it as the experience of a lifetime—their ultimate sexual fantasy coming true.

Ideas like this may be inadvertently fueled by the fact that some women when raped actually achieve orgasm, even though they may strongly resist and are traumatized by the attack.[34] True prevalence data are hard to come by because rape victims understandably are embarrassed to admit that they achieved orgasm during such a dis-graceful violation. Clinical reports place the rate of the victim expe-riencing orgasm at about 5 to 6 percent, but clinicians also report that they suspect the true rate to be higher. This may well be the case, because research reports document that physiological arousal and lubrication occurs in 21 percent of all cases. Why would that happen? Because in half the cases, the date-raped woman was actually attracted to the perpetrator before the act. Orgasm and the associated contractions are thought to facilitate conception by contracting the cervix and rhythmically dipping it into the sperm pool. This admittedly has a modest effect, as sperm retention is increased by only approximately 5 percent with orgasm.

Clearly, conception does not require orgasm,[35] so we cannot place too much weight on the physiological arousal of some women during rape as a prelude to pregnancy. Nevertheless, the fact remains that rapists generally select their victims and appear to consciously or subconsciously select more fertile women. This selection strategy would explain the purported increased pregnancy rate in rape victims and can be viewed in an evolutionary context. If a man is going to take risks raping a woman, the strategy would be to pick the fertile one and enhance one's inclusive fitness.

There are, of course, risks associated with this particular cheating strategy. The male could suffer physical injury. Worse, he could be detected and beaten. Throughout much of human history rapists have been alienated or killed. In modern times he would be thrown into prison alongside psychopaths and murderers, where as a sex offender he is at high risk for being beaten and raped himself. So evolutionary theory argues that there is a subconscious cost-benefit analysis at work—weighing the potential costs resulting from detection against the benefits of producing a child. Dominant men with resources can already attract mates, so one might expect that the cost-benefit analysis might tip the scales in favor of rape when the perpetrator has relatively fewer resources. In support of this prediction, rapists are indeed more likely than non-rapists to have lower socioeconomic status, to leave school at an earlier age, and to have unstable job histories in unskilled occupations.[36]

We can question evolutionary theory because it can be too all-encompassing; we cannot take it too far in explaining violence. Drug cartels in Colombia and the availability of handguns in the United States contribute significantly to explaining why these countries today have high homicide rates, and yet these influences lie outside the domain of evolutionary theory. I think you would admit that an evolutionary perspective can help explain facts about rape in quite a compelling way.

While women of any age can be raped, we've noted that men are much more likely to rape women of reproductive age.[37] Interestingly, women of reproductive age who are raped experience more extreme psychological pain than younger or older women. This has been interpreted as an evolutionary learning mechanism that focuses these women's attention on avoiding contexts where they could be raped and have their overall reproductive success reduced.[38] At another level, we know that men find it far easier than women to have sex without concomitant emotional involvement. Why? Because they do not need to hang around after the sex act is over. In contrast, from an evolutionary perspective, women need a long-term commitment from their male mate to help rear any child that might result from their union, and so they have more need of an emotional, personal relationship. Finally, men very rarely kill the women they rape; although they have the potential to kill, they want their offspring to survive.

But what about rapes that occur between partners in a marriage or other long-term relationship? Between 10 percent and 26 percent of women report being raped during their marriage.[39] How can this be viewed through evolutionary lenses?

A great deal of research has documented that both physical and sexual violence perpetrated by men in a relationship is fueled by sexual jealousy.[40] Infidelity is very distressing for both males and females, but men and women differ in terms of what causes these distressing feelings. Jealousy is the primary motive for a husband to kill his wife in 24 percent of cases, compared with only 7.7 percent of cases in which the wife kills her husband.[41]

Think about this yourself in your own life. Imagine that you are deeply involved in a serious romantic relationship. Now you discover that your partner has become very interested in somebody else. Now imagine two different scenarios. In the first, your partner has a deep emotional—but not sexual—relationship with the other person. In the second scenario imagine that your partner has enjoyed a sexual—but not emotional—relationship with the other person. Which one of these scenarios would upset you most?

David Buss, of the University of Texas at Austin, who conducted research into this question, found that men were twice as likely to find the second scenario the most upsetting—it's the sexual relationship that bothers them, not the emotional relationship. While men find the sexual infidelity most distressing, women in contrast find the emotional infidelity most distressing. These sex differences were still true for scenarios where both forms of infidelity occurred. These findings on Americans also hold true in South Korea, Japan, Germany, and the Netherlands.[42] Men and women in different cultures differ in just the same way. Relatedly, men have been reported to be better than women in their ability to detect infidelity[43] and are more likely to simply suspect infidelity in their female spouses.[44]

What can explain the replicable sex difference in the green-eyed monster of jealousy? The explanation is that men are more distressed about infidelity because they could end up wasting resources and energy in raising a child genetically unrelated to them. Women, on the other hand, are concerned about infidelity because it means they may lose the protection, emotional support, and tangible resources provided by their partner. In both cases, resources are again the driving force behind our intense emotional feelings, but in subtly different ways.

These findings on jealousy now render for us a perspective on why male sexual jealousy can fuel so much physical and sexual aggression in partner relationships. Men who force sex on their spouses are found to have higher levels of sexual jealousy than men who do not.[45] Men may use violence as a mechanism to deter future defection by their female partner.[46] A woman will think twice about having another dangerous liaison if it results in her being battered nearly to death.

Yet this gives us even more food for thought at the evolutionary dining table, where resources and reproduction are the vittles. Why would a male partner rape his female partner in response to an infidelity? You might say it's simply an act of revenge. But lurking under the surface of this social argument may be a deep-rooted evolutionary battle that influences violence and crime—sperm wars.

If a woman did have sex with another man, from an evolutionary standpoint her partner will want to inseminate her as quickly as possible. His sperm will then compete with sperm from the unknown rival in a battle to access the woman's egg. Furthermore, by getting his sperm into her reproductive tract at regular intervals during a potentially prolonged period of suspected sexual infidelity, he puts off the chance that any foreign sperm will be successful in getting to that prized egg. At regular intervals he can top off his sperm in her cervix by injecting 300 million warriors. Half of these will end up in a flow-back that comes out of the vagina and onto the bed sheets, while the rest have further work to do, beginning their arduous journey for the next few days toward the egg in competition with someone else's sperm.[47]

In the genetic cheating game there's no stopping men. Women certainly have a hard time of it. They get raped by strangers. They get raped by friends. They get raped by their partners. Yet women are not always the victims. We'll see that they have their own subtle and conniving ways of waging war to promote their selfish genetic interests.

MEN ARE WARRIORS, WOMEN ARE WORRIERS

Let's start with men as warriors. We all know that men are more violent than women. It's true across all our human cultures, in every part of the world. The Yanomamo are not the only group whose men gather together to conduct killings in other villages. There has never in the history of humankind been *one example* of women banding together to wage war on another society to gain territory, resources, or power.[48] Think about it. It is *always* men. There are about nine male murderers for every one female murderer. When it comes to same-sex homicides, data from twenty studies show that 97 percent of the perpetrators are male.[49] Men are murderers.

The simple evolutionary explanation is that women are worth fighting for. They are the valuable resource that men want to get their hands on. Women bear the children, worry about their health, and make up the bulk of the parental investment. This is also true throughout the animal kingdom. Where one sex provides the greater parental investment, the other sex will fight to access that resource. Evolutionary theory argues that poorer people kill because they are lacking resources, an argument shared in common with sociological perspectives. And the reason men are overwhelmingly the victims of homicide is because men are in competition with other men over those resources. Men who murder are also about twice as likely to be unmarried as non-murdering men of the same age.[50] They have a greater need to get in on the reproductive act, and are willing to take warrior risks. For men one of the underlying causal currents for violence is competition for resources and difficulties in attracting females into a long-term relationship.

Let's also not forget warrior men in the home context. Violence can be used to dominate, control, and deter a potentially unfaithful spouse. Just as lions who take over a female from another male will kill the young and inseminate the lioness, aggression toward stepchildren is a strategic way of motivating the unwanted brood to move on and not take up resources needed for the next generation bred by the stepfather.[51]

Consider also that sex differences in aggression are in place as early as seventeen months of age.[52] Boys are toddler warriors. This might be expected from an evolutionary perspective that says males need to be more innately wired for physical aggression than females, to prepare them for later combat for resources. Seventeen months is a bit too young for sex differences to be explained in terms of socialization differences. Social-learning theories of why males are more aggressive run into trouble with the fact that the gender difference in aggression, which is in place very early on, does not change throughout childhood and adolescence.[53] Socialization theory would instead expect sex differences to increase throughout childhood, with increased exposure to aggressive role models, the media, and parenting influences, but they do not. Consider also that violence increases throughout the teenage years to peak at age nineteen. This is consistent with the notion that aggression and violence are tied to sexual selection and competition for mates, processes that peak at approximately this age.[54]

While male warriors perpetrate most violent offending, females can be aggressive too, in a surreptitious sort of way. On balance, however, women tend to be worriers rather than warriors for reasons that evolutionary psychology can explain.

Women have to be very careful in their use of aggression and sensitive in their perception of it because personal survival is more critical to women than to men. That's because they bear the brunt of child care and their survival is critical to the survival of their offspring. In unison with this standpoint, laboratory studies show that women consistently rate the dangerousness of an aggressive, provocative encounter higher than men do.[55] Women are also more fearful than men of situations and contexts that can involve bodily injury.[56] They are more likely to develop phobias of animals and medical and dental procedures. While they are more averse to physically risky forms of sensation-seeking, they are not averse to seeking forms of stimulation that do not involve physical risk—things like novel experiences through music, art, and travel.[57] Women also have a much greater concern over health issues than men. They rate health as more important and also go to the doctor more often.[58]

Fearfulness of bodily and health injury is therefore the psychological mechanism that evolution has built into women to protect them from death, helping to ensure the survival of their young. Thus, the fact that women are far less physically aggressive than males, in almost all arenas in life and in all cultures across the world, can be explained by an evolutionary principle.[59] Women are more averse to physical aggression than men because of its reproductive impact. Yet what would happen if we lowered the risk of bodily injury from aggression?

In this case a different scenario gets played out. John Archer, of the University of Central Lancashire, has documented that the sex difference in aggression is highest at the most severe levels of physical aggression, is much lower when it comes to verbal aggression, and is negligible with "indirect aggression."[60] Essentially, females are much more likely to engage in aggression when the cost to them in terms of physical injury is minimal. Indeed, Nicki Crick, at the University of Minnesota, has argued that females are more likely than males to engage in this "indirect" or "relational aggression," which takes the form of excluding others from social relationships and group activities and damaging their reputation in their peer groups—gossiping, spreading rumors, humiliating the individual. Ladies, do you recall this from your teenage days or experience it now in your current working life?

So rather than being physically violent, women take a more passive-aggressive strategy. They compete in terms of physical attractiveness—the quality most desired by men, who use it as a guide to fertility—and allow access to the man with the most resources. David Buss argues that women are much more likely to call their competitors ugly, make fun of their appearance, and comment on their fat thighs.[61] Women attempt to ruin their rivals' reputation by saying they have a lot of boyfriends, sleep around a lot, and are sexually promiscuous.[62] Men don't like hearing that from an evolutionary standpoint because if they get together with such a woman, they may end up rearing some other man's offspring. Consequently, such slanderous gossiping is an effective verbal-aggression strategy for women to use that does not run a high risk of physical harm.

We've seen here how violence and aggression is based partly on primeval evolutionary forces from the past. While reciprocal altruism can rule the day, antisocial cheating can also be a successful reproductive strategy, especially when psychopathic cheats migrate from one population to another. I've tried to illustrate how stealing, rape, homicide, infanticide, spousal abuse, and spouse killing can all be viewed from an evolutionary perspective. We've also seen anthropological examples of how different ecological settings could have given rise to either cheating or reciprocal altruist reproductive strategies. Males have evolved to use physical aggression to increase genetic fitness, while women have evolved to be concerned over their own health and that of their progeny, resorting to a safer form of relational aggression to protect their genetic interests. While evolutionary theory cannot, by any means, explain all violence, it at least provides us with a broad conceptual base with some degree of explanatory power.

The seeds of sin are rooted in our evolutionary past, the time when hominids formed social groups that shaped norms for helping behaviors—norms that a minority could break. Genes are the name of the evolutionary game, and therein lies an important implication for us. Talk about evolution, and by necessity we invoke genes. I've argued that antisocial, psychopathic behavior has evolved in some of us as a stable evolutionary strategy. In this ruthless, selfish context, rape is viewed not simply as a mechanism by which men exert power and control over women, as many feminists would argue. It is also the ultimate evolutionary cheating strategy—"love" them and leave them. Inseminate as many women as you can, then leave them to get on with the hard work of raising Cain and reproducing your bad genes. So the next step we will take in tracing the anatomy of violence is to understand the genetic basis to brutishness, and which individual genes stand out as our "usual" suspects.

2

SEEDS OF SIN

The Genetic Basis to Crime

Jeffrey Landrigan never knew his father. He was born on March 17, 1962, to a mother who abandoned him at a day-care center when he was just eight months old. But little Landrigan got lucky. He was adopted into an all-American family in Oklahoma. His adoptive father was a geologist named Nick Landrigan, whose wife, Dot, was a doting mother to both Jeffrey and their biological daughter, Shannon. Well-educated, straight-laced, and respectable, they provided a perfect new beginning for little Jeffrey.

Yet an insidious shadow from the past was cast over this baby that was to effectively seal his fate. By the age of two he was already throwing temper tantrums and displaying emotional dyscontrol that quickly escalated. He began abusing alcohol at the age of ten. His first arrest came when he was eleven, after he burglarized a home and attempted to break open the safe. He skipped school, abused drugs, stole cars, and spent time in detention centers. He was moving rapidly into his criminal career. When he turned twenty he had a drinking bout with a childhood friend who wanted Jeffrey to be the godfather of his soon-to-be child. Jeffrey's response? He stabbed his friend to death outside his friend's trailer. In 1982 he started a twenty-year sentence for second-degree murder.

Incredibly, Landrigan escaped from prison, on November 11, 1989, and headed out to Phoenix, Arizona. It could have been a new life and a clean sheet, yet murder seemed almost destiny for Landrigan. In a Burger King in Phoenix he struck up a conversation with Chester Dyer. Dyer was later found stabbed and strangled to death with an electrical cord, with lacerations on his face and back. Pornographic playing cards were strewn around the bed, with the ace of hearts propped up maliciously on the victim's back. But Landrigan's luck was running out. While exiting the apartment he left his footprint in sugar on the floor. He was consequently arrested, found guilty of homicide, and sentenced to death.

This might have been the last chapter in Landrigan's dramatic, topsy-turvy life. But the strangest twist was yet to come. While Landrigan was on death row in Arizona, another inmate told him of a man named Darrel Hill, a con he had met while on death row in Arkansas. Darrel Hill was Jeffrey's spitting image. Hill turned out to be the biological father that Jeffrey Landrigan had never seen. He was a dead ringer for Landrigan, and looks were not the only eerie similarity.

Darrel Hill had himself started his criminal career at an early age. He too was a drug addict. Like Landrigan he had killed not once but twice. He too had escaped from prison. Landrigan had clearly inherited much more than his father's looks. They could hardly have been more similar.

And that's not all. Jeffrey Landrigan's grandfather—Darrel Hill's father—was also an institutionalized criminal, who was shot to death by police after he robbed a drug store in a high-speed chase in 1961. He died just feet away from his then twenty-one-year-old son Darrel.

What do we make of this? Perhaps Darrel Hill summed it up best when he said:

> It don't take anyone too smart to look at three generations of outlaws and see there's a link of some kind, there's a pattern.[1]

Is there a "killer gene"? Or if not one, then multiple genes that, either on their own or in an intricate conspiracy with the environment, shape killers like Hill and Landrigan? Jeffrey Landrigan was adopted and raised in a safe and nurturing environment, yet despite all the love that his parents gave him—he could not be salvaged. This fascinating natural experiment—in which a baby with a violent heritage was transferred from a life of poverty and squalor into a loving, caring, successful family, yet still became a killer—suggests that there really is a genetic predisposition to violence.

Criminologists for decades have strongly resisted this idea. In this chapter I'm going to not just try to persuade *you* beyond a reasonable doubt, but also explain why social scientists are also opening up their minds to this fascinating and important perspective. To begin with, we'll delve into results from adoption studies that systematically examine cases similar to Landrigan's. In these studies, babies whose biological fathers were criminals were adopted away into noncriminal homes. We'll see that such babies were much more likely to become adult criminals than were babies who were also adopted but whose biological fathers were not criminals.

A second research design that uses identical and fraternal twins renders the same conclusion. Identical twins, who by definition have all of their genes in common, are much more similar to each other on crime and aggression than fraternal twins, who have only 50 percent of their genes in common.

A third but more unusual study comes to the same conclusion: identical twins who were separated at birth are surprisingly similar with respect to antisocial personality, despite being reared in very different environments.

These twin and adoption studies tell us that there is a significant genetic loading for aggression, but they do not tell us which specific genes are involved. So we'll finally turn to research at the molecular level that is now beginning to unmask the mean genes giving rise to aggression.

DOUBLE TROUBLE

About 2 percent of us are twins. Almost all of these twins are fraternal, or dizygotic, twins, who have about 50 percent of their genetic material in common. They develop from two separate eggs that are fertilized by two separate sperm, and effectively they are just like normal brothers and sisters. Much rarer—only 8 percent of all twins—are identical, or monozygotic, twins. These twins have virtually 100 percent of their genes in common because they develop from a single egg-sperm pair-up—a zygote—that basically malfunctions and splits into two.[2] Behavioral geneticists have used this malfunctioning twist of nature to examine genetic influences on antisocial and aggressive behavior. It's the perfect natural experiment for exploring the extent to which any behavioral, physical, or psychological characteristic is influenced by genetics.

Although I've mentioned that fraternal twins share on average 50 percent of their genes, I should qualify this. You actually have about 99 percent of your genes in common with me. Both of us share about 98 percent of our genes with chimpanzees, who themselves are genetically more similar to humans than they are to gorillas. Speaking of monkeys, we even have 60 percent of our genes in common with banana trees. So when we talk about fraternal twins having 50 percent of their genes in common, we are referring to 50 percent of just those small genetic differences that separate all human beings. Similarly, identical twins are not absolutely 100 percent genetically identical, but are 99 percent identical in that remaining 1 percent of genetic variation that differentiates all of us.

How do people go about setting up a twin study in the first place? Laura Baker, my longtime colleague at the University of Southern California, brainstormed with me one lunchtime about a nifty study we could do together. She knew a lot about twins. I knew a fair bit about antisocial behavior in kids. So she thought we could do a twin study on child antisocial behavior. Once we got our grant funded by the National Institute of Mental Health, we set to work. While I was setting up our psychophysiology laboratory, Laura started recruiting the twins. She worked with the Los Angeles Unified School District and sent letters to all parents who had a nine-year-old child in a school in Southern California.[3] We ended up getting 1,210 twins to participate. So then we were good to go.

The caregiver and the twin pair would come in for a full day of assessment—cognitive, psychophysiological, personality, social, and behavioral testing. The parents, the kids, and their teachers would fill out checklists on behavior, including antisocial behaviors. Do they bully other kids? Do they steal? Are they cruel to animals? Do they get into fights? Do they physically attack others? Do they skip school? Do they set fires? All the things that are the hallmark of a troublesome kid and a budding offender-to-be. Now we had our measures of antisocial behavior in 1,210 children.

So how do we work out if antisocial behavior in nine-year-olds is under genetic control? We look at how similar the identical twins are to each other, and compare that to how similar the fraternal twins are. Remember that identical twins are more genetically similar than fraternal twins. So if genes play some role in shaping antisocial behavior, you'd expect pairs of identical twins to be more similar in their level of antisocial behavior than fraternal twins. We use sophisticated statistical techniques—multivariate genetic analysis using structural equation modeling—to compute estimates of the heritability of this behavior.

What did Laura and I find? Heritabilities that ranged from .40 to .50. That means that 40 to 50 percent of the variability among us in antisocial behavior is explained by genetics. It did not matter who rated the child's behavior. If it was the teacher, the heritability was 40 percent. If it was the parent, it was 47 percent. If it was the kids themselves, heritability was 50 percent.[4] So no matter who makes the assessment, about half of the variation in antisocial behavior among kids is under genetic control. Half of the answer to why some of us are antisocial while others are not is due to genetics.

Our findings became even more dramatic when we combined our different measures of antisocial behavior. No measure is perfectly reliable. You know how parents, teachers, and kids can disagree on things. How can we derive a more reliable measure of antisocial behavior? By averaging the three informant sources to get a "common view" of what the child really does. When we did that, we found that 96 *percent* of the variance in this combined view of antisocial behavior is heritable. There is no contribution at all from the shared environment, and only a 4 percent contribution from the non-shared environment.[5] Once we have a more reliable measure of antisocial behavior the genetic influence goes way up. We must be very cautious not to overestimate the importance of genetic factors, but all in all there is no question that antisocial behavior is heritable—and significantly so.[6]

Twin studies also tell us that aggression and violence are heritable. In our study we measured reactive and proactive aggression. Reactive aggression is a case of someone hitting you, and you hitting them back—a sort of "defensive" or retaliatory aggression where you stand your ground. That form of aggression had a heritability of 38 percent. Proactive aggression, on the other hand, is meaner and crueler—you use force to get things from others. That had a somewhat higher heritability of 50 percent.[7] Again, the influence of the shared environment was minimal for both forms of aggression, and indeed was even *nonexistent* for boys.

Dozens of other twin studies have found the same effect in children, adolescents, and adults—males and females alike. Indeed, a meta-analysis of 103 studies compared heritability of aggressive behavior with rule-breaking, nonaggressive behavior.[8] Nonaggressive antisocial behavior was 48 percent heritable, while aggressive behavior was 65 percent heritable. Yet again, shared environmental influences were small for nonaggressive antisocial behavior (18 percent) and minimal for aggressive behavior (5 percent). Genetics and the non-shared environmental influences rule the roost when it comes to aggression. We also know that genetic influences are strongest for criminal careers that start early, occur across many settings, are persistent and severe,[9] and involve callous, unemotional symptoms like lack of remorse.[10] This is exactly the form of antisocial behavior that later gives rise to adult violence.

PLACING IDENTICAL PEAS INTO DIFFERENT PODS

One problem with twin studies is something called the equal environments assumption. Identical twins may be treated more equally by parents, teachers, and even peers than fraternal twins are. So an argument can be made that, sure, identical twins may be more alike on antisocial behavior than fraternal twins. But it's not because they are more genetically similar—it's because they are more *environmentally* similar.

This problem is circumvented in studies of identical twins reared apart. These are powerful studies for establishing heritability. Naturally they are very rare. However, one such study has been conducted on antisocial behavior in children and adults, consisting of thirty-two sets of monozygotic twins who were separated shortly after birth and reared apart.[11] The result? Statistically significant heritabilities of 41 percent for children and 28 percent for adults.

These findings from a large sample are striking. But perhaps more dramatic are findings from a case study of just eight monozygotic twin pairs reared apart where it was known that one twin was convicted of a crime.[12] The critical question was this: How many of the other eight twins from these pairs were also criminals? Of the eight, four had also committed one or more crimes, indicating clear evidence for the role of genetic factors. Because they were reared apart, you cannot say that the similarity is due to having the same upbringing—it's more to do with genetics.

One of the four concordant cases consisted of a pair of female Mexican monozygotic twins separated at nine months. They were brought up by parents with very different personalities.[13] Their environmental upbringing was also very different. One twin was brought up in a town, while the other twin grew up in the desert. Nevertheless, quite independently and as if by magic, just after reaching puberty both twins left their homes, took to the streets, and started to commit juvenile crimes. Both were separately institutionalized several times for their offenses. Recidivistic female crime is unusual, and when packed in the form of identical twins reared apart, it is even more extraordinary. Here we see the powerful influence of the genes that these two girls shared in common. It's a case of dark genetic forces overshadowing the power of the environment.

Studies of twins reared apart represent an important research strategy for understanding the genetics of crime. While the eight case studies are not methodologically strong they do illustrate the usefulness of this twin approach. Together with the finding of a methodologically stronger research study that utilized thirty-two pairs of twins reared apart and observed the same findings,[14] they add yet another important strand of support for a genetic predisposition to crime and antisocial behavior.

BUT WHAT ABOUT THE ENVIRONMENT?

Well might you ask. If you're an advocate of the importance of the environment, all this genetic stuff is disconcerting. But here's some good news. Genetic studies inform us about environmental influences just as much as they tell us about genetic influences. Twin studies tell us that about 50 percent of the variance in antisocial behavior is explained by *environmental influences*. The genes-versus-environment battle comes out as a tie.

But you know from when you were a teenager yourself that there are different types of environmental influences. Which one is more important in shaping antisocial behavior in children? The influences that come from within the family? Or the influences that come from outside the family? What would your guess be on who is more important in shaping kids' behavior—their home and parents? Or influences outside of the home?

It turns out that parents don't count as much as you would like to think. When Laura and I examined this, the familial home influences accounted for on average 22 percent of the total variance in antisocial behavior. In contrast, environmental influences outside the family accounted for 33 percent of the variance.[15] Even at nine years of age, children are being influenced—even pushed and shoved—in directions dictated by their peers rather than their parents.

This may sound hard to believe, but our study is no fluke. When you look at the results from overarching reviews of all genetic studies of antisocial behavior—over a hundred of them—you get the same result.[16] The same is true for a wide range of behavioral and personality measures, so it doesn't apply just to antisocial behavior. Indeed, Tom Bouchard, a leading behavioral geneticist at the University of Minnesota, has argued that shared environmental influences on adult personality are almost zero.[17] Yes, zero—no influence at all.

If like me you are a parent, it's sobering news. Do you want to believe that all your valiant caregiving efforts are worth almost zero? We face enormous cognitive dissonance—we do not want to believe findings like these because it means all our best efforts have been a waste of time.

It's frankly very upsetting. Parents want their children to be like them, and they put a lot of work into raising their children. And lo and behold, they turn out just the way their parents had wanted. So parents naturally believe that of course their efforts made a difference. But what if it's genetics? Parents silently and passively contribute half of their genetic material to the child. They cannot see their DNA and how it influences their child. They can, however, see all their socialization efforts, and if their child turns out well, their conclusion that their efforts really counted is reinforced. In our desire to believe we make a difference, we may not want to believe that our perceptions on how important we are as parents are wrong.

Taken together, the astonishing depth and breadth of twin studies is one factor that is beginning to change criminologists' minds about genetics. Slowly but surely, more pirate ships have appeared on the research horizon flying a genetic Jolly Roger. One ship you can ignore, but not an armada. Yet explaining the sea change that is occurring in social scientists' minds goes far beyond this fleet.

ADOPTION STUDIES—BACK ON THE LANDRIGAN TRAIL

Twin studies may underestimate the extent to which genes shape anti-social behavior because the error in the measure of antisocial behavior gets counted as non-shared *environmental* influences. But as we have seen, they may also overestimate genetic influences due to breakage of the equal-environments assumption. We need a pointer to get back onto the right path.

We leave environmental territory and get back onto the genetic trail. Recall that Darrel Hill left his son Jeffrey Landrigan at birth, and we saw an eerie likeness in their adult violent behavior. Now let's magnify this case study hundreds of times—by studying together hundreds of Jeffrey Landrigans—to see scientifically if there is a father-son linkage. A linkage even though the offspring never grew up with his true parent, but was instead brought up in a different home, a different environment, with a different way of living.

In the adoption design, offspring are separated from their criminal biological parents early in life and fostered out to completely different families. This is the experimental group. The control group consists of babies also fostered out soon after birth, but their biological parents do not have criminal records. If the offspring with criminal parents grow up themselves to become criminals at a higher rate than adopted children whose biological parents were not criminals, this would indicate a genetic influence stemming from their biological criminal parents.

That is precisely what has been found. In a landmark adoption study of crime, my colleague Sarnoff Mednick demonstrated that the adopted-away offspring of criminal parents in Denmark were more likely to become criminals as adults than the adopted-away offspring of noncriminal biological parents.[18] You can see these findings illustrated in Figure 2.1.

Mednick grouped the adoptees based on the number of criminal convictions of their parents. The adoptee controls, of course, had parents with zero convictions. Some adoptees had parents with one conviction, some two, and so on. What you see plotted in Figure 2.1 is the number of criminal convictions in the *adoptees* as a function of the degree of criminality in their biological parents. You can clearly see that

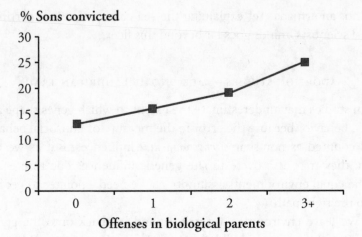

% Sons convicted

Offenses in biological parents

Figure 2.1 The increase in homicide rates in adoptees as a function of
the degree of criminal offending in the biological parents

the more convictions the biological parents had, the more offending
there was in their adopted-away offspring. It's a very clear demonstration that one's genetic heritage predisposes one to crime. It's also a reliable finding—almost every other adoption study on crime has observed
the same finding, and there are more than a dozen of them.[19] The findings are replicated time and time again across independent research
laboratories in different countries.

Which isn't to say there aren't caveats. Adoption agencies, for
example, try to place babies into adopting families who are similar
to the true biological parents—a process termed "selective placement." Furthermore, there could be differences in the length of
time the baby is with their natural mother. If antisocial mothers are
neglectful of their offspring before adoption, this negative bonding
experience—an environmental process—might account for the later
antisocial behavior. Mednick, however, was careful to control for
these factors. His findings could not be explained away by selective
placement of adoptees into adoption homes of a similar socioeconomic status, or the age at which the infant was taken away from the
mother. Other studies have similarly controlled for methodological
confounds like these.[20]

Of course, twin and adoption studies, like all other studies, have
their methodological weaknesses. Critics of the conclusion that there is

a genetic contribution to crime will eagerly latch onto such limitations. Their objections may seem to disqualify the conclusions, but it's a false alarm. These studies represent different people, time, places, measures, and designs.[21] All these differences should often lead to the expectation of divergent, different results—yet very tellingly they all converge on the same intrinsic finding.[22]

Let's apply this principle to the current context. Participants in more than a hundred genetic studies of antisocial behavior have ranged in age from nineteen months to seventy years. They cover the period from the Great Depression to the present. They represent many different Western nations, including Australia, the Netherlands, Norway, Sweden, the United Kingdom, and the United States. They use a wide variety of measures of antisocial behavior. They are made up of twin studies, adoption studies, and sibling designs. They also include large-scale studies that represent the general population and use advanced quantitative modeling techniques. They include studies conducted in the past fifteen years, and the findings from yesteryear stand up in studies done today.[23] Taken together as a whole, these studies converge on a simple truth that even the strongest critics of genetic influences in violence are finding harder to resist—genes give us half the answer to the question of why some of us are criminal, and others are not.[24]

ACNE AND XYY

What specific seeds account for sin? It's a big question, and it has always been controversial. In the past the most sensationalized link between violence and genes has been the case of XYY.

Normally we each have twenty-three pairs of chromosomes, with each chromosome being a bundle of many genes. One of these chromosomes is the sex chromosome—X or Y. Each parent gives one chromosome to each pair, which determines if we end up as an XY (male) or an XX (female). But on rare occasions there's a mistake. Instead of one Y chromosome pairing with one X, two Y chromosomes pair with one X. The result is a male who receives an extra male chromosome—XYY.

Soon after the XYY condition was first discovered, in 1961, there were rumblings that it might be linked to violence. In 1965 the prestigious science journal *Nature* published research findings from blood tests of Scottish prisoners in a special security hospital for the mentally disabled, which showed that 4 percent had an additional Y chromosome.[25] While it may not sound dramatic, this rate is forty times higher than the 1-in-1,000 rate of XYY reported for the general population.

A year later, in July 1966, while England was busy trying to win soccer's World Cup, a man named Richard Speck killed eight nurses in a dormitory in Chicago. He held them in their dorm at knifepoint, leading them out of the room one by one to rape and strangle them to death. One of the nurses, Corazon Amurao, surreptitiously slipped under a bed during the ordeal. While Speck thought he was raping the last nurse on a dorm bed, Corazon was huddled under the bed, terrified that her turn would be next. But Speck miscounted how many victims he had in the room and left. He was eventually caught. Corazon Amurao positively identified him in an identity parade and he was charged with the homicides.

A sensational twist in the dramatic coverage of this crime was the claim that Speck was XYY. At least superficially there was reason to suspect this possibility. XYY males are taller, averaging about six feet. They also have a history of learning disability and have IQs somewhat lower than average. It was also thought that XYYs had acne and that severe acne might be a marker for XYY—a mark of Cain.[26]

Speck was six feet one inches tall, and was not that intellectually sharp, as indicated by his miscounting of his victims and the struggles he had in school—he repeated the eighth grade and dropped out before he was sixteen. He also had a pockmarked face due to acne scarring. In a blaze of publicity just before his appeal against his conviction it was reported that Speck was an XYY.[27] This came shortly after a few high-profile scientific publications reported on the XYY-crime link, including Mary Telfer's report in *Science* that XYY was overrepresented in men in criminal institutions in Pennsylvania.[28]

It turned out that Speck was not XYY at all. To be sure, his face was pockmarked, as you can clearly see in Figure 2.2. Yet even before the trial began, Eric Engel, a Swiss neuroendocrinologist at Vanderbilt University, had performed a chromosome analysis on Speck and found him to be a completely normal XY male.[29] But the erroneous newspaper reporting fueled the public belief that XYY might be a cause of violence. It became almost folklore.

The link between XYY and violence in particular was debunked in a definitive study by Sarnoff Mednick and his colleagues in an influential paper published in *Science*.[30] They took a population of 28,884 men

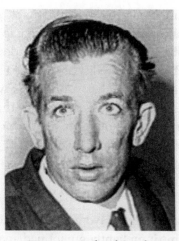

Figure 2.2 Richard Speck

born in Copenhagen and conducted a sex chromosome screen of the 4,139 who were over six feet tall. Twelve were found to be XYY. They then checked crime convictions of these twelve and compared them to normal XY males whose crime records were also checked. The result did indeed demonstrate that XYY is associated with crime in general, with a crime rate of 41.7 percent in the XYY group versus 9.3 percent in the controls. However, the rate of violent offending in the XYY group was 8.4 percent compared to 1.8 percent in the controls—a fivefold increase, which, while very large, was *statistically* nonsignificant due to the small sample.

Social scientists lapped up the findings. Criminology textbooks routinely reported this study as proof that there was no genetic basis to violence. Some even erroneously used this finding to scotch the whole idea of a genetic basis to crime in general. But let's get the facts clear.

While it is true that technically there is no statistically sound evidence to link the XYY syndrome to violence, this does not embarrass the notion of a heritable basis to crime, for four reasons. First, although XYY males do not commit more violent offenses than controls, they do commit more petty property offending.[31] Second, while the XYY syndrome represents a genetic abnormality, many criminologists misunderstand it. The XYY karyotype is not a heritable condition that is passed on from parents to offspring. It stems instead from random chromosomal mutations at the time of conception. Consequently, XYY research has no bearing at all on the issue of whether crime and violence is *heritable*.

Third, even if the XYY syndrome were a heritable genetic disorder and failed to show a relationship with crime, such a failure does not invalidate the significant findings of many twin and adoption studies that do show a relationship between heredity and crime. Fourth, recent studies with larger sample sizes show that young boys with XYY are indeed rated as more aggressive and more delinquent than controls.[32] As we shall see, there are many genes other than those on the Y chromosome that likely play a role in criminal behavior.

A MEAN MONOAMINE

For social scientists, the ugly Hydra head of the genetics of crime seemed to have been triumphantly guillotined and buried forever. But legend has it that when one of Hydra's heads was cut off, several more grew in its place. The intellectual battle over whether genes play a role in violence was just warming up.

Han Brunner was a doctor in the University Hospital in Nijmegen in the Netherlands who was approached one day in 1978 by a woman wanting genetic counseling.[33] Many of her male relatives seemed to have significant behavior problems. The problem was in their eyes, she said—it was the way they looked at you, frightening and aggressive.[34] Her ten-year-old son was showing signs of behavior problems and she also had two daughters. Might they be carriers of some genetic defect that results in aggression?

Han Brunner went on a systematic investigation, traveling around the Netherlands to track down this extended family across four generations. His research was fastidious. He even visited shelters that housed some of the woman's relatives. He interviewed them and took blood samples for genetic analyses. Fifteen years after that woman's visit, Brunner and his colleagues published their findings in *Science*. What they turned up was astonishing and almost eerie.

The fourteen male relatives that he studied showed a history of violence and impulsive aggression. It was almost a rerun of the Jeffrey Landrigan–Darrel Hill three-generation clan. In the four-generation family tree that Brunner drew up, only the male offspring of females were affected. That had to mean that whatever the genetic abnormality, it had to be carried this time on the X chromosome—the one transmitted by women. When Brunner genotyped the families he found an astonishing abnormality. These males had a defective gene—the MAOA gene, which normally produces the enzyme monoamine oxidase A. He sequenced this gene, analyzed it in detail, and found a mutation in it that resulted in no functional MAOA at all. All the affected members had this mutant form of the MAOA gene.[35]

MAOA is an enzyme that metabolizes several neurotransmitters involved in impulse control, attention, and other cognitive functions, including dopamine, norepinephrine, and serotonin.[36] Mutations in the normal MAOA gene lead to deficient production of the MAOA enzyme. It wasn't just that it was low in the affected family members, it was virtually nonexistent. A total lack of MAOA has profound effects. It disrupts the normal function of other neurotransmitters, resulting in a wide range of disorders—including attention-deficit/hyperactivity disorder, alcoholism, drug abuse, impulsivity, and other risky behaviors. Han Brunner also found that the lack of MAOA in his affected family members resulted in lower IQ. We know that low IQ is a very well-replicated risk factor for crime and violence.[37] Put this together with impulsivity, inattention, and drug or alcohol abuse, and you can see why impulsive aggression is not an unexpected outcome.

I saw Han in 2011 at a meeting in Amsterdam, and his perspective from the time of the publication of his work in *Science* was interesting. He sensibly recognized the controversy and was aware of the potential misuse of medical genetic research. So upon publication he couched his findings very cautiously. He used words like "abnormal behavior" instead of "aggression," and "associated" instead of "causes" in the title of his publication in *Science*. Despite this, the media were again blasting out the message of a new gene for crime. Han is at pains to maintain that there is no one gene for crime, that the genetic abnormality he discovered is extremely rare, and that the environment is critically important also.[38] He has put that message forward ever since I have known him, and yet it is consistently misinterpreted by social scientists who want to discredit his work, and the media who want to sensationalize it. Despite the onslaught and criticisms, however, another Hydra head was soon to pop up and go back into battle to persuade social scientists of the potency of the genetic argument.

THE WARRIOR GENE RIDES AGAIN

Han Brunner's novel finding in 1993 was given a big boost in 1995 by Jean Shih, a colleague of mine when I was at the University of Southern

California. Jean and her research team were investigating the effects of knocking out the MAOA gene in mice. You can knock out or deactivate a gene in mice by replacing it with an artificial DNA sequence. Once in a while Jean's team would come into their lab in the morning and notice a dead mouse. It did not take them long to work out that mice with deletion of the MAOA gene had become ferociously aggressive and were attacking other mice.[39] Jean had found a gene linked to aggression—and it happened to be the same gene that Han Brunner had found to be abnormal in his Dutch family.

The third Hydra head of the genetics perspective came at the beginning of this century. It proved to be a turning point not just on the genetics of crime, but almost the genetics of pretty well everything else worth talking about.

Two scientists—Terrie Moffitt and Avshalom Caspi, at Duke University—paved the way for research in this field with their seminal paper in *Science* in 2002. Widely regarded as one of the most important research papers in the social and behavioral sciences, it demonstrates something we will focus on much more in a later chapter—that genetic and biological factors *interact* with social factors in predisposing someone to later antisocial and violent behavior. So yes, individual genes are important—but in a specific social context.

Terrie—or Temi, as I have known her since I first met her in Tuscany when she was still a graduate student—had set up a major longitudinal study on antisocial behavior in Dunedin in New Zealand. Gold was struck near Dunedin in 1861 and in the ensuing gold rush it became the largest city in New Zealand, and still remains the second-largest city on the south island. Avshalom, Temi's husband, struck gold with the Dunedin data in his brilliant analysis of a gene that regulates the enzyme MAOA and how a variation of that gene combines with child abuse to produce antisocial behavior.

Although we share genes in common, there are variations in any gene, with different sequences of DNA at any specific location. These "genetic polymorphisms" give rise to differences among us—such as having different blood types, blue eyes versus brown eyes, or straight hair versus curly hair. One such genetic polymorphism results in different levels of MAOA. It's quite easy to genotype an individual, either from a blood sample or, even less invasively, from a saliva sample. About 30 percent of us have a variation in the MAOA gene that gives rise to relatively low levels of this enzyme, resulting in disturbances in neurotransmitter levels. The rest of us have relatively normal levels of MAOA. Caspi and Moffitt repeatedly assessed over a thousand children from Dunedin on antisocial behavior from age three to twenty-one. They also knew which ones had experienced no maltreatment from age three to eleven years, which had some maltreatment, and which were severe maltreatment. What they found was that low levels of MAOA were associated with later antisocial and violent behavior, particularly when the children had been severely abused.[40]

It was a dramatic discovery because it highlighted the complexity of understanding the genetic and biological basis of antisocial and violent behavior—something we'll return to later on. The New Zealand findings also brought more weight to bear on the earlier human findings from the Netherlands and the animal findings from the United States. Different research methodologies were beginning to converge on the same conclusion—low MAOA is to some extent associated with violent and aggressive behavior.

Yet new molecular genetic findings such as these come and go like lightning bolts out of the blue. Does this one replicate? Largely speaking it does. Four years after the original finding of Caspi and Moffitt, a meta-analysis that pooled results from five studies confirmed the original effect,[41] and it has since been linked to antisocial personality disorder.[42]

While these studies have shown that the low-MAOA gene is especially related to antisocial behavior in those with a history of abuse, studies are also beginning to show direct links between this gene and antisocial personality characteristics—irrespective of whether subjects have been abused.[43] Both men and women with the low-MAOA gene report higher levels of lifelong aggression.[44] Men with a rarer genetic abnormality of the MAOA gene that results in excessively low levels of MAOA have twice the level of serious delinquency and adult violence of normal controls.[45] Furthermore, the link goes beyond self-reports or psychiatric interviews. Those with the low-MAOA gene also show more aggressive behavior in a laboratory setting.[46] There's no single gene for crime or violence, but initial research does highlight some partial role played by this gene.[47]

Another chimeric Hydra head arose, again in New Zealand, in August 2006, but this time the battle was uglier and even more controversial. Researchers reported that the Maori had twice the level of the genotype conferring low levels of MAOA compared with Caucasians in New Zealand. The researchers were immediately quoted in newspapers as saying that this difference

> goes a long way to explaining some of the problems Maori have. Obviously, this means they are going to be more aggressive and violent and more likely to get involved in risk-taking behaviour like gambling.[48]

The headline of MAORI VIOLENCE BLAMED ON GENE helped not one bit. In the furor that followed, scientists, politicians, journalists, and pretty well everyone else dived into the hot and at times hostile debate that ensued.

The researchers who had presented the finding countered that they had been badly misquoted, and in a clarification argued:

> The extrapolation and negative twisting of this notion by journalists or politicians to try and explain non-medical antisocial issues like criminality need to be recognized as having no scientific support whatsoever and should be ignored.[49]

At the same time, they also argued that the low-MAOA genotype—which had come to be known as "the warrior gene"[50] based on research on aggression in monkeys[51]—was evidence of positive natural selection for the Maori. They hypothesized that the Maori have been well recognized as fearless warriors and historically had embarked on long, dangerous canoe voyages in their migration from Polynesia to New Zealand. They were also the survivors of warfare with other island tribes. They consequently argued from this "warrior gene hypothesis" that evolutionary forces may have resulted in the doubling of the frequency of the low-MAOA gene in Maori.[52] Put another way, this gene may have conferred a "survival of the fearsome" advantage on an indigenous group that now makes up 15 percent of the New Zealand population.

Some argued that the suggestion does a great disservice to the Maori people.[53] Others raised ethical concerns about the harm such speculation can do, including diverting attention from the poorer social and economic conditions of the Maori.[54] The authors of the warrior-gene hypothesis counter that it is both unethical and unscientific to ignore genetic difference in the Maori, a difference that could have potentially important medical and treatment implications for understanding disease disparities.[55]

There is no question that we all must be extraordinarily cautious in interpreting any genetic differences between ethnic groups, especially with respect to crime and violence. At the same time, the evolutionary argument put forward is not entirely implausible. Counterpoint: While the base rate of the low-MAOA gene is about 34 percent in Caucasian males and 56 percent in the Maori, it is 77 percent in Chinese males. Yet the homicide rate in China, at about 2.1 per 100,000, is less than that of the United States—the Chinese are not exactly known for their fearless, warrior-like tendencies.[56] We'll return to the ethical issues on the biology of violence, but for now let's turn away from the debate on genes and violence in the Maori and back to a more established body of evidence that does not rest on ethnic-group differences.

Importantly, let's consider that the *type* of aggression we are talking about may make a difference. The MAOA warrior gene may well be especially important in predisposing people to hot-blooded, emotional, and impulsive forms of aggression—rather than cold-blooded, regulated aggression. Han Brunner documented that the men in his Dutch kindred study tended to display more impulsive forms of aggression that often occurred in response to anger, fear, or frustration.[57] Consistent with this interpretation was research done in Los Angeles that found that UCLA students with the low-MAOA gene not only had more aggressive personalities, but showed greater interpersonal hypersensitivity—their feelings were more easily hurt.[58] They also showed a greater brain response to being socially excluded, suggesting that they were indeed more easily upset by personal slights.

Those with the warrior gene are more hypersensitive to criticism, which in turn results in increased impulsive aggression.[59] Australians with the warrior gene not only exhibit higher levels of antisocial personality, but also show an abnormal brain response to processing emotional stimuli.[60] No, I'm not going to say it's all due to Australians' being the offspring of 160,000 convicts shipped out from England in the eighteenth and nineteenth centuries. I believe instead that this indicates that the low-MAOA gene has an across-the-board linkage with crime. By and large it cuts across cultures.

JIMMY "THE FUSE"—EXPLOSIVE BRAIN CHEMISTRY

I've so far discussed one particular gene—the "warrior gene"—because it has quite strong scientific support as contributing to antisocial and aggressive behavior. Yet other genes are also involved. The 5HTT gene,[61] the DRD2 gene,[62] the DAT1 gene,[63] and the DRD4[64] have all appeared on the gene landscape as linked to antisocial and aggressive behavior. What do these particular genes do? They regulate two important neurotransmitters in the brain—serotonin and dopamine.

But before going further, let's gain another perspective on this aspect of the anatomy of violence. From the genetic makeup of the brain it's only a brief step to the chemistry of violence. The essence of the molecular genetic research we have been touching on above—identifying specific genes that predispose individuals to crime—is that genes code for neurotransmitter functioning. Neurotransmitters are brain chemicals essential to brain functioning. There are more than a hundred of them and they help to transmit signals from one brain cell to another to communicate information. Change the level of these neurotransmitters, and you change cognition, emotion, and behavior. Genes that influence neurotransmitter functioning can therefore result in aggressive thoughts, feelings, and behaviors.

Take dopamine, for example. Dopamine helps produce drive and motivation. It is critically involved in reward-seeking behavior. Aggressive behavior can be rewarding, and in animals dopamine receptors help code for this rewarding property of aggression.[65] When dopamine is experimentally increased in animals it fuels aggression, while blocking dopamine decreases aggression.[66] You can think of it as the accelerator in the car that helps move us forward to things that we want.

There is a very different story to tell on serotonin. The serotonin-transporter gene is one of the most intensively researched genes in my fields of psychology, psychiatry, and neuroscience.[67] There are two versions of this gene—the short-allele version and the long-allele version. About 16 percent of us have the short-allele version.[68] This version makes our brains overrespond to emotional stimuli[69] and can result in us letting off steam when we get overheated. It is thought to be associated with low serotonin because those with the short-allele version have reduced levels of serotonin in their bloodstream.[70]

So, do violent offenders have low serotonin levels? The research on this started with a seminal study in 1979 of military personnel. Fred Goodwin was an exceptional scientist and director of the National Institute of Mental Health. He first asked his military-personnel participants how many fights and assaults they'd gotten themselves into. They were then put on bed rest and fasted overnight. Instead of getting a breakfast when they woke up the next morning, they got a spinal tap—a needle through their spinal cord—to obtain a sample of their cerebrospinal fluid. From this fluid Fred and his team assayed levels of serotonin.

What they found was dramatic and created a watershed in the research field on the biology of violence. Resting levels of serotonin explained a full 85 percent of the variation in the incidents of aggression in the lives of these military men.[71] That relationship is enormously high, and perhaps the finding was a little too good to be true. Subsequent studies have shown that the relationship between aggression and low serotonin levels is not as strong as originally thought—it explains more like 10 percent of the variation in aggression—but it is still relatively strong and has been extremely well replicated in adults,[72] especially in relation to those committing impulsive violent acts.[73]

Why would low serotonin result in violence? Serotonin is a mood stabilizer, which has an inhibitory function in the brain. It is thought to be one of the biological brakes on impulsive, thoughtless behavior. It innervates—or lubricates—a part of the brain called the frontal cortex, which, as we will see in the next chapter, is critically important in regulating aggression. The less serotonin you have, the more rash you may be. Brain-imaging research has shown that people given a drink that reduces serotonin by depleting tryptophan—an amino acid critical for serotonin production—are more likely to retaliate when they are made an unfair offer in a game.[74] Without serotonin they get upset more easily when annoyed. Combine a low serotonin predisposition to an unfair social situation that irritates you, and blowing your fuse is just around the corner.

That may explain what happened to James Filiaggi—known as "Jimmy the Fuse." Jimmy was of Italian ancestry with a pretty good upbringing—not all that different from me and Lombroso. In fact, like me, he used to be an accountant. But throughout Jimmy's life he showed signs of a very short temper—the reason for his nickname. As a boy, Filiaggi bit off the end of his brother Tony's finger, and took a piece out of his schoolteacher's hand. He also attacked a nun, which, not surprisingly, resulted in expulsion from school. And yet he was also a smart kid who graduated with honors and went into finance. One night Filiaggi got really upset during an argument with his estranged wife, and, feeling threatened by him, she called 911. He shot her in the head.

Facing the death penalty, Jimmy's defense team brought in another Italian, Emil Coccaro—a dynamic worldwide authority on serotonin and aggression. A spinal tap and consequent biochemical assay conducted by Emil convincingly demonstrated that Filiaggi had extremely low levels of serotonin.

And that was not all. Coccaro also found that Filiaggi had very high levels of dopamine, the neurotransmitter that promotes reward-seeking behavior and drug abuse. Filiaggi had the worst of both worlds—a combination of reward-seeking together with reduced inhibition. His foot was pushing down hard on the accelerator for rewards, but never applying the brakes for inhibition. It was that chemical cocktail that likely sent Filiaggi off the rails with the estranged wife he wanted back. Most of us have biochemical brakes on our behavior. Filiaggi didn't.

As Filiaggi faced the death penalty for his crime, could these neurotransmitter abnormalities perhaps be used as a mitigating factor to persuade the court that due to a biological makeup beyond his control, he was prone to impulsive, aggressive behavior? No. Filiaggi was executed by lethal injection. We will consider whether such mitigation could and should happen in a later chapter, but for now imagine whether in your mind it might make a difference if you were on the jury.

The links between brain chemistry and violence in humans are complex, as the case of Jimmy the Fuse illustrates. But let us also not forget that the environment is critically important. Filiaggi's low serotonin did not act in isolation in predisposing him to aggression. It also required a suitably provocative social context to result in his hot-blooded violence. In contrast to the link between the *short*-allele version of the serotonin transporter gene, which links to impulsive, hot-headed aggression, the *long* allele, as one of my former graduate students, Andrea Glenn, has persuasively argued, is associated with more cold-blooded and planned psychopathic behavior in those with low responsivity to stress.[75]

We have a long way to go in understanding the neurochemistry of violence, not just with respect to soothing serotonin but also to reward-driving dopamine. Yet there are some provocative links. For example, one study has observed that the dopamine transporter DAT1 gene, which is linked to violence, is also associated with the number of sexual partners you have.[76] As we just saw in the previous chapter, this suggests from an evolutionary perspective that violence—while self-damaging in many ways—may still be adaptive in terms of reproducing one's genes. There is a common genetic mechanism that plays out at the neurotransmitter level linking the two topics of hot sex and heated violence, and in the next decade we are sure to find out much more on this neurochemical component of the anatomy of violence.

THE END OF THE BEGINNING

Scientific inquiry is just beginning to scratch the surface in understanding the specific genes that create violence. We have reasons to be humble about our conclusions, and yet proud for how far we've come. Twenty years ago, molecular genetics was a fledgling field of research. Now it is a major enterprise providing us with a detailed look at the structure and function of genes. A major beginning step was the Human Genome Project—one of the most important international research projects of our time. It began in 1990, and by 2000 researchers had come up with a working draft of the human genome. It turns out that we have far fewer genes than was originally thought—about 21,000—roughly the same number that mice have. Although human genes are mapped and are available to all of us on the Internet, a lot remains unknown. For example, about 98 percent of our DNA is "junk" DNA,[77] meaning that it does not encode protein sequences—we don't yet know what it's there for or what it does.

Take that as a crucial caveat. The knowledge base that I have quickly sketched out above on what genes are connected to violence is going to change enormously over time. Nevertheless, the basic message—that genes have a strong influence on our behavior—will remain intact. We stand on the threshold of unlocking many untold secrets of our genetic makeup with all the medical benefits and ethical conundrums that come with that knowledge. Behavioral genetics is a shadowy black box because, while it tells us what proportion of a given behavior is genetically influenced, it does not identify the specific genes lurking in there that predispose one to violence. Molecular genetics is poised to pry open that black box and shed light on the dark figures of violence.

As researchers illuminate the role that "junk" DNA has in the transcription of protein-coding sequences and the regulation of gene expression,[78] we'll uncover yet more knowledge on the genetics of violence. Critical here will be uncovering what environmental influences interact with what genes in causing crime—a new development that is finally beginning to excite social scientists who previously held the genetics of crime in distrust.

With this chapter on genetics we have reached the end of the beginning. Scientifically, the Human Genome Project has completed the task of setting the stage of gene-behavior discovery, and we will move on to a more complete elaboration of what genes shape crime and violence. This also marks the end of the beginning of our investigation into the seeds of sin. But before ending, let's return briefly to our beginning. Darrel Hill, on death row, summed it up succinctly when he said:

> I don't think there can be any doubt in anyone's mind that he (Jeffrey Landrigan) was fulfilling his destiny . . . I believe that when he was conceived, what I was, he became . . . The last time I saw him he was a baby in a bed, and underneath his mattress I had two .38 pistols and Demerol; that's what he was sleeping on.[79]

Placing that gun and drugs under his baby boy's pillow foreshadowed what was to come. Like father, like son—whether it is violence, drugs, or alcohol. Landrigan was seemingly doing little more in life than acting out the sins of his biological father.

3.

MURDEROUS MINDS

How Violent Brains Malfunction

Randy Kraft was a man with a murderous mind. But you would never have guessed it from meeting with him. A computer consultant with an IQ of 129, close to mine, he grew up in Southern California just south of where I used to teach, at USC. Randy, like me, was brought up the son of respectable, hardworking parents. He was the youngest in his family, and again like me he had three older sisters. He grew up in a middle-class, conservative area in a rather normal, even uneventful home life that matched my own. A smart schoolkid, he was placed in accelerated classes, just like I was. He attended Westminster High School and the prestigious Claremont Men's College, an elite liberal-arts college, where he gained a degree in economics. Randy had a lot going for him.

You can see for yourself from Randy's Web page, where he reminisces about his childhood, that his life was pretty much all "apple pie and Chevrolet" back in the good old '50s and '60s. Randy talks affectionately about his home life, replete with happy memories of bowling with his dad and preparing strawberries and whipped cream with his mother. He reflects on the excitement of witnessing with his father the pale, eerie light thrown from a Nevada nuclear test-site explosion, and his first school dance with a girl, at age thirteen. Randy ruminates on his home, set against the backdrop of strawberry fields in rural Orange County. He clearly loved helping his dad make a morning fire from garden rubbish, and he paints a multimodal, colorful scene of sounds, smells, and textures:

> Today, when I look back, I can smell the distinct, sweet odor of a damp grass fire, and hear the frenetic crackle of the struggling flame, and see the ribbon of white smoke curling far into the blue morning sky. And there is Dad in his old style undershirt and baggy pants, piling more onto the fire with the pitchfork, and I'm helping him.[1]

It could be something you or I might write about our home lives. Except when you look back at your life there is no smell of fresh blood. You do not hear the frenetic cries of your struggling victims, echoing through the deathly dark night sky. You do not see their loosened and disordered underwear, their pants undone in a violent assault, or feel the fire flaming your passion during the heated rapes. The whitening of your victim's face as you strangle them until they turn a pale shade of blue. The wetness of their lap as their pelvic muscles relax after death, releasing urine from their bladder.

You would not have experienced this. Randy repeatedly did. It's a different scene from the one that Randy sketched, and one he insists he never acted in. Yet he is on death row in San Quentin, having killed an estimated sixty-four times between September 1971 and May 1983.

The very likeable Randy would socialize in the evenings with his adult and teenage victims-to-be, share beers with them, and take them cruising around in his car. Then, after drugging them with a mix of tranquilizers and beer, he would playfully torture them, rape them, and then dump their bodies out of his car—earning him the nickname "the Freeway Killer." Some he would strangle, some he'd shoot. All were teenage boys and young adult men.[2]

Randy could still be killing today if not for a bit of bad luck on May 14, 1983. It was one o'clock in the morning. He'd been out having a good time. After a drink or two he was driving at a steady 45 m.p.h. in his Toyota Celica on the Interstate 5 portion of the San Diego Freeway just south of L.A. Although he wasn't speeding, Randy's driving was just a little bit erratic. He then made an illegal lane change that ended his killing career.

A California Highway Patrol car had been following him. It put on its lights and hailed him with its public-address system. Randy dutifully pulled over, just ahead of the police. Rather than wait for the cops to come to his car, he walked back to them with a bottle of Moosehead beer.

Kraft admitted that he'd had three to four beers that night, but said he was not drunk. The cops checked him out with a sobriety test. This was one test in his life that Randy failed. They had to charge him with driving under the influence of alcohol.

That meant having to take him in and impound his vehicle. Sergeant Michael Howard walked ahead to Randy's car. It was only then that he became a little suspicious. There was someone slumped in the passenger seat. You have to give the California Highway Patrol some credit. They normally allow a sober passenger to drive the car back home—that way the driver would not have to pay the impound fee. Maybe this other guy could help Randy out.

Thinking the passenger was asleep, Sergeant Howard politely knocked on the window, but there was no response. That was a bit odd. He opened the door and shook the passenger. Still no response. Most peculiar—maybe he was drunk, too. He then lifted the jacket lying on the passenger's lap, and that's when he noticed that his pants were undone, his penis and testicles were sticking out, and there were ligature marks on his wrists.

The paramedics were brought to the scene but it was too late. The dead body belonged to Terry Gambrel, a twenty-five-year-old U.S. Marine. He had drunk the equivalent of two beers and had ingested some Ativan—but not enough to kill him. The responsibility for that lay at Randy's door. He had strangled the Marine to death.

Randy was up the creek without a paddle. The well-mannered, meticulous, soft-spoken, hardworking computer consultant was none other than the Freeway Killer, soon to be renamed the Scorecard Killer. The nickname evolved from the fact that in the trunk of his car, lying inside his briefcase, was a long, two-columned list of coded names like "England," "Angel," and "Hari Kari." This was Randy's hit list. Like myself in my accounting days, Randy liked to keep orderly numerical lists. Some entries appeared to have been double killings because their code names were "2 in 1 Hitch" or "2 in 1 Beach"—perhaps two hitchhikers or two killings down by the beach. Many of the coded entries made a lot of sense. "Euclid" referred to the ramp where Kraft dumped the body of his victim Scott Hughes, "EDM" referred to the initials of another victim, Edward Daniel Moore, while "Jail Out" was a reference to Roland Young, whom Randy killed just hours after Young's release from jail.

After every sexual score with these men, Randy wrote them up. According to his notes, the Scorecard Killer had murdered sixty-four young men in a twelve-year period, getting away scot-free all those years, until that fateful night when he was caught for little more than a traffic violation. A trivial mistake by an otherwise meticulously detailed murderous mind—a mind and brain that we'll examine soon in our search for understanding of the functional neuroanatomy of violence.

Unlike Kraft, the vast majority of murderers kill only once. Such was Antonio Bustamante. A different killer with a different background, Antonio was born in Mexico and came to the United States at the age of fourteen. Like many Mexican-Americans he had a strong connection with his family. Although they were poor, Antonio grew up to be a law-abiding teenager and young adult.

But then an insidious change took place. He got caught up in drugs. He stole to support his habit. His criminal career then took off as his identity as an industrious, law-abiding immigrant ended. He became impulsive, increasingly argumentative, and got into more fights. For the next two decades he was in and out of prison. His heroin addiction meant he was constantly in need of money.

In September 1986, three years after Randy Kraft's arrest, Bustamante burglarized a home. He did not find cash, but did uncover traveler's checks. Things were looking good until he was surprised by the eighty-year-old occupant, who had returned from a nearby grocery store. Bustamante was six feet two inches tall and weighed in at 210 pounds. You'd think that it would not be too hard to get away from an eighty-year-old man, but Bustamante's fight-or-flight system decided to fight instead of take flight. Bustamante beat the defenseless old man to death with his fists. According to the prosecution, blood was splattered everywhere in the apartment.

Bustamante was a messy and disorganized killer. He'd left his fingerprints everywhere at the crime scene. He hadn't even bothered to clean himself up. When he went to cash the traveler's checks they had blood on them. In an even more remarkable oversight, he was still wearing his bloody clothes when arrested by the police.

Two distinct types of killers: the cool, calculating Kraft, and the bungling, bullheaded Bustamante. Divergent home backgrounds. Different ethnic backgrounds. Dissimilar criminal backgrounds. Distinct modi operandi. A very disparate number of victims. If you could look inside the minds of these men, what would you see? Would the brain scan of a murderer look like yours? Where exactly in the brain would the difference be? How would the brain functioning of serial killers like Randy Kraft differ from those of less memorable but more common-variety one-off killers like Antonio Bustamante? And how do any of us—who presumably have not killed—fit into the picture?

Not that long ago, such questions were the province of pulp fiction. In Jonathan Demme's movie *Silence of the Lambs* the serial killer Hannibal Lecter scolds FBI agent Clarice Starling for trying to dissect him with a paper-and-pencil questionnaire, what he termed a "blunt little tool." But today brain-imaging technology is giving us a much sharper instrument to probe the anatomy of violence. It's giving us tangible visual evidence that there is something wrong with how such killers' brains function. While these studies are still coming of age and have their limitations, they not only provide a basis upon which future research may build, but also raise provocative and important questions about free will, blame, and punishment that we'll return to in chapter 10.

But before getting to these complex ramifications, let's look at the scientific evidence showing that murderers have a mind to crime; we can now bear witness to that fact by studying their brain functioning.

THE BRAINS OF MURDERERS

We've come a long, long way in our understanding of the brain. Aristotle thought the organ was a radiator to cool blood. Descartes thought it was an antenna for the spirit to communicate with the body. The phrenologist Franz Gall believed that bumps on the skull revealed an individual's personality. Now we know that this three-pound lump of gray matter is behind everything we do—seeing, hearing, touching, moving, speaking, tasting, feeling, thinking, and of course book reading. And if all actions and behaviors stem from the brain, then why not violent behavior? Why not homicide?

Before 1994, I'd never done a brain-imaging study of murderers. Neither had anyone else. It's not too surprising, given the difficulty of recruiting and testing a substantial number of the minuscule proportion of us who commits homicide—less than one in 20,000 in any one year in the United States.

But one reason I emigrated from England to California in 1987 was that in addition to the good weather, there were plenty of murderers who could be recruited into my research studies. Credit for recruiting the unusual sample I studied goes to my colleague Monte Buchsbaum, who was just down the road from me at the University of California in Irvine. We identified the subjects through referrals from defense attorneys. Because California has the death penalty, their clients would die unless mitigating circumstances like brain abnormalities could be documented. We were able to build up a unique and sizable research sample.

So, complete with shackles and chains, and flanked by guards, our forty-one murderers trooped into the brain-scanning facility. They looked pretty formidable, intimidating, and ominous. Yet in reality they were very cooperative. We forget that for 99.9 percent of their lives, murderers are just like me and you. That's why they always come across as your next-door neighbor. Tragic actions in a few fleeting moments set murderers apart from the rest of us. As we shall see, their brain functioning also sets them apart.

The technique we used to scan their brains was positron-emission tomography—PET for short. It allows us to measure the metabolic activity of many different regions of the brain at the same time, including the prefrontal cortex—the very front part of the brain, which sits right above your eyes and immediately behind your forehead. We used the continuous performance task to activate or "challenge" the prefrontal cortex. The subject had to press a response button every time they saw the figure "o" flashed on a computer screen. This went on for thirty-two minutes. Believe me, it's very boring. But the task requires sustaining attention for a long period, and the prefrontal cortex plays an important role in maintaining vigilance. It's this part of the brain that is active in you now and that has gotten you to reach this point in the book. After the task, the murderer was taken to the PET scanner, which measured glucose metabolism occurring during the earlier task, rather than afterward in the scanner. The higher the glucose metabolism, the more that part of the brain was working during the cognitive task.

What did the study of forty-one murderers and forty-one age- and sex-matched normal controls reveal? Our key finding is illustrated in

Figure 3.1, in the color-plate section, which shows the brain scan of a normal control on the left and the brain scan of a murderer on the right. It shows a horizontal slice through the brain, so you are looking down on it with a bird's-eye view. The prefrontal region is at the top, and the occipital cortex—the back part of the brain, where vision is controlled—is at the bottom. The warm colors—red and yellow—indicate areas of high glucose metabolism while cool colors like blue and green indicate low brain functioning.

If you look at the normal control, on the left, you can see strong activation in the prefrontal cortex as well as the occipital cortex (at the bottom). The murderer, on the right, shows strong activation in the occipital cortex, just like the normal control. There's nothing wrong with his visual system. In stark contrast to the normal control, however, the murderer shows a striking lack of activation in the prefrontal cortex. Overall, the forty-one murderers showed a significant reduction in prefrontal glucose metabolism compared with the controls.[3]

Why should poor prefrontal functioning predispose one to violence? What can help us to form a bridge between a bad brain and bad behavior? And what happens after impairment to the prefrontal cortex? These questions can be answered at different conceptual levels.

1. At an *emotional* level, reduced prefrontal functioning results in a loss of control over the evolutionarily more primitive parts of the brain, such as the limbic system, that generate raw emotions like anger and rage.[4] The more sophisticated prefrontal cortex keeps a lid on these limbic emotions. Take that lid off, and the emotions will boil over.

2. At a *behavioral* level, we know from research on neurological patients that damage to the prefrontal cortex results in risk-taking, irresponsibility, and rule-breaking.[5] It's not far to go from these behavioral changes to violent behavior.

3. At a *personality* level, frontal damage has been shown to result in a whole host of personality changes. These include impulsivity, loss of self-control, and inability to modify and inhibit behavior appropriately.[6] Can you imagine these types of personality traits in violent offenders?

4. At a *social* level, prefrontal damage results in immaturity, lack of tact, and poor social judgment.[7] From here we can imagine how a lack of social skills can result in socially inappropriate behavior and poorer ability to formulate nonaggressive solutions to fractious social encounters.

5. At a *cognitive* level, poor frontal functioning results in a loss of intellectual flexibility and poorer problem-solving skills.[8] These intellectual impairments can later result in school failure, unemployment, and economic deprivation, all factors that predispose someone to a criminal and violent way of life.

It's not just one level of analysis but five—five reasons we might expect that poor prefrontal functioning could predispose a person to violent behavior. It's not surprising, therefore, that poor prefrontal functioning is the best-replicated correlate of antisocial and violent behavior.[9]

Fact or artifact? Is there a true relationship between poor prefrontal functioning and homicide, or is it explained instead by some methodological artifact? We think fact. Group differences in brain functioning could not be explained away by group differences in age, sex, handedness, history of head injury, medications, or illegal drug use prior to scanning. Furthermore, the murderers could do the task—their performance was just as good as the controls', possibly because the behavioral occipital cortex was *more* activated in the murderers than in the controls.[10] The murderers likely recruited this visual brain area into action to help them perform the visual task and to compensate for their poorer prefrontal functioning. Prefrontal dysfunction in murderers is fact, and not artifact.

BUSTAMANTE'S BUST HEAD—AND MONTE'S TESTIMONY

Our study constituted the first brain-imaging evidence to show that the brains of a large sample of murderers are functionally different from those of the general population. Nevertheless we must be cautious. Violence is enormously complex, and prefrontal dysfunction doesn't apply to all murderers.

To illustrate this further, let's return to Antonio and Randy and delve further into their murderous minds. Antonio Bustamante, as you will recall, was an impulsive criminal who had for years been spiraling downhill until he finally hit rock bottom in an unplanned, impulsive killing of a defenseless old man during a botched burglary. As the prosecution attorney Joseph Beard argued, it was a vicious and needless attack motivated by greed and money. He inevitably sought the death penalty.

Bustamante had been charged by the police no fewer than twenty-nine times prior to his arrest for homicide. His crimes included theft, breaking and entering, drug offenses, strong-arm robbery, and unlawful flight to avoid prosecution. His background and pattern of offending was typical of many lifelong recidivistic criminals. He was your typical thug.

With one curious exception. Looking closely at his records, I see that his offending did not start until he was nearly twenty-two. That's simply not typical of your recidivistic violent offender, whose antisocial behavior typically starts much earlier—often in childhood and certainly by early adolescence. And yet by all accounts Bustamante was a well-behaved teenager. So what gives?

The defense team, led by Christopher Plourd, looked over his history and the circumstances of the homicide. Something seemed strange to them too. Bustamante had been very messy and disorganized in stealing and cashing the traveler's checks. There was blood all over them. He'd left his fingerprints everywhere at the crime scene. He was still in his bloody clothes when he was arrested. Does this sound like a well-oiled, efficient killing machine to you? Probably not. Maybe this particular killing machine had a screw loose.

Plourd discovered that his client had suffered a head injury from a crowbar at the age of twenty. By all accounts Bustamante's personality changed radically afterward, transforming him from a well-regulated individual into a recklessly impulsive and emotionally labile renegade. Believing that this history of head injury was significant, Plourd had his client's brain scanned. It was at this point that Monte Buchsbaum, a world-leading schizophrenia expert and brain-imaging researcher, became involved. He testified at trial that Bustamante was suffering from dysfunction to the prefrontal cortex.

Antonio Bustamante was one of the forty-one murderers whose brains we had scanned, and his scan was telling. If you were sitting on the jury, what would you yourself think? Could the injury have turned Bustamante into a monster of a man unable to regulate and control his actions and emotions? Would you buy the neurological evidence that damage to the orbitofrontal cortex impairs decision-making and releases the brakes on emotion regulation, and that the brain scan provided objective evidence for this?

Take a good look at Figure 3.2, in the color-plate section and you can bear witness yourself. You can see the brain impairment to Antonio Bustamante, on the right-hand side. The orbitofrontal cortex is at the top. It's a cool-colored green compared with the big blotch of red in the normal control on the left. Bustamante's brain is not normal. At least, that's what the jury believed—they spared Bustamante the death penalty.

The prosecution was flabbergasted. As prosecution attorney Joseph Beard said:

> I'd never seen anything like this before. I didn't even know what a PET scan was. One of them was labeled "Bustamante" and the other was labeled "Normal." They were obviously different. The shapes were different, the colors were different. . . . I don't think it's an excuse. From my perspective, its hocus-pocus. . . . I'm not sure that they had the wherewithal to say that someone hitting him with a pipe 20 years before dramatically changed an altar-boy into a killer.[11]

And that hocus-pocus PET scan still hangs on the wall of Joe Beard's office as a reminder of the brain excuse that defendants increasingly ply in capital cases. A reminder of how pretty-colored pictures of the brain can be used to sway jurors' perspectives on innocence versus guilt—on life in prison versus the death penalty.

Yes, the causal direction of the relationship between prefrontal dysfunction and violence is certainly open to question. Imaging does not demonstrate causality. There is only an association, and many possible counter-explanations. We'll never know what Bustamante's brain scan looked like the day before the homicide. We'll never know if Bustamante's poor orbitofrontal functioning *caused* him—in one way or another—to morph from an altar boy into a killer who beat an old man to death.

Nevertheless, let's try to put the pieces together just as any detective or doctor would. Antonio Bustamante was as good as gold growing up, right until early adulthood. Then, at age twenty, a crowbar from hell struck the altar boy. Medical records from that time attest that it resulted in a very significant head injury. This injury likely increased Bustamante's impulsivity and lowered his threshold for more accidents. Not long after the crow-bar injury, he was involved in a serious automobile accident that resulted in yet more head injuries.[12]

For the two decades that followed, Bustamante was incessantly in trouble with the law. He also had more bar brawls, which very likely resulted in further head injuries. It's not exactly Jekyll and Hyde, but it's not far off. At the relatively late age of twenty-two, he clocks up the first offense in his life—just *after* the crowbar and automobile incidents that resulted in head injury. Bustamante suddenly switches from good to evil, tumbling into a turbulent world of drugs and crime, eventually ending up at the house of his victim—and homicide. I think the order of events is telling.

Let's put this Jekyll-and-Hyde transformation together with the medical fact that the area of the brain most susceptible to damage from head injury is the orbitofrontal cortex. Combine this with the well-known neurological fact that damage to the orbitofrontal cortex frequently results in disinhibited, impulsive behavior, poor decision-making, and a lack of emotional control.[13] Blend this with Bustamante's PET scan, revealing reduced orbitofrontal functioning. Consider that his crime was impulsive, not planned. While it was vicious, it was also very unsophisticated. His homicide was followed by disorganized, thoughtless actions. He made no attempt to cover his tracks.

You don't have to be a Sherlock Holmes to deduce that it was the head injury at twenty—well beyond his control—that likely caused his poor prefrontal functioning and the later impulsive, violent offending. Even the plodding Doctor Watson with his nineteenth-century medical knowledge would likely have come to the same conclusion. But is this scenario true of all killers?

THE BRAIN OF A SERIAL KILLER

In striking contrast to Antonio Bustamante we have our other killer, Randy Kraft. You'll recall from Randy's early life history that we see nothing extraordinary. He grew up as an all-American boy in conservative Orange County in Southern California—not exactly the greatest risk factor for violence.

Would Randy have the same prefrontal impairments that we saw in Bustamante? Think about it. The selection of the victim. Working out how to orchestrate the evening beginning with friendly drinks. Being able to booze and schmooze without losing executive control over the situation. Timing the point to strike. The escalation to drugging the victim. Ensuring that he is well bound and cannot escape. All those bodies to get rid of. All that mess to clear up. Working on murder into the early hours only to show up for work the same morning and put in a hard day's computing.

How did he do it? You can see for yourself in Randy's brain scan. Take a good look at Figure 3.3, in the color-plate section, and focus on the three scans in a row. On the left you have the normal control, on the right you have a single murderer, and in the middle you have Randy Kraft—labeled "Multiple Murderer." Check out the difference between Randy's brain and the single murderer. What stands out is that he does not have reduced frontal functioning. Instead, that part of the brain is lit up like a Christmas tree.

To me, Randy is the exception that proves the rule. Here we have a man capable of killing approximately sixty-four people in a twelve-year period without getting caught. You have to have good prefrontal functioning to pull that off. He had an excellent ability to plan, to regulate his actions, to think ahead, to consider alternative plans of action, to sustain attention, and to keep on task. It's exactly what you need to be a successful serial killer. He's an exception in that he differs from other killers in his brain profile. He proves the rule that a *lack* of frontal functioning results in a *lack* of ability to plan, regulate, and control one's impulses, resulting in not just homicide but early apprehension.

Let's look further into Randy's mental makeup, and piece together why he succeeded in staying so successful in slaying while other killers are caught more quickly. To begin with, in stark contrast with Antonio Bustamante, who had *twenty-eight* arrests before his homicide, Randy Kraft had *almost nothing* in his criminal record before he was apprehended. It was almost as clean as a whistle, and what little there was is illuminating. Let me elaborate.

This story starts in the summer of '66. It was the summer of Speck—the summer that Richard Speck was killing nurses in Chicago. It was also the summer of a historic first that I will never forget. It was the one and only time that England won the World Cup in soccer. I was twelve and Randy was twenty-one. He was also never to forget that time, but for a different reason. It was his first police bust.

Randy was taking a stroll at Huntington Beach just south of L.A.

and propositioned a young man on the beach. Unfortunately for him the young man was an undercover police officer. Randy was charged with lewd conduct but nothing came of it, even though it was duly recorded. That's because he was told something that many first-time offenders are told: "Just don't do it again."[14]

I suspect this was a double message for Randy. The message said: (1) watch out, the police are about, and (2) smarten up your act, and you can beat the cops. Remember that this was five years before Randy's first known homicide. It was a scare that smartened him up in a way that his well-functioning prefrontal cortex could register. Poor frontal functioning results in poor social judgment, loss of self-control, and an inability to modify behavior appropriately. It was *good* frontal functioning that helped Randy to learn from his mistakes and adjust his careless behavior accordingly. Once bitten, twice shy.

And yet Randy still wanted sex. What's a man to do? Well, one adaptive strategy is to move from adults to adolescents—lower-hanging fruit that yields easier and more satisfying pickings, and a new sensual exploration of younger flesh. Given that there was also less chance of getting caught by an undercover vice officer, this is what Randy decided to do.

There are likely many victims in the four years since that initial sting that we'll never know about. The only one who lived to tell the tale was Joey Fancher. It was March 1970, and young Joey was just a wayward thirteen-year-old from Westminster, not far from where Randy was living in Long Beach. Joey had skipped school to race up and down the Huntington Beach boardwalk on his bicycle. There Randy clapped eyes on him. He gave Joey a cigarette, and perhaps having a sense of the kind of kid Joey was, asked him a question. Had he ever had sex with a woman? No. Would he like it? Yes! So off the two sped on Randy's motorbike, back to his apartment under the pretext of making young Joey's adolescent dreams of lovemaking come true.

The bike ride itself might have been a buzz for the boy as he clung on to this cool beach dude, but a bigger buzz awaited him. Once in the apartment Randy brought out the next enticement—a bit of dope. The boy felt woozy with the cannabis, so Randy—the benevolent host that he was—brought Joey just the thing that would wipe away that wooze. Four little red capsules with some Spanish sangria to wash it down. Now the boy was all Randy's, to fulfill his wildest wishes with. Kraft forced the disoriented boy to give him oral sex. Joey resisted, but would years later tell a jury, Kraft "put his hands on my head and forced me. I couldn't do nothing. Period. It was like I was a rag doll."[15]

Joey retched with the ejaculate in his mouth. Kraft then took him to his bedroom, placed him on the mattress, and sodomized him. You'd think that after taking a break to go to the bathroom Randy might have gotten the better of his overflowing emotion and backed off just a bit. Instead, he beat the boy mercilessly and sodomized him yet again. Joey the rag doll was passing in and out of consciousness in a drug-drenched haze. He could still feel the intense pain of the anal penetration. He wept with the physical and psychological torture. He vomited from the alcohol-drug mix. Randy made one more trip to the bathroom. This time he came back out and nonchalantly told the boy he was going off to work—as simple as that. Randy just left the apartment, as cool as a cucumber.

Herein lies the tragic moment. If the correct action had been taken, Kraft would have been removed from circulation. He would never have been able to continue his pedophilic impulses. But it was not to be. Joey got out of the house, crossed Ocean Boulevard in a haze, and was almost hit by a car. He just managed to make it across the road to a bar and appeal for help. A customer called 911 and Joey was taken to a hospital to have his stomach pumped to discharge the drugs and alcohol. Two police officers then returned with Joey and his family to Randy's apartment, where Joey had left his new shoes. There they found a hoard of seventy-six photographs, largely of men in various stages of orgasm.

You'd *really* think something would have happened, but it didn't. Joey was not much different from many other sexually abused children. Too ashamed of what had happened to him, Joey could not bring himself to tell the police and parents about the wretched rag-doll rape and beating at the hands of Kraft. It was too humiliating. Plus, the police had done their inspection without a search warrant. They did not charge Randy.

For his troubles Joey ended up that night getting a beating from his grandfather—who mercilessly used a board with a nail in it—for cutting school and almost losing his new shoes. This was on top of the intense pain from his bleeding and a torn rectum that took two weeks to heal, while he kept his lips firmly sealed on the rape.

As for Randy, I can imagine him carefully contemplating at the end of that evening how close he had come to conviction for pedophilic rape and assault. His prefrontal cortex was recognizing once again that he must be much more careful. The under part of the prefrontal cortex specializes in learning from experience and fine-tuning decision-making based on past experience.[16] Randy was contemplating how to proceed. Dead men tell no tales. From now on, he would leave no witnesses, and to our knowledge he made his first killing the following year.

Let's look back at Randy's brain in Figure 3.3 and compare it this time to the normal control. You can see more activation in the very middle—the thalamus—as well as excellent activation of the occipital cortex at the bottom and the temporal cortex at the side-middle area. You don't see as much activation in either the normal control or the one-off killer.

But we did see this in someone else who had a brain scan very much like Randy's. That scan is shown above Randy's in Figure 3.3. Take a look at this one and compare it to the three you see below it. Which one would you say it most approximates? It's not a perfect match, but it does seem more similar to Randy's than the others. Note the plentiful prefrontal activation at the top, the bilateral thalamic activation in the very middle, the occipital activation at the bottom, and the temporal lobe activation at the sides.

What's interesting about this brain scan is that it's my brain scan. As you noticed earlier, it's hard for me not to see parallels between Randy's life and mine, and the parallels go on. We both have flat feet and we both love tennis. Randy was one of the four top seeds in the Westminster High varsity tennis team. I was not as good, but I captained the tennis team at my college at Oxford University.

Randy also had an elder sister who was a primary-school teacher, just as I did—we were both influenced by that sisterly connection. At university I very much wanted to be a primary-school teacher and I was accepted for postgraduate teacher training at Brighton. I particularly wanted to teach eight-year-olds, because during university breaks I took children on holiday for a charitable trust. I had different age groups but felt I could connect with eight-year-olds. Randy also wanted to be a primary-school teacher and spent a semester working as a teacher's aide with third-graders aged eight and nine. Neither of us sustained our career goal. We've both been caught drunk in our cars in Southern California by the police, albeit under different circumstances. And we both have the same brain functioning.

Am I a serial killer? I've never been caught and convicted for homicide. Nor any other offense, for that matter, with the exception of smuggling moon cake from Shanghai into Melbourne in 2000, for which I was fined about $175. Might I have a brain predisposition to be a serial killer? Maybe. Does this similarity in scans demonstrate that brain imaging is *not* diagnostic? I'd like to believe so.

Clearly there are "normal" people like myself—and perhaps yourself—with "abnormal" brain scans. And by the same token, there are "abnormal" violent individuals who have quite normal brain functioning. We cannot use brain imaging as a high-tech tool to tell who's normal, who's a one-off killer, and who's a serial killer. It's just not that simple. Yet at the same time we are beginning to gain important clues as to which brain regions—when dysfunctional—could give rise to violence.

So there we have them. Bustamante, Kraft, and Raine. Three different individuals with different yet somewhat similar backgrounds and brains. We've seen that the prefrontal cortex is a key brain area that is dysfunctional in murderers. And while I'd like to emphasize that fact, the exception presented by Randy Kraft gives us pause. While we cannot read too much into one case study, such fascinating individuals do, as we're about to see, generate interesting hypotheses for further testing.

REACTIVE AND PROACTIVE AGGRESSION

Analyzing Randy's brain made us reflect upon an important distinction in violence research—between "proactive" and "reactive" aggression. This distinction has been around for a long time in the work of Ken Dodge, at Duke, and Reid Meloy, in San Diego. The basic idea is that some predatory people—the proactives—use violence to get what they want in life.

Randy was proactively aggressive. He carefully planned his actions, drugging his victims, having sex with them, and then impassionately dispatching them. Like a good computer specialist, he was methodical, logical, calculating, and an able trouble-shooter of problems. Proactively aggressive kids will bully others to get their money, games, and candy. There's a means to an end. Proactives plan ahead. They are regulated, controlled, and driven by rewards that are either external and material or internal and psychological. They are also cold-blooded and dispassionate. They'll carefully plan the heist they have been thinking through, and they'll not think twice about killing if need be. Quite a lot of serial killers fit this bill—like Harold Shipman, in England, who killed an estimated 284, most of them elderly women; Ted Kaczynski, the Unabomber, whose terror campaign was conducted with mail bombs; Peter Sutcliffe, who bumped off thirteen women in the north of England; and Ted Bundy, who carefully killed about thirty-five young women, many of them college students.

Flip the aggression coin and the other side to the Randy Krafts of the world are "reactive" aggressives. These more hot-blooded individuals lash out emotionally in the face of a provocative stimulus. Someone has insulted them and called them names. They've lent money and it has not been returned. They've been verbally threatened. So they hit back in anger.

Take Ron and Reggie Kray, two identical twins who grew up in east London and operated in the swinging '60s, the same time that Randy Kraft was operating in Southern California. Reggie Kray's killing of Jack "the Hat" McVitie was an example of reactive aggression. It went like this.

McVitie had said mean things about Reggie's schizophrenic twin brother, Ron. True, Ron Kray was fond of his food, and yes, he enjoyed exploring the boundaries of his sexuality. But there are more subtle ways of expressing these facts than to call him "a fat poof," as Jack "the Hat" did. Jack also owed the Kray twins a hundred pounds, which did not help things. Adding injury to insult, one night walking out of a Chinese restaurant, Reggie bumped into McVitie, who said, "I'll kill you, Kray, if it's the last fucking thing I do."[17] Now, that's not nice.

Reggie decided that that was going to be Jack McVitie's last supper. Later that night Reggie pushed a knife into McVitie's face and stabbed him to death in an explosive fit of pent-up anger. Reggie would have blown Jack's head off, but his .32 automatic jammed twice, so he had to use a knife instead. Reactive aggression is much more emotional and unregulated. So in this context, although they were both murderers, Kraft and Kray were more like apples and oranges.

Given this proactive-reactive subdivision, I decided to categorize our forty-one murderers into proactive, predatory killers and reactive, emotional killers. We scanned all sources for all the information that we could dredge up on our subjects—attorney records, preliminary-hearing transcripts, court transcripts, national and local newspaper stories, reports and interviews from psychologists, psychiatrists, and social workers, and of course rap sheets. We even interviewed some of the previous prosecution and defense attorneys for more information on the killings. In the end, we classified twenty-four murderers as "reactive" killers and fifteen as "proactive" killers.[18] In a number of the homicides there were elements of both proactive and reactive aggression, so they were left unclassified.[19] Think of a revenge killing, for example. Someone gets really upset by an insult, and in response they set about carefully getting their own back. They are indeed reacting to a slight, but they plan their sweet revenge carefully and thoughtfully, and obtain satisfaction in doing so—a psychological gain. They are not unlike terrorists who react to a sociopolitical, ideological insult by carefully planning a counterattack.

The results of our reactive-proactive comparisons are illustrated in Figure 3.4, in the color-plate section. Here you're looking down on the brain and the prefrontal cortex is again at the top. This time the subregion you see is called the ventral—or underneath—prefrontal cortex. The reactive, hot-blooded murderer has low prefrontal functioning in the ventral subregion. In contrast, the predatory, cold-blooded killer has just as much prefrontal activation as the normal controls. Like Randy Kraft, they've got the goods to make a cold, calculated kill. In contrast, the hot-blooded killers are not so hot when it comes to prefrontal regulatory activation.

We see here—even at a visual level—that homicide is nuanced. Yes, there is a cerebral basis to violence. And yes, the prefrontal cortex is one of the culprits. But even among the tiny proportion of us who kill there are differences. Our group of predatory, proactive killers features the same regulatory brain control as Randy. The brain anatomy of murder is color-coded on a reactive-proactive aggression spectrum.

PREFRONTAL CONTROL RELATIVE TO LIMBIC ACTIVATION

Wait a bit. If these predatory killers have relatively normal prefrontal functioning, what made them killers in the first place?

Let's plumb the depths of the murderous mind. Deeper down in the brain, well below the civilized upper crust of the prefrontal cortex, we arrive at the limbic system, site of the emotions, and the more primitive parts of our neural makeup. Here the amygdala fires up our emotions and stimulates both predatory and affective attack.[20, 21] The hippocampus modulates and regulates aggression and when stimulated sets in motion predatory attack.[22, 23] The thalamus is a relay station between the emotional limbic areas and the regulatory cortical areas. The midbrain when stoked up gives expression to full-blooded affective emotional aggression.[24]

We combined these regions to get an overall measure of subcortical activation in the reactive murderers, the proactive murderers, and the normal controls. We found that *both* murderer groups showed *higher* activation of these subcortical limbic regions than the controls, especially in the more "emotional" right hemisphere of the brain. Below the façade of the boy-next-door that many cold-blooded killers are able to portray, there's a lot bubbling under in that deeper subcortical cauldron of brain functioning.

What exactly is going on here? We can think of these deeper limbic emotion-related brain regions as partly being responsible for deep-seated aggression and rage, which both groups of killers have in common. The difference, however, is that the cold-blooded killers have sufficient prefrontal regulatory resources to act out their aggression in a relatively careful and premeditated fashion. They feel as angry as anyone, but instead of getting mad, they get even. In contrast, while the hot-blooded killers also have a mass of angry feelings simmering away, they don't have sufficient prefrontal resources to express their anger in a controlled and regulated fashion. Someone gets their goat, they see red, and they blow their lid. Before you know it, blood flows.

This seeming paradox of good frontal regulatory control and increased limbic activation in predatory, proactive killers can be exemplified by a number of serial killers. Take Ted Bundy, who may have killed as many as a hundred women and girls, mostly college students. His homicides were the epitome of planning. With his arm in a sling to make him look vulnerable, Bundy would politely ask a young woman to help him carry something to his car. Using his beguiling charm, good looks, and debonair manners, he would lure her to a safe place where with demonic fury he would tear into her—biting her buttocks, gnawing her nipples, and bashing her head in a sexual orgy that ended in a brutal beating and killing. Despite all the planning and forethought that carefully preceded his attacks, once that stealthy lion had stalked his prey, he unleashed with ferocious fury the ultimate attack. The emotional limbic cauldron was overflowing into an unbridled, unregulated killing.

The study I did with Monte, like all initial findings, requires replication and extension. Another study of eleven impulsive murderers also using the continuous-performance task replicated our findings of reduced prefrontal activation.[25] Yet because these studies are so hard to conduct, the reality is that virtually no other research group has been able to build upon and extend our initial findings on murderers.[26] For many researchers, linking the brain to homicide is a bridge too far. Nobody can cross it.

I take our instrumental proactive murderers as a model for serial killers, on whom we know very little scientifically. If I could perform brain scans on a significant group of serial killers, I might expect a brain profile similar to our proactively aggressive killers—a hotbed of seething limbic activation bubbling under the good prefrontal functioning that allows them to carefully plan their actions. Yet even within this pack of serial killers, make no mistake—there will inevitably be several shades of gray lurking in the etiological shadows.

THE FUNCTIONAL NEUROANATOMY OF MURDEROUS MINDS

We've seen that the prefrontal cortex is critical in regulating and controlling both behavior and emotion. We've also seen that excessive subcortical activity may fuel the heightened emotion that we see in our violent offenders. We could stop there in our mapping of the mind of the murderer. We have the essence here in a nutshell. Yet as I readily acknowledged above, the scientific reality would be that we are being overly simplistic. We get back to the complexity of homicide, psychopathy, and criminal offending, and the inevitability that any attempt to explain and understand such behavior through functional neuroanatomy—the workings of the brain—is going to be enormously complex. Here I'll give you just a piece of the exciting neuroanatomical action that is taking place today in our probing of the murderous mind.

Moving from the front of the brain, where we have been focusing, to the relatively less explored posterior part, we'll start with the angular gyrus—area 39 in the map created by the German anatomist Korbinian Brodmann in 1909. The angular gyrus lies in the inferior, or lower, half of the parietal lobe, above the superior temporal cortex, and in front of the visual cortex. It is consequently in a prime position in the brain, lying at the junction of three of the four major lobes—the parietal, the temporal, and the occipital cortices. It connects and integrates information from many modalities—visual, auditory, somatosensory, vestibular—in order to perform complex functions. It lies on the surface of the brain. Find the top of your ear with your fingers and move them up a couple of inches—1.5 inches behind that spot is about where the angular gyrus lies.

We imaged the angular gyrus in our murderers and found significantly lower glucose metabolism in this structure than in those of the controls. In Sweden, researchers also found reduced cerebral blood flow in this area of the brain in impulsive, violent criminals.[27] Other researchers have argued for angular gyral dysfunction in violent offenders as well.[28]

How might dysfunction of the angular gyrus translate to violence and offending? The angular gyrus is one of the latest areas of the brain to develop, and so, not surprisingly, the abilities it governs are complex and sophisticated. Unlike the visual cortex, which comes online immediately for the newborn infant, the angular gyrus subserves functions that include reading and arithmetic, abilities that as we know do not start early in life, but develop much later in childhood. So, for example, reductions in glucose metabolism in the left angular gyrus have been associated with reduced verbal ability,[29] while damage to this region in neurological patients results in problems with reading and arithmetic[30]—complex functions that involve integration of information across multiple domains. Writing ability is also affected in a subtle way. For example, letters may be missing or duplicated, or be widely spaced. Punctuation is off, and capital letters may be disregarded.

So if the angular gyrus is not functioning well, then a child's reading, writing, and arithmetic are going to suffer—the three R's that are the foundations of scholastic performance. What do we know about violent offenders? They do poorly at school. If you do poorly at school, you'll have a problem getting a job. You won't get as much money as you'd like. You'll then be more likely to use violence to get what you want in life—things you cannot get because of your educational failure. The root cause may be brain-based, but the path to violence may well lie along school and occupational failure—a social/educational process.

The hippocampus and its surrounding area, the parahippocampal gyrus, is another brain region that is disturbed in offenders. The hippocampus lies just behind the amygdala and its Latin name means sea horse. We've touched on this area above in connection with our sample of murderers, and other researchers are also finding that offenders have functional disturbances in this brain region. One study on antisocial, conduct-disordered boys from London showed reduced function of the hippocampus during an attention task.[31] In Sweden, the neuroscientist Henrik Soderstrom found reduced hippocampal functioning to be associated with higher psychopathy scores in violent offenders.[32] In the United States, Kent Kiehl has argued that the parahippocampal gyrus contributes to symptoms of psychopathy.[33] Researchers in Germany led by Jürgen Müller also found reduced parahippocampal functioning in adult psychopaths,[34] while Daniel Amen in California found the same finding in impulsive murderers.[35]

We need to ask why hippocampal impairment would make an individual more likely to offend. For one thing, it makes up part of the emotional limbic system. We know in turn that psychopaths and other offenders have abnormal emotional responses. The hippocampus is also part of the neural network that forms the basis for the processing of socially relevant information, and it is involved in recognizing and appraising objects. Disruption to such a system could in part relate to the socially inappropriate behavior shown by some violent individuals, as well as the misrecognition and misappraisal of ambiguous stimuli in social situations that can result in violent encounters.[36]

The hippocampus is critical for learning and memory. It's one of the first areas to go in people with Alzheimer's disease. With my longtime colleagues Rolf and Magda Loeber in Pittsburgh I studied the ability of schoolboys to remember both verbal material and nonverbal, visuospatial material. The result? Boys who had been persistently antisocial from the age of six to sixteen as rated by their parents and teachers did more poorly on these hippocampal memory tasks than controls.[37]

We also know that the hippocampus plays a role in fear conditioning, and as we'll see in a later chapter, antisocial and psychopathic individuals have a particular deficit in this form of learning. Psychopaths are fearless individuals, as are many other violent offenders. It's worth noting that researchers from Italy and Finland have found a structural abnormality in the hippocampus of psychopaths, which plays an important role in fear conditioning and emotional responding.[38]

Yet there's more to the hippocampus than memory and ability. It is a key component in the limbic circuit that *regulates* emotional behavior,[39] and it has been implicated in aggressive, antisocial behavior in both animals and humans. In animals, it regulates aggression through its connections to deep structures in the middle of the brain, including the lateral hypothalamus and what's called the periaqueductal gray, structures important in controlling both defensive rage attack and predatory attack.[40] So a poorly functioning hippocampus will be of little help to either an offender who is beginning to fly off the handle in the first stage of an argument, or one who is seeking revenge.

Another brain area that is believed to be dysfunctional in offenders is the posterior cingulate, lying more toward the rear of the head and deep inside the middle of the brain. This region has been found to be poorly functioning in adult criminal psychopaths,[41] conduct-disordered boys,[42] and aggressive patients.[43] Because this brain region is also important in the recall of emotional memories[44] and the experiencing of emotions,[45] a disturbance to this area will likely result in a disturbance in emotion, including causing anger. We also know that the posterior cingulate is involved in self-referential thinking—the ability to reflect back on oneself and understand how one's behavior can affect others.[46] So if a psychopath fails to understand how his actions can harm others, this could help explain his thoughtless, antisocial acts and his failure to accept responsibility for his actions.

A NEW EXCUSE FOR WIFE ABUSE?

Killing is one thing. Striking your wife across the face is another. The trouble with research like mine on murder is that killing is very rare. What about more common acts of serious violence like spousal abuse?

Of course, I'm not saying that spousal abuse is trivial by any means, but it's far more common than homicide. Are spouse-abusers different from killers in brain functioning? Or can we discern similar patterns in these common-variety offenders? To help answer that question, let's take a trip to Hong Kong.

It's a fantastic place. I took my family there when I was on sabbatical at Hong Kong University. People were so sweet and polite. The very first morning that I took my two young boys, Andrew and Philip, to Victoria Kindergarten in the Fortress Hill area, we were stopped in the street by a young woman. She asked if she could help hold the boys' hands. Well, why not? So off we all marched, hand in hand to preschool, where she duly said good-bye to the boys, thanked me, and vanished into thin air amid the bustling streets.

Strange, isn't it? Maybe she was a nutcase, but I don't think so. She was a smartly dressed professional. To her, my two-year-old tots were cute curiosities, decked out in their red school blazers, gray trousers, satchels, and mixed Asian-and-Caucasian faces. It was typical of the graciousness, courtesy, and respect for the family and children that Hong Kongers have.

Yet lurking beneath that civilized façade lies the cruel visage of domestic violence. I did a survey of 622 Hong Kong undergraduate students. They were not all rich kids by any means, but they were largely from the privileged classes. You don't expect much to have gone on in their homes in their formative years. But I nevertheless asked them how their parents dealt with conflicts before the kids were eleven— *before* they could turn into troublesome teenagers. Sixty-two percent had parents who would insult or swear at them, 65 percent had parents who would do or say something just to spite them, while 48 percent were slapped or spanked.

No big deal, you'll say, if you remember being on the receiving end of a good spanking or two as a child. Surely this happens in the best of homes. But let's get beyond the simple stuff. Fifty-one percent went on to admit that their parents would hit them with an object. Forty percent were physically beaten. Six percent had actually been *choked,* while 5 percent had been deliberately burned or scalded. Seven percent had even been threatened with a knife or gun. In all cases it was their own parents perpetrating the abuse. So how often did *your* parents choke and burn you or put a gun to your head before you turned eleven?

Serious domestic violence was pretty rampant even in the homes of these educated, better-off undergraduates. True base rates are likely a lot higher, since people forget what really happened after ten years. Plus, you never want to admit—even to yourself—that you had parents bordering on the sadistic and inhuman. Some of these kids were having the living daylights beaten out of them—some repeatedly—behind closed doors. And these are the better-off kids. Heaven knows what was happening—and still is going on—to kids from much poorer homes in Hong Kong.

And where the kids are getting beaten, the wives are being bashed. Today it's hard to believe, but until about 1980 spousal abuse was hidden under the carpet at home.[47] A man who gave his wife a belting was not considered a criminal; such treatment was part and parcel of everyday married life. Even after the recent criminalization of spousal abuse, wife battering is still rife. The prevalence of spousal abuse each year is approximately 13 percent in the United States, with an estimated 2 million to 4 million victims a year.[48] It accounts for about half of all female homicides and is a leading cause of injury to developing fetuses.[49] It's a shocking, disgraceful, criminal offense, and yet it's all too common and frequently tolerated in some households.

Let's face up to these spouse-abusers. If we can look beyond their eyes and into their brains, do these men also have a dysfunctional cortex? They batter women, but is that because they have battered brains?

Tatia Lee is a brilliantly creative clinical neuroscientist at Hong Kong University with a penchant for sailing into uncharted waters. She conducted some of the very first brain-imaging work on lie detection, and she was just a couple of doors down from my office during my time there in 2005. Together with her graduate student, we teamed up to test our ideas on spousal abuse. We recruited twenty-three men referred by police to social-welfare departments and psychology practices for physically abusing their wives. Our main hypothesis was that such men may overrespond to emotional stimuli, and that that may in part be a cause of their abuse. We measured their reactive and proactive aggression and also gave them two verbal and visual emotion tasks.

The verbal task is called the emotional Stroop task. The subject is first presented with the name of a color, like "blue." They then see an emotionally negative word like "kill," which is either printed in blue or another color, and have to judge whether the color of the word "kill" was blue or not. The same thing is done with nonemotional words, like "change." We then measure how long it takes them to respond. People who take longer to respond to the emotional word than to the neutral word are showing a cognitive bias to negative affect stimuli— meaning that the negative emotional nature of the word has hijacked their brain's attention and slowed down their responses.

In the visual task, the subjects viewed neutral pictures like a chair and also emotionally provocative pictures—things like a man holding up another man in a robbery with a gun to his head, or a man holding a woman from behind with a sharp knife across her throat. In both of these verbal and visual tasks we scanned their brain using functional magnetic resonance imaging (fMRI). Our research resulted in fourfold findings.

First, spouse-abusers were strongly characterized by reactive aggression—where the individual responds aggressively in the face of provocation. In contrast, once we controlled for this, the spouse-abusers showed no proactive aggression. They were not using aggression in a planned, premeditated, manipulative fashion.

Second, in the emotional Stroop task, the spouse-abusers were slower in responding to emotional words. Negative emotional stimuli were capturing their attention much more than normal.

Third, in functional brain scans during the emotional Stroop task, our spouse-abusers showed much greater activation of the emotional amygdala to negative-emotion words, together with less activation in the regulatory prefrontal cortex.

Fourth, when batterers saw pictures of visually threatening stimuli, they showed hyper-responding in widespread brain areas covering the occipital-temporal-parietal regions. These regions are exceptionally sensitive to the recognition of objects[50] and to spatial perception.[51] They indicate that batterers experience greater visual arousal when exposed to threatening stimuli.

Putting these four findings together, a pernicious pattern unfolds. Spouse-abusers have a reactive aggressive personality that makes them more likely to lash out when provoked. Emotional words inordinately grab their attention. They are less able to inhibit the distracting emotional characteristics of stimuli, resulting in impaired cognitive performance. When presented with aggressive stimuli their brains overrespond at an emotional level and underrespond at a cognitive control level. Spouse-abusers are constitutionally different from other men.

These neurocognitive characteristics of batterers may partly contribute to their abusive behavior. Some researchers have documented that batterers do not listen to reason, and instead emotionally react out of all proportion to a situation.[52] Excessive attentional processing to a visual stimulus like a frown or a scolding voice may distract the batterer's attention and make him misinterpret the social interchange. It could contribute to the racing thoughts, irrational behavior, and escalating negative emotion that characterize wife-batterers.[53]

To my knowledge, these are the first physiological studies of *any* kind to show brain abnormalities in spouse-abusers when reacting to emotional stimuli, and the first to demonstrate hyperreactivity to threatening stimuli. Our findings challenge an exclusively social perspective on spousal abuse and suggest instead a neurobiological predisposition to battering. Historically, the prevailing clinical perspective has been that spousal abuse is a conscious, deliberate, and premeditated use of power to subjugate and control the female partner for selfish instrumental gain.[54] An alternative hypothesis that Tatia and I suggest is that spousal abuse has a significant brain-based reactively aggressive component.[55]

Is this a newfangled excuse for wife-abuse? I'm not exactly saying that abusers are not to blame. And I'm not saying that all abusers are like this. But I do think we need to recognize that there's more to domestic violence than the traditional feminist perspective cares to admit. Feminists argue that the cause of spousal abuse lies in a patriarchal society that sanctions men's using physical power to control women. We argue instead that neurobiology nudges some men to overreact at home and that we need to consider a contribution by the brain to spousal abuse. Why? Because traditional treatment programs to treat spouse-abusers based on the feminist perspective simply do not work.[56] We need to incorporate neurobiological perspectives into domestic-abuse treatment programs if we genuinely want to eradicate this completely unacceptable behavior of men toward women.

THE LYING BRAIN

So far we have been talking about people who are characterized by the media as brutes, monsters, and villains. We have been discussing despicable deeds that include murder, child rape, and wife-battering. And you may be sitting there dispassionately reflecting on how this other half lives, and what exactly makes these mean men tick.

But what about you? What's ticking away inside you when you perpetrate an antisocial act? Oh, so you're not antisocial? You really think that? Well, not perhaps antisocial at the level that we have been discussing so far, but let's turn to two arenas that will be much more familiar to your daily experience than murder and spouse-battering. You're not perhaps so wonderful after all.

Let's start with lying. And please do not protest your innocence any further, because as Mark Twain rightly put it, "everybody lies—every day, every hour, awake, asleep, in his dreams, in his joy, in his mourning."[57] You do lie—honestly you do. So how do we probe your antisocial mind? What instruments can we use to detect when people are telling whoppers?

"Oh, Agent Starling, you think you can dissect me with this blunt little tool?" Hannibal Lecter in the classic thriller *The Silence of the Lambs* had a point, and Clarice Starling, the FBI agent interviewing him, should have known better. The paper-and-pencil questionnaire tools she was using on the serial killer Hannibal "the Cannibal" Lecter in his prison cell have been traditionally used by forensic specialists to probe the minds of murderers. But they have been ineffective in revealing much that is fundamentally wrong with psychopaths like Lecter. After all, psychopaths have been known to tell a white lie or two about themselves, so do you really think they will tell the truth in a simple questionnaire? We need something far sharper than a blunt pencil and paper questionnaire to learn when people are fibbing.

A big fat sixty-ton magnet of the type used in MRI does not sound very sharp, but it's not a blunt tool. When it comes to discerning truth from fiction, it's as sharp as a razor. My academic friends Tatia Lee, at Hong Kong University, Sean Spence, at the University of Sheffield,[58] and Dan Langleben, at the University of Pennsylvania, are a triumvirate of pioneering scholars who each independently stumbled onto a sublime truth about lying—the prefrontal cortex is critical.

Tatia Lee took normal individuals—just like you—and put them into a scanner. She then gave them tasks during which they had to either tell the truth or lie. Sometimes they lied about themselves, just as we do in life. So a question might be, "Were you born in Darlington?" "Yes," I would say. "No," you would say. We are telling the truth. And while that is happening, Tatia collects data on what the brain is doing. Then she reverses the situation. "Were you born in Darlington?" "No," I say. "Yes," you say. This is an autobiographic lie—similar to when you sometimes lie to your friend about whether you are free to meet up tonight or not.

In another task, subjects were given a simple memory task to complete in which a three-digit number—like 714—was quickly followed by either the same or a different set of numbers. The subject had to say whether the sets of digits were the same or different. Sometimes they were instructed to tell the truth, while at other times they had to deliberately lie and feign memory impairment—just like some people feign injury after a claimed accident to financially gain from medical insurance.

It did not matter what the task was, Tatia found that lying was consistently associated with *increased* activation in the prefrontal cortex as well as areas of the parietal cortex.[59] At just the same time as Tatia was doing her work in Hong Kong, Sean Spence[60] and Dan Langleben[61] independently found essentially the same pattern of findings in England and the United States, results that span three different continents and cultures. In stark contrast, telling the truth was not associated with *any* increase in cortical activation.

What's going on here? The bottom line to deceit is that this anti-social act is a complex executive function that requires a lot of frontal lobe processing. Telling the truth is actually very easy. Telling tall tales is much harder and requires much more processing resources and brain activation. Deception involves theory of mind. When I lie to you about where I was at eight p.m. on Wednesday, January 27, I need to have an understanding of what you know about me—and what you do not know. Was I really celebrating my birthday with my family? I need to have a sense of what you think is plausible, and what is not. For this "mind reading" we need to recruit a number of brain regions that form connections between the frontal cortex and subregions in the temporal and parietal lobes.

Yesterday it was paper-and-pencil tools. Today it's becoming brain-imaging paraphernalia. By combining brain-imaging method-ology with machine learning—equally new sophisticated statistical techniques—Dan Langleben and Ruben Gur, at the University of Penn-sylvania, have been demonstrating accuracy rates upward of 88 percent in detecting deception. The disconcerting question is, How much lon-ger will our lying minds remain stubbornly private to the latest inves-tigative lie-detection tools? The current view is that lie detection based on functional imaging is not sufficiently developed for use in courts of law,[62] although that could conceivably change in the future. For now, however, let's turn to another antisocial arena that we frequently find ourselves caught up in and conflicted by—making moral decisions.

COMPARING YOUR MORAL BRAIN WITH THE ANTISOCIAL BRAIN

You know cannabis is illegal, but you've taken it anyway. You know you should not download movies from the Internet, but you persist in breaking copyright laws. And now you are reporting your taxes and wondering if you should nudge up those tax-deductible charitable con-tributions a hundred or two.

We've all had those moments of being torn between right and wrong—between heaven and hell. The devil and the angel are bat-tling it out hell-for-leather inside our hot heads, beating out the even-tual choice with hammer and tongs. You've wondered what on earth to do.

But you've never wondered what's going on inside your brain during these moments, have you? That's what a lot of social scientists and philosophers have been pondering for over a decade. And now we have some fairly clear-cut answers.

It goes like this. We slot you inside a brain scanner and present you with a series of moral dilemmas using visual scripts. We'll start with what is called a "personal" moral dilemma—one that's really up close and personal. This one could almost have been plucked from a page in the life of Phineas Gage, a railway worker whom you'll meet in a later chapter. You're standing on a footbridge looking down on a railway track. Below you, farther back along the track, is a runaway trolley that is about to plow into a group of five unsuspecting railway men working farther ahead on the track. Standing next to you is a rather corpulent gentleman.

Here's the deal. If you do nothing, five innocent men are going to die right before your eyes. Alternatively, you can push the big bloke off the bridge. He's a goner, but his big body will block the runaway trolley and save the lives of five men. What do you do?

You only have two choices. You are out there on that bridge, hearing the death rattle of the oncoming trolley and envisioning the gory carnage that will occur. No, you're not allowed to throw yourself off the bridge instead—saint that you are. You're just not big enough to block the trolley. Calling out to the railway workers won't work either.

Put this book down and reflect on your decision—to do nothing or to push the man off the bridge.

It's difficult, isn't it? And we can push and pull our minds in different directions. Are you really going to stand idly by and let five innocent men die? Look, the obese guy is likely to die early from heart disease anyway—why not give his life a dignified and worthwhile ending by saving five innocent men?

Then again, isn't it sort of wrong to kill? But at the same time it's five for one—surely you cannot ignore those odds? This dilemma is damned difficult—it's very personal and involves a high degree of conflict.

Josh Greene, an amazing philosopher and neuroscientist at Harvard, published the first study to describe what happens at a neural level during personal moral dilemmas like this.[63] Compared to more "impersonal" moral dilemmas that do not bring you face-to-face with someone else, your brain shows increased activation in a circuit that comprises the medial prefrontal cortex, the angular gyrus, the posterior cingulate, and the amygdala. This makes sense, as these brain areas contribute to complex thinking, and the ability to step outside of yourself and evaluate the bigger social picture.

But let's get back to how you actually processed the dilemma. I'm not as interested in exactly what decision you came to as I am in how you *felt*. Wasn't it awkward? Didn't you feel uncomfortable? You may have even physically squirmed in your seat a bit just as one undergraduate student did in my class earlier this week when I described this dilemma. This is where that amygdala and other limbic activation comes in, contributing to the emotional "conscience" component of moral decision-making alongside some subregions of the prefrontal cortex.

What your actual answer was is not entirely uninteresting either. About 85 percent of you felt you could not bring yourself to push that man off the bridge. About 15 percent, however, would have sacrificed him. These numbers are obtained in large-scale surveys of moral dilemmas. In contrast, if you put the same question to patients who have lesions to the ventral prefrontal cortex—people who as we'll later see are more psychopathic than the rest of us—that "push-him-off" rate triples to about 45 percent.[64]

If these same patients with ventral prefrontal lesions are with other villagers hiding in a cellar from invading troops above, and if their baby starts crying, they are three times more likely to smother their baby to prevent the enemy from finding and killing everyone. This is a high-conflict dilemma. They are making a utilitarian moral decision—the greater good of the greater number.

Don't worry too much if you chose to push the man off the bridge or smother your own baby. The seventeenth-century English philosopher Jeremy Bentham, who espoused utilitarianism, would have been proud of you. It does not necessarily mean you have a frontal brain lesion or that you are a psychopath—although you may have a slightly different way of thinking about life than others.

Josh Greene was not able to image the ventral prefrontal cortex back in 2001 when he conducted his groundbreaking study, due to what we call "susceptibility artifact," but many other studies have replicated and extended Greene's findings and shown activation of this region during moral-dilemma tasks.[65, 66] The ventral prefrontal region is critical for making "appropriate" moral decisions—or at least passive decisions that result in no harm to others.

We'll come back to morality very soon, but here I want to recap where we stand with our murderous minds. I've been arguing that the prefrontal cortex and limbic system are misfiring in violent offenders. We also found that our murderers had poorer functioning in the angular gyrus. We've seen that other studies of antisocial individuals reveal abnormalities in the posterior cingulate, the amygdala, and the hippocampus, while others document abnormal functioning in the superior temporal gyrus in violent offenders,[67] psychopaths,[68] and antisocial individuals.[69]

Let's now compare this hit list of brain areas in antisocials to the hit list activated when normal people contemplate a moral dilemma. What are the areas most commonly activated across studies in moral tasks? They are none other than the polar/medial prefrontal cortex, the ventral prefrontal cortex, the angular gyrus, the posterior cingulate, and the amygdala.[70] There is an undeniable degree of overlap.

Let me make the point visually for you. Figure 3.5, in the color-plate section, puts together these two sets of findings—the antisocial brain and the moral brain—to create a neural model of morality and antisociality. The top scan slices the brain right down the middle from front to back—you can see the nose on the left. The middle scan slices the brain head-on. The bottom slice is a bird's-eye view looking down on the brain. Brain regions implicated in both offending and moral decision-making are colored yellow. Areas found to be abnormal only in offenders are colored in red, and areas linked only to moral-judgment tasks are colored in green.

You can see that there are substantial areas of overlap between antisocial/psychopathic behavior and making moral judgments. Brain regions common to both include the ventral prefrontal cortex, the polar/medial prefrontal areas, the amygdala, the angular gyrus, and the posterior superior temporal gyrus.

It's not a perfect match by any means. Furthermore, while the posterior cingulate is activated during moral judgment, evidence implicating this region in antisocial behavior is sparse to date, although studies have indeed found abnormalities in the posterior cingulate in psychopaths,[71] impulsively aggressive patients,[72] and spouse-abusers.[73] Nevertheless, there are commonalities we cannot ignore. Some parts of offenders' brains critical for thinking morally just don't seem to be functioning very well.

JOLLY JANE'S VOLUPTUOUS BRAIN

We have been learning what brain areas are activated when normal people make moral decisions. But what happens in the brains of psychopaths when given the same moral dilemmas?

Historically, psychopaths have been viewed as "morally insane." On the outside they seem normal, and can even be very pleasant, sociable, and likeable. Ted Bundy is a classic example of a serial killer who had a charismatic personality that allowed him to lure young female victims into his deadly trap.[74] Yet when it comes to having a sense of morality, there is something missing in psychopaths. Here we'll take a closer look at what this "moral insanity" is like from a real-world case. What exactly is broken in the brains of psychopaths at the moral level?

My sister Roma was a nurse. My wife, Jianghong, is a nurse. My cousin Heather is a nurse. So allow me to pick the case of a nurse for our discussion of a breakdown in the moral brain. "Jolly" Jane Toppan cheerfully killed at least thirty-one people in Massachusetts during a six-year period, from 1895 to 1901. Like Randy Kraft, she was not caught for several years. Nicknamed "Jolly Jane" by hospital staff and patients due to her gregarious and happy demeanor, she became one of the most successful private nurses in Cambridge.

Jolly Jane liked to live life to the full. Like many serial killers, she enjoyed experimenting in her modus operandi and exploring her life-or-death power over others. Like many modern-day female offenders, she particularly took pleasure in experimenting with drugs—but in an unusual way. One of her greatest excitements in life was to see life itself slowly sucked out of the patients she cared for. She would first inject them with an overdose of morphine. She would then sit patiently with them, gazing into their eyes almost like a lover, observing the moment when their pupils contracted and their breath shortened.[75] Just when they were about to sink into a coma, Jane would revive them with a jab of atropine—an alkaloid extracted from deadly nightshade. It blocks the activity of the vagus nerve. This causes the contracted pupils to dilate, the slowing heart to beat rapidly, the cooling body to sweat, and shaking spasms to overcome the patient. Eventually they would die, but not before Jane had her high from observing their eyes dilate and watching their bodies contort in a slow death.

As with Randy Kraft, the only insight we have into what else Jolly

Jane would get up to during these murderous moments comes from the dramatic testimony of the one individual to survive an attack. Amelia Phinney was a thirty-six-year-old patient hospitalized with a uterine ulcer in 1887. Jolly Jane attentively floated around her like Florence Nightingale. The good nurse gave her patient a drink purportedly to help her pain—to Amelia it tasted bitter. Then Amelia felt her throat dry up, her body turn numb, and her eyes become heavy. She felt herself sinking into sleep.

At that point she became aware of something unusual—Jane was pulling back the bedsheets and getting into bed with her. Jolly Jane stroked her hair, kissed her face, and cuddled up to her. After a period of carnal embraces, Jane jumped onto her knees to peer deeply into her patient's pupils. She then gave Amelia another drink—presumably atropine to reverse the physiological symptoms of morphine. At that critical point, Jane abruptly disengaged. Amelia was aware of Jane dashing quickly out of the room—presumably because she heard someone approaching.

So Amelia Phinney lived to tell the tale, but not immediately. To this patient, the experience was so utterly bizarre that it must surely have been a dream during her ill state. Like Joey Fancher, who only testified a long time after his attack, in the court case of Randy Kraft, Amelia kept her bizarre story to herself. It came to light fourteen years later, after Jolly Jane's arrest, in 1901.[76] As with Randy Kraft, a serial killer who could have been caught, she continued on her killing spree.

Unlike many other female serial killers, who frequently kill for monetary gain, Jane was not profiting from her murders. The killings did, however, give her what she herself termed "voluptuous delight"— a shorthand nineteenth-century term for a sexual turn-on. Today she would be called a lust serial killer—which is very unusual for a female. Yet while Jane needed her sexual turn-ons, as a nurse aren't there other ways of getting such worldly pleasures? How could she morally justify her actions given the awful loss of innocent life?

It seemed almost motiveless malignity. It doesn't morally make much sense. And in fact, this is essentially how Jane herself sums it all up:

When I try to picture it, I say to myself, "I have poisoned Minnie Gibbs, my dear friend. I have poisoned Mrs. Gordon. I have poisoned Mr. and Mrs. Davis." This does not convey anything to me, and when I try to sense the condition of the children and all the consequences, I cannot realize what an awful thing it is. Why don't I feel sorry and grieve over it? I cannot make any sense of it.[77]

Jane could never understand herself. Nor could those who knew her. After her arrest a deluge of letters were received attesting to the fact that she was a compassionate, dedicated, and caring professional. She could not have committed these heinous deeds. If you look at her picture, in Figure 3.6, and peer into her eyes, can't you too see a gentle, kindhearted, motherly nurse?

Jane racked her mind for the cause of her crimes. She could gaze longingly into the eyes of her dying victims and experience her voluptuous delight while watching their agony. She knew what she was doing. She knew she was killing. Jane was utterly perplexed when at her trial in 1902 she was found not guilty by reason of insanity. To her mind, she could not possibly be insane because she knew full well what she was doing.[78] She truly could not make sense of it.

But I feel I can. And I literally mean *feel*. Jane knew cognitively what was moral behavior and what was not. Of course she could tell right from wrong at a thinking, cognitive level. But she did not have the *feeling* of what is moral. She could not empathize *emotionally* with the human suffering that resulted from her actions. She couldn't grieve or even feel sorry for her victims. I strongly suspect it was because she had a defective amygdala and ventral prefrontal cortex. She lacked the feeling for what is moral.

Figure 3.6 Jane Toppan

That moral feeling, centered on the amygdala and prefrontal cortex, is the emotional engine that translates the cognitive recognition that your act is immoral into behavioral inhibition. It holds you back from committing an immoral act, even though a part of you wants to move forward to gain your voluptuous pleasure. I submit that this emotional brake on immorality functions much more poorly in psychopathic individuals like Jolly Jane.

Jane could look into her victims and literally see them suffer. But what she could not see was her functional brain scan. There I believe she would have seen the faulty emotional wiring of her immoral brain that contributed to her killings. Jane died at the age of eighty-one, just before the outbreak of World War II, so of course I cannot test my theory. Yet if we look back at Jane's history, we can at least recognize the many social and psychological trappings of a psychopathic personality.

Jane was born into a desperately poor family of Irish immigrants. Her mother died when she was only a year old, and she clearly suffered from the type of maternal deprivation and breakage of the mother-infant bonding process found in the backgrounds of psychopaths.[79] Add to this a poor father who was mentally ill and could not care for his family, and a grandmother equally destitute and unable to care for the children. Jane was institutionalized until the age of five, and passed off as an Italian orphan because of the shame of being Irish. She was "adopted" into a home where she was treated as a servant girl.[80] With an early environment like this, the seeds of psychopathy grow rapidly.

The young Jane duly went on to exhibit the psychopathic traits of being sociable and charming, developing the reputation of being the life and soul of the party. She evidenced pathological lying and deception, weaving fanciful stories of her father living in China, her sister marrying an English lord, and the czar of Russia offering her a nursing position. She was a stimulation-seeker who also committed acts of petty theft against other nurses and patients. She conned and manipulated her hospital superiors. Among her victims were her own stepsister and her stepsister's husband. She was essentially superficial, with her surface joviality hiding a more disturbed, deep-seated personality disorder.

All of these characteristics are features of psychopaths,[81] and psychopathy provides a fertile ground for serial killers. While Jane gave detailed confessions on thirty-one murders, she claimed before she was locked away, in 1902, "It would be safe to say that I killed at least 100 persons."[82] Unless you are like Randy Kraft, who used a scorecard to keep an accurate tally, it's easy to lose count.

WHAT'S WRONG WITH JOLLY JANE'S PSYCHOPATHIC BRAIN?

So Jane was a psychopath. But would she have the type of brain functioning that might explain her moral insanity? While we cannot scan Jolly Jane's brain, we can scan the brains of her fellow psychopaths today and put them through the same moral dilemmas given to normal people.

This is exactly what my gifted graduate student Andrea Glenn did. We'd discovered that temp agencies are home to higher-than-normal numbers of psychopaths, a point we'll discuss in greater detail in chapter 4. Just like Josh Greene, Andrea confronted our subjects with personal, emotional, and moral dilemmas that involved harm to other people: Should you smother your crying baby to save yourself and other townspeople hiding from terrorists who would otherwise hear the sound and kill you all? We also presented subjects with less emotional, impersonal moral dilemmas: Should you keep money you found in a lost wallet?

Andrea found that individuals with high psychopathy scores showed reduced activity in the amygdala during emotional, personal moral decision-making.[83] While the amygdala, the neural seat of emotion, shows a bright glow in normal people when faced with emotion-provoking moral dilemmas, this emotional candle is barely flickering in highly psychopathic individuals.

Findings demonstrate that amygdala functioning is disrupted during moral decision-making in psychopathy and seems to be at its core. Without such amygdala activation, individuals may not think twice about conning and manipulating others. Just like Jane, they happily live out their immoral lives without feeling guilt or remorse. So when Jolly Jane manipulated others, stole their possessions, or thought about killing someone for frivolous reasons, she did not have that amygdala activation firing inside her to hold her back—no sense of shame.

Indeed, Jane's emotions were almost dead. Like a pathological stimulation-seeker, she was so removed from her natural feelings that she had to go to very extreme lengths to register a tangible feeling of "voluptuous delight." Consider the killing of Elizabeth, her sister-in-law. Jane confessed that she had deliberately prolonged her life so that she could witness more of her suffering: "I held her in my arms and watched with delight as she gasped her life out."[84] Cuddling and groping in bed with Elizabeth in the moments of her sister-in-law's death was just about the only way Jolly Jane could apparently be truly happy, and experience some sense of emotion in her life.

We know the amygdala is centrally involved in responding to cues of distress in others, thus guiding individuals away from antisocial behavior,[85] and we also know from work by the leading psychopathy researcher James Blair that psychopaths are less capable of recognizing negative emotions—including fear and sadness—in others' faces. So when Jane with her malfunctioning amygdala peered with intense curiosity into her victim's hapless eyes and felt their bodies, I think she was trying to register an emotion in the face of her victim. Was her patient experiencing fear? Was it sadness? Or perhaps it was pleasure? Jane's emotional brain and amygdala were desperately stumbling around, trying to work it all out. That voyeuristic experience piqued her curiosity while, simultaneously, she was devoid of any natural feeling that could give her cause for moral concern at what she was doing.

Andrea Glenn found that the medial prefrontal cortex, the posterior cingulate, and the angular gyrus were also dysfunctional in psychopaths during moral decision-making and were particularly associated with interpersonal features of the psychopath—superficial charm, lying and deception, egocentricity, and manipulation. These brain areas are also part of the neural circuit of moral decision-making and are involved in self-reflection, emotion perspective-taking, and integrating emotion into social thinking.[86] In turn, we can certainly see that Jane's social thinking was very disturbed. She could not take the emotional perspective of her victims. Try as hard as she might, she could not reflect and understand emotionally even her own behavior—she could not integrate emotion into social thinking. This partly explains her perplexing, psychopathic behavior. And given our brain-imaging findings on psychopaths, I suspect that Jane's aberrant behavior can be explained by a fundamental failure in the neural circuitry of morality. That's what I believe was egregiously wrong with Jane's psychopathic brain.

PIECING THE BRAIN TOGETHER

We have seen in this chapter that the violent brain functions very differently from yours. If we had to pick the area of primary difference, it would be the prefrontal cortex. We've seen how impulsive, reactive aggression can result from a lack of normal regulatory and inhibitory functioning. We have witnessed this in the reactive, impulsive, hot-blooded homicide committed by Antonio Bustamante. Being more regulated and controlled, proactively aggressive murderers do not have that same degree of prefrontal dysfunction, but like their reactively aggressive counterparts they do have a mass of limbic activity bubbling over in their brains that fuels violent, aggressive outbursts follows their careful planning.

We've also seen that there is not one but multiple brain areas which, when dysfunctional, can predispose one to violence. It's not just the dorsal and ventral regions of the prefrontal cortex that are dysfunctional, but also the amygdala, the hippocampus, the angular gyrus, and the temporal cortex. Yet future research will show it's even more complicated. The antisocial brain is a patchwork of dysfunctional neural systems and we are only just on the threshold of putting together these pieces to better understand it.

We've seen that poor brain functioning is not restricted to rare forms of violence. We've witnessed a frontal-limbic imbalance in relatively common forms of violence like domestic abuse—the overactivation of the amygdala combined with under-activation of the regulatory frontal cortex. Increasingly, the scope of functional brain imaging research is seeping into our personal lives. We are detecting a network of brain areas that unite in shaping the moral decisions we make on a daily basis—brain areas that are just not functioning normally in "morally insane" psychopaths and serial killers like Jolly Jane Toppan. These individuals lack the *feeling* of what is moral, and that partly accounts for their inexplicably egregious behavior.

But let's return to our point of departure. What do we really make of the horrific homicides perpetrated by Randy Kraft? We've seen how highly regulated and controlled this computing consultant was. Surely Randy had enough prefrontal control to keep his carnal desires in check. Randy was a heartless, cold-blooded killer—and I mean heartless almost literally. In our next stop through the body in this anatomy of violence we will leave the brain and travel to the heart of the matter—to the cardiovascular and autonomic nervous system.

4.

COLD-BLOODED KILLERS

The Autonomic Nervous System

Imagine committing a heinous crime that benefits you but brings harm to others. Putting a knife into a hateful husband who beats you. Strangling a belligerent boss at work. Breaking into a house at night and robbing it. Taking revenge on the man who stole your girlfriend. Embezzling millions of dollars from your company. Worse still, abducting, torturing, raping, and killing innumerable strangers, one by one.

Think hard about it, putting yourself into the actual situation. You've been drinking late at night on campus and your passion and mind have gotten out of control. Your girlfriend seemed bored with you, started making eyes at other guys, then gave some lame excuse to go. She jilted you right there at the bar. You'd really wanted sex with her that night and now you feel frustrated and angry.

You are walking back to your dorm and it's late at night. Then, not far ahead, you see a pretty student. You increase your pace to catch up, but keep a safe distance and walk softly so as not to make too much noise. As you come to the part of the path that breaks away from the buildings and moves into the trees, shrubs, and bushes, you catch up with her. You look quickly over your shoulder and no one is there. You grab her from behind. You place one hand on her mouth and shove her into the bushes and onto the ground. You take out a knife and threaten to take her life unless she performs specified sex acts with you. You rape her. You can hear and feel her heart beating, loud as thunder in her terror, and it turns you on. Then, with one hand over her mouth, you take the knife and stab her through the heart as you gaze into her eyes to watch her expression of utter and complete fear, see her pupils contract, feel her body writhe, and hear her breathing shorten.

After committing the crime you attempt to cover your tracks. But the next day the police arrive outside your door. You are arrested. You must create an alibi and stick with it as you are grilled by suspicious authorities, keeping track of the lies, knowing that one false move could send you to the death chamber.

What's going on inside you? What's going on inside an actual perpetrator? I want to argue in this chapter that you and a real-life criminal radically differ—or at least I hope you do. It's likely that you would perspire and your heart would quicken when you initially contemplated raping that girl, or during the interrogation. You may have been slightly nauseated just reading what I asked you to envision. Even the thought of it probably evoked negative emotions like disgust. But many violent offenders barely break a sweat when they violate the law, no matter how grave the transgression.

You have a conscience that was prickled at just the *thought* of committing the act, let alone the actual commission and completion. Others do not. You have a heart, while others are heartless. I'll argue here that your conscience is predicated on the good functioning of your autonomic nervous system, a part of the body sometimes referred to as the "visceral" nervous system due to its key role in emotion. The most important breakthrough in our understanding of this region of the anatomy of violence is that the nervous system of some offenders is simply not as "nervous" as the rest of ours. It confers on them a fearless, risk-taking, conscience-free personality that can result in criminal, violent, and even psychopathic behavior. They are biologically different from us. At the heart of this autonomic predisposition to violence, to which we first turn in this chapter, is the heart itself.

It may seem obvious to say that bomb-disposal experts and Theodore Kaczynski have something in common. Ted Kaczynski, otherwise known as the Unabomber, started off as a professor at the University of California, Berkeley, before embarking on his deadly career of violence, from 1978 to 1995. During this time he killed three people and injured twenty-three others with bombs he sent in the mail or placed on planes. His first target was Northwestern University, and then he moved on to the University of Utah, Vanderbilt, UC Berkeley, the University of Michigan, and Yale. He left bombs on an American Airlines flight and sent a mail bomb to the president of United Airlines. His success is marked by his years-long evasion of one of the FBI's most expensive investigations in its history.

His big mistake came when he published a 35,000-word manifesto in *The New York Times* and *The Washington Post.* He had threatened to kill again unless it was published, and both the FBI and the attorney general acquiesced. The manifesto ranted against industrialized society, leftism, and scientists, and how they were controlling society and restricting freedom. In a flash of bad luck for Kaczynski, his estranged brother picked up the newspaper and happened to recognize some unusual words and phrases, such as "cool-headed logicians." These phrases were reminiscent of letters he had received from Ted. Sections of the search warrant that allowed agents to search Kaczynski's remote cabin in Lincoln, Montana, document that even then, many FBI experts didn't believe that he was the author of the Unabomber manifesto. But all of that doubt was put to bed when FBI agents dropped in on him in 1996 and happened to notice a live bomb on the table together with the manifesto.

So, outside of the obvious, what would Ted Kaczynski have in common with a bomb-disposal expert? One key trait: in dealing with deadly contraptions both need nerves of steel and a certain degree of fearlessness. One British Army bomb-disposal expert working in Bosnia reflected on his job in this way: "It sounds dangerous but . . . I've not been in any situations where I felt in danger."[1] He is able to put that fear aside. Furthermore, both bomb-disposal experts and serial killers are intelligent. Kaczynski was a child mathematics prodigy who went to Harvard University at the tender age of sixteen. After gaining his PhD in mathematics from the University of Michigan, he was welcomed into a professorship at UC Berkeley. His IQ was above the genius level[2]—he scored 167 at age eleven.[3] Despite the despicable nature of Kaczynski's acts, he, like a bomb-disposal expert, was an intelligent and, in many ways, a highly rational individual.

But digging a bit deeper into the biology that our "test subjects" share, we find something else in common—a low resting heart rate. Some people kill in such a manner that we call them cold-blooded killers without thinking too much about the term. Yet what if this description turns out to be more literal than figurative?

In the anatomy of violence, the heart is a central organ orchestrating the tendency to antisocial and violent behavior. As we so often do in biology, let's start with animals. Rabbits who are aggressive and dominant indeed have lower resting heart rates than subordinate, nonaggressive rabbits.[4] Furthermore, when dominance in these rabbits is experimentally manipulated, heart rate goes down as dominance goes up. The same relationships have been found throughout the animal kingdom in macaques, baboons, tree shrews, and mice.[5]

Yet the idea that a low heart rate can raise the odds of someone becoming antisocial and violent may strike you as something too simple to believe.[6] In an age of powerful and sensitive diagnostic tools like functional brain imaging, there is something crude about linking violent behavior with a biomarker that is so astonishingly simple and easy to measure. Does this claim for the biology of crime and violence stand up to serious scientific scrutiny?

In my first research as a PhD student, at York University, in England, I found that a low resting heart rate characterizes antisocial schoolboys.[7] I found the same result when I moved to Nottingham University.[8] Maybe it was a fluke? So when I moved to the University of Southern California my colleagues and I conducted a meta-analysis of the heart rate–antisocial relationship. This involved us taking into account *all* studies we could find that had investigated this issue in child and adolescent samples.[9] We found forty publications, involving a total of 5,868 children. Pooling all studies gives you a much clearer view of the true picture.

What stood out clearly was that antisocial kids really do have lower resting heart rates.[10] We also looked at heart rate during a stressor—for example, while the subject was waiting for a medical exam. In the laboratory, kids would be asked to do a difficult mental arithmetic task like counting backward in sevens from 1,000. If you don't think it's all that stressful, just try it! In these cases involving stress, the overall differences become even larger.

In our meta-analysis, resting heart rate explained about 5 percent of the differences between subjects in antisocial behavior. That might not seem like much to you, but put into medical context, the relationship is strong.[11] It's much stronger than the relationship between smoking and lung cancer, or the effectiveness of taking aspirin to reduce the risk of death from a heart attack, or antihypertensive medication and reductions in strokes. Each of these are important and powerful relationships in the medical world, and in each case they are dwarfed by the strength of the heart rate–antisocial relationship.[12]

In fact, to get something as strong as the heart rate–antisocial relationship, you have to turn to the effect of nicotine patches in reducing smoking, or the ability of SAT scores to predict later college GPAs. If we now turn to resting heart rate during a stressor, this seemingly innocuous biomarker suddenly explains 12 percent of the variation that exists among us in antisocial behavior. This is as strong as the ability of mammograms to detect breast cancer, the accuracy of home pregnancy test kits, and success of sleeping pills in improving chronic insomnia. It's hard to ignore these medical relationships. It's equally hard to ignore the relationship between heart rate and antisocial behavior. It's clinically meaningful and significant.

It is not that low heart rate characterizes only one subgroup of antisocial kids. It applies to young as well as older children, and to girls as well as boys. So boys with low heart rates are more antisocial than boys with high heart rates. Girls with low heart rates are more antisocial than girls with high heart rates.[13]

However, heart rate may partly explain the gender differences in antisocial behavior. If you take your pulse using your watch, count the number of beats in one minute, and compare it with your opposite-sex sibling or partner's pulse, you will likely find that if you are female, your heart rate is several beats a minute higher than your male counterpart's. Males in general have lower heart rates than females; it's a robust finding.[14] There is the same sex difference in antisocial behavior. The sex difference in heart rate is in place as early as age three, with boys having a heart rate that is 6.1 beats a minute lower than girls.[15] This sex difference in heart rate starts just before sex differences in antisocial behavior begin to emerge.[16] The strong and replicated sex difference in heart rate provides one intriguing clue as to why men commit more crime than women—they have lower heart rates.

Let's shift from comparing genders to comparing generations. Twin studies have repeatedly found substantial heritability for resting heart rate.[17] They have also found that the offspring of criminal parents have low resting heart rates.[18] Given the fact that there is significant heritability for childhood aggression and adult antisocial behavior, and given that there is transmission of antisocial behavior from parent to child, low heart rate may be one of the heritable mechanisms that account for the transmission of antisocial behavior from one generation to the next.

A lot of studies have measured heart rate and antisocial behavior concurrently, at the same point in time. But a stronger design would be to assess heart rate early in life—and then show that it's related to antisocial behavior at a later age. That's called a prospective longitudinal design. Five such longitudinal studies from England, New Zealand, and Mauritius have indeed confirmed that low heart rate in childhood—as early as age three—is a *predictor* of later delinquent, criminal, and violent behavior.

Now, it's important to note that these studies do not demonstrate causality, and nobody is arguing that we can tell who exactly in a classroom of kids is going to become antisocial on the basis of heart rate alone. But it's a factor, and by teasing out the temporal ordering of the variables in question through research that follows young children into adulthood, we move one step further in support of the causal model that low heart rate early in life raises the *odds* of someone becoming a future offender.

Could it be that social factors cause both crime and low heart rates—giving the false impression that low heart rates cause crime? David Farrington, of Cambridge University, one of the world's leading criminologists, examined this issue in establishing the best independent early predictors of convictions for violence. He found that out of forty-eight predictors (family, socioeconomic position, academic attainment, and personality—everything from low social class to low IQ to impulsivity), only *two* were related to violence independent of all other risk factors: low resting heart rate and poor concentration.[19] Indeed, low heart rate was even more strongly related to measures of violence than having a criminal parent—one of the best social predictors of later crime.[20] These findings led Farrington to conclude that "low heart rate may be one of the most important explanatory factors for violence."[21]

Let's look at this relationship coming from the other direction. While a low heart rate raises the odds that someone will become antisocial, a high heart rate actually *reduces* the odds of later crime. I conducted a study of English schoolboys who were antisocial at age fifteen but who desisted from adult crime at age twenty-nine. I then matched them against seventeen antisocial adolescents who had become criminal by age twenty-nine and also with seventeen non-antisocial, non-criminal controls. The ones who desisted from crime had significantly *higher* resting heart rates relative to both criminal and control groups, indicating that a high heart rate protects against adult crime.[22]

On the treatment side, medications like stimulants that raise heart rate reduce antisocial behavior.[23] Studies are also showing that heart rate may help predict which children will benefit from therapy—and which won't. One study from Germany found that children who before treatment had low heart rates were *less* responsive to behavior therapy.[24] Interventions may be more effective in antisocial children with normal or high heart rates in whom the causes of their antisocial behavior may be more environmental than genetic. Knowledge of resting heart rate may not just help predict which children are more at risk for later criminal behavior, it may also provide invaluable knowledge in treatment programs.

Again on the medical side, one of the big problems is that it's nearly impossible to find a biomarker that is diagnostically specific to just one psychiatric disorder. For example, there are many biological correlates of depression, but they are also found in patients with anxiety and other mental illnesses. An unusual and important feature of the low heart rate–antisocial relationship is its diagnostic specificity. While other psychiatric conditions, including alcoholism, depression, schizophrenia, and anxiety disorders, have if anything been linked to a *higher* resting heart rate, no psychiatric condition other than conduct disorder—i.e., antisocial and aggressive behavior—has been associated with a *low* resting heart rate.[25]

The above studies have largely focused on violent criminals, psychopaths, and conduct-disordered children. But how much of a transgressor do you have to be to have a slower heart rate?

I was pondering this issue on the sabbatical I took with my family at Hong Kong University. In Hong Kong it is rare for pedestrians to cross the road on a red light, even when the coast is clear. But there are always a few who do. Whenever I would take my boys out to the park, we'd inevitably come upon a crossing and they would see some adults breaking the rule. They would point at them and call them "naughty penguins"—after Pingu, a cartoon they watched about an adventurous but mischievous little penguin. So it occurred to me—do naughty penguins also have low resting heart rates?

With the help of eight undergraduates, I collected heart-rate data on 622 Hong Kong students and asked them about their habits, including how many times they ever crossed the road on a red light. We found a difference. It was not big, just two beats a minute, but it was statistically significant and in the right direction. Naughty penguins really do have lower resting heart rates! Of course, this minor infraction is just the tip of the antisocial iceberg, but it indicates that low heart rate covers the whole spectrum of antisocial acts down to the smallest transgression.

Taking all these points together, it's hard to deny that a true, replicable relationship exists between low cardiovascular arousal and violence. When one line of scientific evidence supports a hypothesis, it is persuasive. But when many separate lines of evidence from different perspectives converge on the same conclusion, the argument becomes truly compelling.

Indeed, this body of evidence has raised the intriguing possibility that low heart rate could be considered a *biomarker* for the diagnosis of conduct disorder.[26] Currently, conduct disorder and almost all clinical disorders like schizophrenia are defined not in terms of biology, but in terms of symptoms that are obtained in an interview with a clinical practitioner. So the clinical symptoms of conduct disorder are things like lying, stealing, fighting, and cruelty to animals. These are all behavioral in nature and rely on subjective verbal reports from caregivers of the children themselves. There are two good reasons that biomarkers are not included in psychological diagnoses. First, they are not found to be diagnostically specific—they apply also to other disorders. Second, in everyday practice it's not that easy for a doctor to scan a patient to assess brain functioning—to say nothing of the extra financial burden scanning would present.

Heart rate is different on both counts. It *is* diagnostically specific, and it is extremely cheap and quick to assess. Think of it yourself. What happens first when you go to your doctor's office? You have your blood pressure and heart rate taken. Adding an objective biomarker to a subjective diagnosis is the holy grail that psychiatry and clinical psychology are searching for in all mental illnesses. Of course, not everyone with a low heart rate becomes a violent offender. My heart rate in my mid-twenties was 48 beats per minute, and the same will be true for a number of you. Yet at an admittedly imperfect level, low heart rate is a telltale sign of transgressors.

GETTING THAT AROUSAL BOOST IN LIFE

So low resting heart rate represents one of the best replicated, most easily measured, and most promising biological correlates of antisocial and aggressive behavior. But *why* does it predispose someone to antisocial behavior? Even with simple biological measures, unfolding the "mechanism of action"—how low heart rate produces antisocial and aggressive behavior—is highly complex. Let's examine a few of the prevailing explanations.

One is fearlessness theory.[27] A low heart rate is thought to reflect a lack of fear.[28] Although we talk about "resting" heart rate, the term is misleading. In research studies, subjects are brought into a novel environment, met by strangers, and have electrodes slapped on them. This is less like "resting" and more like experiencing a mild stressor. Timid, anxious children will have higher heart rates. Those lacking fear will have lower heart rates.

As outlined above, there are some particularly fearless individuals such as bomb-disposal experts who function perfectly well in society and also have particularly low heart rates.[29] After all, it takes nerves of steel to defuse a bomb. By the same token, antisocial and violent behavior requires a degree of fearlessness. If a boy lacks fear, he is more likely to get into a fight because he is not afraid of getting hurt. Similarly, punishments like prison do not motivate many offenders to desist from violence because this punishment does not hold fear for them.

Fearlessness theory receives support from research showing that low heart rate provides the underpinning for a fearless, uninhibited temperament in infancy and childhood,[30] and that the more uninhibited a preschooler is, the more aggressive he or she will be later in life.[31] Adolescents with low heart rates are also better able to stand stress, indicating that such individuals are more insensitive to social stressors, including socializing punishments.[32]

Another theoretical explanation of the low heart rate–antisocial behavior connection lies in empathy. Children with low heart rates are less empathic than children with high heart rates.[33] Children who lack empathy are less able to put themselves into another person's shoes and to imagine what it must feel like to be bullied and hit. Those with low empathy may be more aggressive because they have no concern for the feelings of others. Certainly children lacking empathy are more antisocial and aggressive.[34]

Another explanation for how low heart rate produces antisocial and aggressive behavior is stimulation-seeking theory. This theory argues that low arousal represents an unpleasant physiological state, and that those who display antisocial behavior seek stimulation to increase their arousal levels to an optimal level.[35] We all have an optimal level of arousal at which we can operate effectively and comfortably.[36] Think of times you come back home and really need some stimulation—you turn on the TV, brew some coffee, turn up the music, get on your cell phone, or go out and party. You are bored and need a buzz. Yet there are other times you instead come home and leave the TV off, turn off your cell phone, and retreat into your own quiet space. The day has been too much and you're over-aroused.

The same need you have applies to kids with chronically low levels of arousal. Preschool boys with low heart rates not only are more antisocial and hyperactive, but they also choose to watch videotapes depicting intense anger more often than kids with more normal heart rates.[37] In my own research, resting heart rate at the age of three characterizes both stimulation-seeking behavior at that same age[38] and aggressive behavior at the age of eleven. Kids with chronically low levels of arousal may get an arousal boost in life by beating someone up, shoplifting, joining a gang, or getting involved in drugs. The harsh reality is that breaking any rule is fun for most kids—just think back to the days when you were a teenager. Living on the edge may not be what parents want for their teenagers, but for the kids themselves it's exciting and gives meaning to life. Perhaps it's not too surprising, then, that resting heart rate is at its lowest in life during adolescence, when stimulation-seeking[39] and antisocial behavior are at their highest.[40] And that craving for an arousal boost in adolescence may be part of the reason violence peaks in the late teenage years.

If you have ever experienced this craving for stimulation, as I did when I was a kid, you get into a state of really just not knowing where to put yourself. You experience an intense feeling of restlessness and emptiness that can peak in a sense of agitation, and a real need to

release some type of hard-to-describe, built-up tension. I have that feeling right now. You want to move around. Once you can find something
to do in order to "shift gears," you feel better.

These feelings are exactly what a significant number of serial killers
report experiencing prior to their homicides. The intense tension and
restlessness. The need to go out in search of a victim. The consequent
excitement of the abduction, torture, rape, and killing. And then the
sense of relief and release from tension.

Why would that be? I suspect it's explained partly by having physiological under-arousal and a stimulation-seeking personality. The important message I really want to convey is a simple medical fact: low heart
rate is a significant risk factor for antisocial behavior. Of course, it is not
the only process within the autonomic nervous system that has gone
awry in antisocial and violent individuals. To put that statement into a
societal context, we need to take a trip to Mauritius.

SHARED EARLY TEMPERAMENTS, DIVERGENT ADULT OUTCOMES

Mauritius is one of the most beautiful tropical islands in the world and
a destination for those seeking a luxury holiday with its consequent
peace, quiet, and harmony. It's also not a bad place to do research. In
the past twenty-five years I have had to drop in on the island thirty-nine
times. One could, of course, research violence in Detroit, but on balance I slightly prefer Mauritius. "It's so delicious," as the advertisements
say along the road going from the airport to La Pirogue beach hotel,
where I always stay. The sun, the palm trees, the beaches, the volcanic
mountains, and some of the warmest and most gracious people I have
ever met make for an exotic mix.

Mauritius is a small island in the Indian Ocean near the Tropic of
Capricorn, lying to the east of Madagascar. It extends thirty-eight miles
from north to south, and twenty-nine miles from east to west. Part of
the African continent, it is a multiracial democratic nation that gained
independence from British rule in 1968 and became a republic within
the Commonwealth of Nations in 1992. With a population of 1.28 million as of July 2009, it is the third most densely populated country in
the world. At the initiation of our longitudinal study, in 1972, Mauritius
was a developing country, but now it is developed and widely viewed as
a model African country.

Mauritius is also a wonderful melting pot of cultures, and the country is again notable in that ethnic tensions are rare. So where is the malevolence in Mauritius? Let's put the previous idea of low arousal and stimulation-seeking into a research context that we undertook there.

Why Mauritius, you might ask. Back in 1967, the WHO—World Health Organization—wanted to learn more about children who were at risk for the development of clinical disorders later in life. It recommended that a study should be conducted in a developing country, that the study should utilize three-year-old children, and that biological methods should be used to identify children at risk for later mental-health problems.[41] Initially, the WHO had targeted India as a possible site,[42] but a medical director from Mauritius successfully argued for the geographical advantage of his country. Mauritius was a small island with low emigration—factors that would permit subjects to be contacted more readily over time than in India.

The Mauritius study was set up in 1972 by Peter Venables, from York University in England, and Sarnoff Mednick, from the University of Southern California. Peter was to become my PhD supervisor five years later, while Sarnoff would eventually lure me to the United States eleven years after that. I became the director of the study in 1987 when Peter retired. The sample was a birth cohort consisting of 1,795 three-year-old children all born in one of two towns—Vacoas and Quatre Bornes, both in the middle of the island and conveniently situated. The research laboratories were in Quatre Bornes.

The study began like this. Families came to the research unit. Mothers sat down with their three-year-old children, and new toys were placed around them. Would the child leave the secure home base of his or her mother and explore the toys? At one extreme, some children would not leave and sat clinging to their mother—they were stimulation-avoiders. Some would come and go from their mother, using her as a "safety net" for exploration. Yet others would freely explore the toys and the new physical environment—the stimulation-seekers or explorers. Children were also placed in a sandbox and rated on their engagement in social play with other children. Their friendliness to the experimenter and their willingness to chitchat was also assessed. These four separate behavioral indicators formed a measure of stimulation-seeking.[43]

Eight years later, when aged eleven, the children were rated by their parents using a checklist of child behavior problems that included aggression—items like "fights other children," "attacks others," "threat-ens others."[44] I found that high-scoring stimulation seekers at age three—the top 15 percent—were more aggressive at age eleven. To be sure, not all stimulation-seekers became aggressive. But to some extent, the early behavior of young children predicted later aggression. Mauritius may be heaven, but like anywhere else, devils roam. Two children in our study illustrate that while arousal and temperament predict aggression, further complexities must be recognized. One little boy, called Raj, and one little girl, called Joëlle, had nearly the lowest heart rates and the highest levels of stimulation-seeking and fearlessness. They fell into the top sixth percentile of their respective gender on these measures when aged three. So how did these two under-aroused stimulation-seekers turn out later in life?

Raj turned out to be not just a stimulation-seeker in adulthood, but also a vicious, psychopathic thug who loved riding motorbikes and terrifying and manipulating people. He was the most psychopathic individual in our entire sample of 900 males, with multiple criminal convictions ranging from theft to assault to robbery. In discussing his social relationships and how he came across to others, he admitted, "There are many people scared of me, most of 'em. I've got to be dan-gerous."[45] He actively enjoyed making people uncomfortable. Like many psychopaths, he took pride in his ability to control and regulate people, especially through his reputation of aggression and violence, which gave him status and power within his peer group. In discussing how his friendships were formed, Raj commented, "I want friends out of fear."[46] When someone expresses desires like that, you get the sense you are dealing with a man who knows no fear himself, yet craves fear in his friends.

Raj's lifelong fearlessness from age three to age twenty-eight gave rise in part to his aggressive behavior, which in turn allowed him to obtain rewards and status from those who feared him. It was reinforced so strongly that it became his modus operandi. When asked about his girlfriend, he mused for a while and then laughed. "Yeh . . . I think she's scared of me too!" he said.[47] It speaks to the callousness and cold-bloodedness that is typical of psychopaths. Just as we saw earlier in our evolutionary perspective on violence and cheating, psychopathy can be a successful reproductive strategy, with power and control over others bringing resources that translate into greater reproductive fitness.

Raj's authority over others through threats and violence pervaded even his intimate social relations. His ability to make people frightened likely enhanced his enjoyment of sexual relations with his girlfriend, in a similar way to the enjoyment that sadistic rapists obtain in terrifying, dominating, and controlling their victims.

Yet was he really that fearless? Surely something must have scared him, sometime. What if he met others like himself?

> Nothing can frighten me. They want to fight with me? I beat them up—that's it, that's all. Ye know what I mean? I just cut their face, ye know what I mean?[48]

He really had neither a sense of fear nor a concern for others. Because he lacked the empathy needed to appreciate others' pain, there was no empathy holding him back from mutilating people's faces. He lay at the extreme of psychopathy—at the extreme of fearlessness.

Did he sometimes feel sorry for the victims of his violence? Did he have a sense of conscience? Raj's reply: "No, 'cos it's them that searched for it."[49] Psychopaths are always more than willing to blame others to justify their actions. They apply the "just deserts" principle to defend their heinous actions. Others get what they deserve because of how *they* behave. This gave Raj free license to do almost anything. Life for renegade Raj and other psychopaths is essentially *jeux sans frontières*— games without boundaries. They are playing out a life full of fun and excitement. This mind-set can make for a nasty piece of work—a callous, unemotional, heartless, cold-blooded, stop-at-nothing psychopath. And it's caused by low physiological arousal, fearlessness, and stimulation-seeking early in life.

The little girl, Joëlle, also turned out to be a fearless stimulation-seeker later in life, but in a very different way. She went on to become Miss Mauritius and obtained her excitement in life though very different avenues.

As an adult, her prevailing memory of herself as a child was one of a thirst for discovery. To try everything out, to explore the world, and to put herself forward. When asked about her memories, Joëlle said, "I wanted to discover so many things about life. The most important thing for me was to express myself."[50] She too wanted to act on the environment, but in a different way from Raj. The desire for discovery, to experience the world, and to give full expression to one's fearless, stimulation-seeking potential need not always result in criminality.

Joëlle went on to live a fulfilled, aggression-free life because despite the biological and temperamental predispositions for an antisocial lifestyle, she was a kind, generous, and sensitive person. She had other factors that protected her from the extreme outcome of a psychopath, and perhaps being a girl, combined with all the genetic and environmental baggage that comes along with a woman's world, made a difference.

In broad terms, the difference between Raj and Joëlle is not unlike that between Ted Kaczynski and our fearless bomb-disposal expert. Biology is not destiny. The same biological predispositions can result in very different outcomes. At the same time, these early biological warning signs can give us a sense of potential problems on the road ahead. Indeed, when it comes to understanding outcomes for violence through the autonomic nervous system, our notion of conscience is key.

CONSCIENCE CONQUERS CRIME

Have you ever thought of killing someone? No? Well, aren't you a Goody Two-shoes.

Seventy-six percent of "normal" men have had at least one homicidal fantasy. For normal women the rate is a bit lower, at 62 percent.[51] Who do you want to kill? Men think about killing co-workers, while women want to kill their family members, especially stepparents. That latter fantasy fits our evolutionary account of homicide—you kill those not genetically related to you. Why do you want to kill? The most common reason is a lover's quarrel, but apparently 3 percent of you have fantasized about killing someone just to experience what it is like to kill someone.[52]

Alfred Hitchcock had a good sense of the surprising range of violent thoughts throughout American society. In his movie *Strangers on a Train* there is a cocktail party scene where a woman imagines a killing:

> I think it would be a wonderful idea. I can take [my husband] out in the car and when we get to a very lonely spot, knock him on the head with a hammer, pour gasoline over him and over the car, and set the whole thing ablaze.[53]

And she laughs.

I hope I never meet some of you, and yet I imagine you have not killed anyone. Why? Because when you really think hard about it, when you put yourself right there in the situation of doing it, you can't follow through. Something's holding you back. I know I can't follow through, no matter how much I've wanted to kill some of my critics. This thing we call a conscience kicks in. It's made up of gut reactions and feelings generated in part by our autonomic nervous system and pulls us back from the brink. And it goes beyond heart rate. What we're talking about here is a symphony of classical conditioning and autonomic reactions that inspire or dissuade us from taking antisocial actions.

How can we measure something as abstract as "conscience"? Well, sweat is a good place to start—specifically something known as classical conditioning as measured by skin conductance. Let's take a quick trip to the laboratory, the kitchen, and then back to the laboratory again.

In the laboratory, skin conductance[54] is measured with small electrodes. We place them on the distal phalanges—the tips—of the first and second fingers of the hand. We then pass a very small electrical current across these two electrodes—so small you would never feel it. The more you sweat, the better the current will be conducted. These very tiny electrical changes—as small as .01 microsiemens (one hundred millionths of a siemen, a unit of conductance)—are amplified so that they can be seen and measured by computer software.

Variations in the size of a subject's sweat response to a simple tone played over headphones reflect differences in the extent to which the subject allocated attentional resources to process the tone.[55] When you pay attention to a sound, the prefrontal cortex, amygdala, hippocampus, and hypothalamus are activated.[56] Some of these "lower" brain areas—the hypothalamus and brain stem—stimulate sweating.[57] So people sweat a bit more when thinking or listening to something. Although the sweat response is a peripheral autonomic measure, it is nevertheless a powerful measure of central nervous system processing.[58] The bigger the skin-conductance response, the greater the degree of attentional processing.

Let's get back to the vexing question of quantifying exactly what a "conscience" is. What ultimately gives us that sense of right and wrong in life? I believe the answer lies in biosocial theory.[59] We can think of a conscience as essentially a set of classically conditioned emotional responses. Criminals and psychopaths show poor fear conditioning—in part because they are chronically under-aroused. Because of this lack of fear conditioning, they lack a fully developed conscience. And it is that lack of conscience—a sense of what is right and what is wrong—that makes them who they are.[60]

It goes like this. Classical conditioning involves learning an association between two events in time. When an initially neutral event (the conditional stimulus) is closely followed by an aversive event (the unconditional stimulus), that initially neutral stimulus will develop the properties of the aversive stimulus. In the classic case of Pavlov's dogs, a bell was paired with the later presentation of food. Food to hungry dogs automatically elicits an unconditional response: salivation. After a number of pairings of the bell with the food, the bell by itself came to elicit the salivation. The dogs learned a relationship between the sound of a bell and the later presentation of food. They conditioned.

Now from the lab to the kitchen. Young children are not too different from Pavlov's dogs. Take the scenario of a small child stealing a cookie from the kitchen. Punishment by the parent, like scolding or a slap, elicits an unconditional response—the child is upset and hurt. After a number of similar learning trials, the sight of the cookie—or just the thought of stealing the cookie—will elicit an uncomfortable feeling, a conditioned response. It is that discomfort that keeps the child from engaging in the theft. The storage in the brain of similar "conditioned emotional responses" developed early in life in lots of different situations accumulates to form what we call "conscience." And that's what stops you from killing someone.

In this analysis, socialized individuals develop a feeling of uneasiness even thinking about stealing something or assaulting someone. That's because such thoughts elicit unconscious memories of punishment that took place early in life, for mild misdemeanors like theft or behaving aggressively. Haven't you sometimes said when discussing a crime to your friend, "I could never even *think* of doing such a thing"? Now you can understand part of the reason. You rarely if ever contemplate such events because even the *thought* of such acts generates previously conditioned emotional responses that produce discomfort in you. Criminal thoughts then get rubbed out of your cognitive repertoire—they are off your radar screen.

There's another side to this that I find interesting. There are some offenses that have an almost unnatural feel about them—they don't seem all that criminal. Think about cheating on your taxes, for example. Imagine pumping up your yearly charitable contributions from $100 to $200. This act does not seem quite as "offensive" as other offenses. I mean, you did give $100 to charity, didn't you? You're not such a bad person, are you? And perhaps the reason it does not seem so bad—and why you might do it—is that there is no convincing analogue of tax evasion in childhood. Parents do not punish us for these "white-collar crimes" but instead focus on more obvious things like stealing and fighting. Consequently, some of us have not developed much of a "conscience" for these acts. That may be why white-collar crimes are committed by people who are supposedly reasonable citizens in society—and why you might think they are not as serious as other criminal offenses.

Plagiarism is another example. It is absolutely rampant in students. The self-report survey I conducted on Hong Kong undergraduates showed that 67 percent had passed off other people's essays as their own work. Similarly, 66.6 percent had copied others' work to meet a course requirement. Despite strict institutional prohibitions against such actions, it goes on unchecked. Perhaps less surprising to you— likely because you have done it too—is that 88.3 percent had bought pirated software or DVDs, while 94.2 percent had illegally downloaded music or movies. Again, there is no convincing childhood analogue of these actions that gets punished, and hence little or no conscience about perpetrating those acts. Parents may pass off their own ideas as their child's when helping them in their schoolwork—and praise their child when rereading that terrific-looking piece of work a few days later. We may even be unknowingly socializing our children into white-collar antisocial habits.

Now to the evidence. A systematic review of all studies conducted on adult criminals, psychopaths, and antisocial adolescents concluded that there is overwhelming evidence for poor fear conditioning in offenders.[61] Nevertheless, living a criminal way of life might cause the poor conditioning, rather than poor conditioning being a causal agent of later crime. While dozens of studies found poor fear conditioning in criminals and psychopaths, none prospectively tested whether poor fear conditioning early in life predicted adult crime. What was really needed was a prospective longitudinal study to prove the point.

FEARLESS TOTS TODAY—RUTHLESS THUGS TOMORROW

Onto the conditioning stage steps Yu Gao, from Beijing Normal University in mainland China. Gao had come to study for her PhD with me at the University of Southern California in 2003. In a collaboration that would span three academic generations, she shed light on the darker developmental question of whether poor fear conditioning predisposes someone to crime.

My own PhD supervisor, Peter Venables, had taken a long look at the fear-conditioning data he had collected in Mauritius and concluded that there was no conditioning. I bought into Peter's conclusions because, well, he was after all one of the world's leading authorities on psychophysiology. You are hardly going to question your own supervisor, are you?

Gao was less gullible and more gutsy. It was an example of where fresh minds give rise to new perspectives, innovation, and progress. We had the help of Mike Dawson, a world-leading authority on fear conditioning. Gao launched herself into the data and with her strong statistical expertise she convincingly demonstrated that fear conditioning had indeed occurred in the three-year-olds. Peter had been too pessimistic—his conditioning paradigm had indeed worked.

Of course, like everything else in life there are differences between us in the degree of fear conditioning. Some condition, and some do not. That's the interesting bit that Gao pounced on. Recall that mothers brought their three-year-old children into the laboratory—1,795 of them in all. Small electrodes were placed on the little fingers of the toddlers to measure skin conductance. Headphones were placed on their heads to deliver the auditory tone stimuli. They sat on their mother's lap for security and comfort. Then the conditioning experiment began.

On some trials, a low-pitched tone predicted that ten seconds later the children would be blasted with an unpleasant loud noise. On other trials, a high-pitched tone would be presented and nothing would happen. The children were not told about the association between the low-pitched tone and the nasty noise. And yet in just three conditioning trials their brains worked it out. As a whole group, the children gave a bigger skin-conductance response to the low tone than to the high tone. They had become conditioned and developed anticipatory fear to the initially neutral tone that had been paired up with the aversive tone.

We sit back and let twenty years go by. The tots are now twenty-three-year-old adults. We search all the court records on the island to see which children grew up to become adult criminals. Out of the 1,795 subjects, 137 had had a conviction. Gao matched each offender with two non-offenders on gender, age, ethnicity, and social adversity—a total of 274. This epidemiological "case-control design" ensures that any group

**Fear conditioning
(in microsiemens) at age 3**

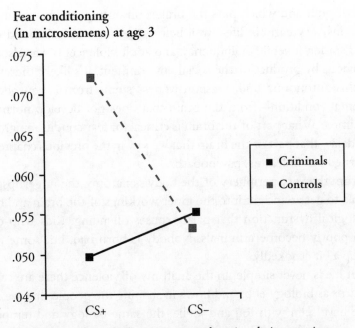

Figure 4.1 Fear conditioning at age three in relation to crime
at age twenty-three. A greater response to the CS+
compared to the CS– indicates fear conditioning.

differences cannot be due to group differences on these demographic measures. Gao then looked at how the two groups fared in their ability to develop conditioned fear twenty years earlier, at age three.

The results were striking. Remember that to show fear conditioning you must show a larger skin conductance response to the low-pitched tone, called the CS+, that predicts the unpleasant tone compared with the high-pitched tone, the CS, that does not predict the aversive tone. Figure 4.1 illustrates the finding. The normal control group showed significant fear conditioning. Their sweat response to the low-pitched (CS+) tone was much bigger than their response to the high-pitched (CS–) tone. Yet the criminals-to-be, back at age three, showed no sign of conditioning at all. They were flat-liners—as a group they did not show *any* fear conditioning. This finding by Yu Gao demonstrated for the first time that an early impairment in autonomic fear conditioning acts as a predisposition to criminality in adulthood.[62]

Gao's research took the field a lot further than before because she documented that a lack of conscience, which normally gives us that sense of guilt and which puts the brakes on outrageous behavior, has its origins very early in life—well before the onset of childhood conduct disorder, juvenile delinquency, and adult violence. It was also not an obvious by-product of the social environment. It's likely, therefore, that this autonomic under-responsiveness stems from a neurodevelopmental condition—from the brain that does not develop normally over time.[63] What part of the brain is critical for fear conditioning? The amygdala—that part of the brain that we saw in the previous chapter to be burned out in fearless psychopaths.

From the very periphery of the body's anatomy, the fingertips, we are able to get an insight into the inner workings of the brain and neurobiological dysfunction that partly causes offending. Kids who condition poorly become criminals. Nobody is born bad, but some may develop a bit crookedly.

Yet life is never simple. In the anatomy of violence there are twists and turns as biology ebbs and flows in shaping the people that we are. As we have seen with Raj and Joëlle, the same biology and temperament may result in different life outcomes. And as we saw with Randy Kraft and Antonio Bustamante in the last chapter, there can be different causes for why two different people both end up as killers. Divergent beginnings, shared endings.

This variability is a real lacuna in our knowledge on the biology of violence. Why doesn't everyone with a slow heartbeat become violent and psychopathic? Can there be two types of adult psychopaths? I believe there can. Rather than showing poor fear conditioning, some psychopaths have surprisingly good autonomic and brain functioning. You likely work with one. One may be a friend or acquaintance. And whether you know it or not, you could even be in a relationship with one. Worse still, you may be one yourself. Let's take a further look.

SUCCESSFUL PSYCHOPATHS

You've had a sense of psychopaths from our discussions of evolutionary cheats and Jolly Jane Toppan. They can be fearless stimulation seekers who are also selfish, charming, and grandiose. As Robert Hare, the world's leading researcher on psychopathy and the creator of the Hare Psychopathy Checklist, succinctly summed it all up in the title of his book, psychopaths are *Without Conscience*.[64] When you lack a conscience you may gain some psychopathic traits. Yet I do not believe that all psychopaths have poor frontal functioning and autonomic under-arousal. Successful psychopaths—those who are not caught and convicted—may be a different beast that we have to contend with.

My interest in successful psychopaths goes back to my accounting days. After I packed in accounting with British Airways for the cloud-capped towers of Oxford University, I was intellectually rich but financially broke. So during my first summer I went back to London and registered at a temporary-employment agency to earn money. It was there, I believe, that I met my first successful psychopath. I had found work as an auditor, and at the company that hired me I met Mike, who was also a temp. I got to know him over drinks in the pub after work. Charming, witty, engaging, and very quickly liked by the permanent staff, Mike was an impressive and professional young man with fascinating life stories and a thirst for adventure, but he soon revealed to me that he was pilfering what he could at work whenever he had a chance, both at this job and, apparently, other jobs. It's not that he admitted to a lot, but I got the sense that he was revealing just snippets of his antisocial lifestyle.

There's nothing more dramatic to say about Mike, except that my memory of him and a few other temps who lived life on the edge stayed with me. I never thought more about Mike until years later as an academic in Los Angeles. I had previously worked with convicted psychopaths in English prisons. I was now working with caught murderers on the verge of execution. I got to wondering whether offenders who were not caught would look the same—biologically, at least—as their caught counterparts. But where would I get "free-range" offenders? Then Mike fleetingly came to mind, along with the answer—temporary-employment agencies.

It was a long shot, but intrigued by the idea, I did a pilot study at the nearby temp agency. I hired temps and paid them to work in my laboratory for three days. The work they did for me? Taking part in experiments. My team and I asked them what crimes they had been committing recently. It sounds a bit naïve. Who would ever tell you about crimes they had committed? And yet before long they were singing like canaries about the robberies, rapes, and even homicides they had committed. My memory of Mike had borne fruit. We quickly got into business, recruiting more temp workers and collecting more data.

To place what I was finding into a research context: the base rate of antisocial personality disorder—lifelong recidivistic offending—is 3 percent in males in the general population. In our temp-agency sample the base rate was an astonishing 24.1 percent—more than eight times the national average.[65] Furthermore, a full 42.9 percent met the adult criteria of antisocial personality disorder[66]—nearly half the sample.[67] Temp agencies were antisocial gold mines, and we started to dig deeper.

Those with "antisocial personality disorder" were perpetrating much more than the mischief I got up to in my youth. Forty-three percent had committed rape. Fifty-three percent had attacked a stranger, causing at the least bruises or bleeding. Twenty-nine percent had committed armed robbery. Thirty-eight percent had fired a handgun at someone. And twenty-nine percent had either attempted or completed homicide.[68] I was realizing that compared with the tigers among my temp-agency recruits, Mike back in England was just a pussycat.[69]

You may wonder why the temps would admit their crimes to us. There are a number of reasons. We obtained a certificate of confidentiality from the secretary of health that protected us from being subpoenaed by any law-enforcement agency in the United States. We could not be forced to reveal our data. In fact, if I did so I would be committing an offense, and could end up as an offender in someone else's study on crime. Our participants therefore were legally protected. Furthermore, they were in a respectable, professional university environment with trustworthy research assistants. Perhaps for the first time in their lives, they could talk about their wrongdoings at length with a professional in full confidence and without risk—even getting into the nitty-gritty of rape and homicide.

Were they fibbing? We think not. There was little or no motivation for such deception, no obvious gain. While some pathological lying cannot be ruled out, we still believe they were antisocial offenders. Put it this way: if they were telling the truth, they were definitely antisocial. If, alternatively, they were lying about their crimes and deceiving us, they were pathological liars and *still* antisocial. In reality, we believe that rates of criminal offending and antisocial personality disorder are *underestimates* of the true base rate in this population, rather than overestimates.

We also found unusually high rates of psychopathic personality as assessed by the Psychopathy Checklist, the "gold standard" instrument for assessing psychopathy.[70] For males, 13.5 percent had a score of 30 or more—the cutoff used to define psychopathy in many prison studies.[71] More than twice that amount—30.3 percent—were above the cutoff of

25 or more that had been adopted in several other studies.[72] For the males whom we focused on in our research, about a third were defined as psychopathic.

How could there be so many more psychopaths in temporary-employment agencies? The answer is that temp agencies are wonderfully safe havens for psychopaths—almost a breeding ground. Psychopaths gain in life by ferociously exploiting others. To begin with, their superficial charm allows them to succeed with their parasitic lifestyle, but ultimately they get caught out by those around them. Once detected, they can pack up and move on to the next social group of victims that they will suck dry. Temporary-employment agencies allow this freedom of movement. They also conduct more limited background checks compared with companies hiring full-time employees. Furthermore, psychopaths are impulsive and unreliable—they only rarely hold down a permanent job. Temporary jobs, in contrast, limit the time that their flaws can be detected by employers. Psychopaths are also stimulation-seekers and love to be on the move for new experiences, and temp agencies give them that freedom, even to move from city to city. Of course, not all people at temp agencies are psychopaths. After all, I was a temp once. But putting all this together, it's no small wonder that we found as many psychopaths as we did.

So now we had our psychopaths. We searched court records to see which ones had been convicted of an offense. Those with a conviction were delineated "unsuccessful" psychopaths. Those without a conviction were the "successful" psychopaths. We did not have many—sixteen unsuccessful psychopaths, thirteen successful psychopaths, and twenty-six controls. But it was a beginning.

Up until this point there had been no empirical research on these individuals except for a seminal, creative investigation conducted by Cathy Widom. From November 1974 to July 1975 she placed an ad in a "counterculture" Boston newspaper that read as follows:

> Wanted: charming, aggressive, carefree people who are impulsively irresponsible but are good at handling people and at looking after number one.[73]

Using a neuropsychological measure, she found that the non-institutionalized psychopaths who responded to her ad did not show the frontal-lobe deficits that one would expect. She went on further to speculate that "autonomic differences found between psychopaths and others may only characterize the institutionalized, unsuccessful psychopath."[74] Teaming up with Joe Newman, a leading psychopathy researcher at the University of Wisconsin, Madison, Widom went on to replicate and extend her original findings.[75] Widom's original study had its limitations. It did not have a control group, and because 46.4 percent had been incarcerated at some point in their lives, they could not be exactly classified as "successful" psychopaths. Furthermore, there were no psychophysiological data to back up her speculative hypothesis.

We, however, did have a psychophysiological laboratory and we set about testing Widom and Newman's ideas. We put all our participants through a social stressor. They were seated in our psychophysiology laboratory. They had electrodes placed on their fingertips to measure skin conductance, and on their arms to measure heart rate. They were acclimated to the setting in what we call a "resting state"—or as near to "rest" as one can get. We made a careful note of their levels of autonomic arousal.

We then sprung on them the social stressor task. They were told that they had to give a speech about their worst faults. They had two minutes to prepare the speech, and two minutes to give it while being videotaped. If the participant hesitated or came to a stop, a research assistant in the room with them would push them to give more details to increase the stress level. The first two preparatory minutes are "anticipatory fear," or what Robert Hare has termed "quasi-conditioning."[76] As in fear conditioning, the question is whether the psychopaths will autonomically respond, both in anticipation of the stressful speech, and also during the speech itself.

The findings are shown in Figure 4.2. The controls show what we all expected—increases in heart rate and sweat rate throughout most of the task. The unsuccessful psychopaths also show what we would expect based on prior research with institutionalized psychopaths, a blunted autonomic stress response—only small increases in sweat rate and heart rate from the resting baseline. The successful psychopaths, in sharp contrast to their unsuccessful counterparts, show significant increases in heart rate and skin conductance relative to their resting state.[77] Essentially there is no difference between the successful psychopaths and the normal controls. Widom's almost prophetic claim, made twenty-three years earlier, had received some initial support.

We also tested our psychopaths and controls on a measure called

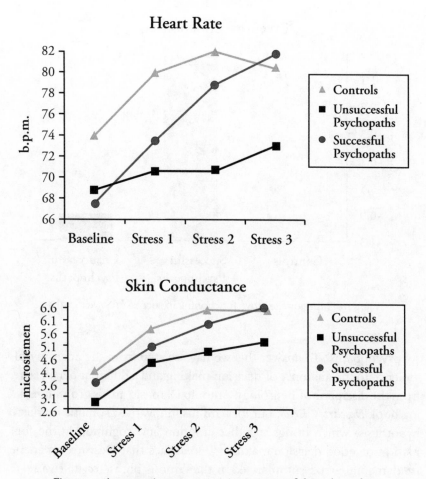

Figure 4.2 Autonomic stress reactivity in successful psychopaths, unsuccessful psychopaths, and controls

"executive functioning." It involves all the cognitive functions that you would like in a successful business executive—planning, attention, cognitive flexibility, and, importantly, the ability to change plans when given feedback that one course of action was inappropriate. How did our three groups do? You can see in Figure 4.3. The controls performed significantly better than unsuccessful psychopaths—that's something you might expect. But take a look at how the successful psychopaths performed. They not only outperformed the failed psychopaths—they also performed significantly *better* than the normal controls.[78]

What are we to make of the surprising findings for the success-

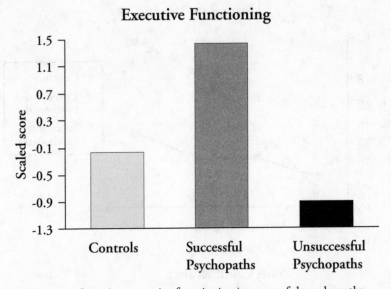

Figure 4.3 Superior executive functioning in successful psychopaths

ful psychopaths? To answer this we have to step from the anatomy of violence to the anatomy of decision-making and a different perspective from the discipline of neurology. Antonio Damasio, in his groundbreaking book *Descartes' Error*, put forward his innovative "somatic marker" hypothesis, which brings together emotion and cognition in the formation of good decision-making.[79] He argues that Descartes made a fundamental error, summarized in the famous phrase *cogito ergo sum*, in believing that there is a fundamental separation of the mind from the body.

Damasio, in contrast, argues for an intimate mind-body connectedness. A good mind makes good decisions, and to do so it has to rely on "somatic markers" produced by the body. These somatic markers are unpleasant autonomic bodily states produced when one is contemplating a risky action or a difficult decision—the pounding heart and the perspiration. These somatic markers have flagged negative outcomes in the individual's past, and are stored in the somatosensory cortex. This input is then transmitted to the prefrontal cortex, where further evaluation and decision-making takes place. If the current situation has been previously linked to a negative outcome, the somatic marker for that past event will sound an alarm bell to the decision-making areas of the brain—no action will be taken. This process may act at either a conscious or a subconscious level and can be thought of as helping to reduce the range of options in decision-making. It is similar to classical conditioning and the anticipatory fear that deters us from conducting an antisocial act previously associated with punishment.

We had always assumed that in order to make good decisions, we need to be removed from our emotions—to be cool, calm, and collected. The revolution Damasio made in cognitive and affective neuroscience was to argue that instead, emotions importantly guide good decision-making. Without emotions and somatic markers, we will not make good decisions.

Now let's turn back to our unsuccessful psychopaths. They have blunted emotions and lack the appropriate autonomic stress response. We can think of that as reduced somatic markers—a relative disconnection between mind and body. That mind-body dualism, according to Damasio, would result in bad decision-making, and certainly incarcerated offenders make many bad life decisions.

Turning to the successful psychopaths, we see that they show intact autonomic stress reactivity and anticipatory fear. They have a mind-body connectedness that allows for somatic markers to help form good decision-making. That translates into superior executive functioning. And I would argue that that is why successful psychopaths are successful.

Recall that we define success here in terms of not being convicted for an offense. Imagine that the successful psychopath is on the street, contemplating robbing a 7-Eleven store. His brain—consciously and also subconsciously—is processing the scene. He's consciously checking up and down the street for specific signs of surveillance—but his subconscious is also forming a gestalt of the whole scene and putting it together. He's about to proceed—but at the last minute he pulls back. There was something about the whole setup that he did not like the look of. He cannot put his finger on it, except that it just did not "feel good."

A somatic marker warning bell had been rung, warning him that previously in a similar situation he was nearly caught. Perhaps it was the same time of day, the same number of people in the shop, the fact that he had also just had a couple of drinks, or a combination of these visual and somatic cues that triggered the warning bell. The heightened autonomic reactivity is giving him an edge over his unsuccessful psychopathic counterpart who does not hear the somatic warning-bell sound and instead ends up hearing the police siren.

So the failed psychopath has reduced autonomic reactivity to cues that signal danger and capture. The successful psychopath has relatively better autonomic functioning and hence is better able to escape detection by the authorities.[80] He also has better executive functioning. But if the successful psychopath does not have the autonomic impairments that haunt failed psychopaths, what made him psychopathic in the first place?

Our original study gives us two initial clues. First, if you look back at Figure 4.2, you can see that in the resting state prior to the social stressor, both psychopathic groups show a low resting heart rate. The successful psychopaths are six beats per minute slower than the control group, and slightly below the level of the unsuccessful psychopaths. So, successful psychopaths have the low resting cardiovascular arousal that we argued earlier may result in stimulation-seeking, a cardinal feature of the psychopath. Second, the successful psychopaths evidenced a psychosocial impairment not shown by the other two groups—being raised by people other than their natural parents or being brought up in a foster home or other institution. Parental absence and a lack of bonding may have helped shape the lack of close social connectedness and the superficiality that typifies psychopathic relationships.

This initial research suggests a methodology for further study of psychopaths in society. Clearly, "success" in these psychopaths—avoiding detection— is relatively modest, and our findings may or may not apply to successful psychopaths who are businessmen, politicians, academics, or terrorists. Nevertheless, we are for the first time obtaining some clues to the makeup of the unstudied psychopaths we know almost nothing about—those that are circulating closely with us in the community.

HOT-BLOODED SERIAL KILLERS

The findings on successful psychopaths may also offer us other insights. It is conceivable that they could give some clues on the makeup of serial killers. What makes someone a serial killer is not just an unanswered question, it is a greatly under-researched one, because it is next to impossible to collect systematic experimental data on a significant number of them. Beyond some very basic facts—that they tend to be white and male, to target strangers, and to use guns infrequently—we really don't know too much about what makes a serial killer.[81]

The predominant tendency is to classify serial killers as "cold-blooded." But might some of their cold-blooded acts be a product of their "hot-blooded" bodies? My speculative working hypothesis is that *some* serial killers have many of the characteristics of successful psychopaths. As a murderer once told me, it's not that easy to kill a person for the first time. But once you have stepped over that threshold, the idea of killing someone else does not carry the same baggage that it used to.

If you manage not to get caught the first time around, you've crossed a major threshold and can make your second killing. You have learned from your first attempt, seen how you nearly slipped up, and adjusted your behavior to become more effective. That's precisely what we found in our successful psychopaths on the executive-functioning task we gave them.[82] You know when to make a move—and when to hold off. What makes for such a capacity? As outlined earlier, it's having a well-functioning autonomic system that provides you with somatic markers—the bodily alarm bells that signal impending capture and the time to beat a hasty retreat.

And perhaps there is a paradox here. I have argued that low heart rate is a well-replicated marker for antisocial behavior—at least in a resting state. What would your heart do if you had just killed someone? I hope you would say it would be beating as fast as a scared rabbit's. Would you feel terrified at what you had done? Very likely. So what would you say a serial killer's heart rate would do after a murder, and how would he feel? I think you would say that his heart rate would be about as normal as could be—he would be cold-blooded. Yet that was not the case for Michael Ross.

Ross was an intelligent serial killer who, just before graduating from Cornell University, started a series of rapes and murders of eight young women in New York and Connecticut. He describes three things that he felt after he had committed homicides:

I remember the very first feeling I had, was my heart beating. I mean really pounding. The second feeling I had was that my hands hurt where I always strangled them with my hands.[83] And the third feeling was, I guess, fear, and the kind of reality set in that there was this dead body in front of me.[84]

Ross is not the exception that you may think. You can imagine that the act of killing someone is utterly revolting, and if you really had to kill someone with your bare hands, the whole experience would be so disgusting that you would vomit. You may imagine that serial killers don't think this way. But you'd be wrong.

That is exactly what happens with some serial killers too. I'm currently working with the Singapore Prison Service. In walking past the building in Changi Prison Complex where they execute murderers, I was reminded of one serial killer they had executed. That was John Scripps—the first Westerner to be hanged in Singapore for committing several homicides. Scripps had all the trappings of a cold-blooded psychopathic killer. After beating senseless his victim Gerard Lowe, an innocent man who had done him no wrong,[85] he proceeded to take his head off:

> Just like a pig, it's almost the same. You cut through the throat and twist the knife through the back of the neck. There ain't much mess if you do it properly.[86]

Scripps was utterly heartless—and yet he threw up. When asked if his victim knew what was going on, he replied:

> He pissed and shit himself. It made a stink. He was shitting himself. Yeah. Right. It wasn't good and I spewed up. He really shit himself, but he couldn't do much about it, could he?[87]

This vomiting reaction is surprising given that the offender was heartless, cruel, and seemingly cold-blooded. One explanation based on twelve case studies of single and multiple murderers suggests that "kindling," or stimulation, of the emotional limbic system can in some cases occur during the killing. This causes hyperactivation of the autonomic nervous system—resulting in nausea, vomiting, profuse sweating, incontinence, or even vertigo.[88] This limbic kindling perspective is very speculative and must be treated with caution. At the same time, it is clear at least that John Scripps was not a man entirely lacking in fear. He was more like one of our successful psychopaths—except that he eventually got caught.

So you may not be as different from a serial killer as you think. Michael Ross demonstrates the autonomic stress reactivity that characterizes our successful psychopaths from temporary-employment agencies. This visceral cardiovascular feedback and heightened emotional awareness constitute the somatic, bodily markers that provide the ventral, or underneath part of, the prefrontal cortex with sound awareness of the social context the person is in. Ross showed the anticipatory fear in a stressful situation that our successful psychopaths showed in the emotional stress task. He had the good executive functions and decision-making ability to plan carefully, stalk his victims, and ensure that the social context was appropriate for what he would do next. Like successful psychopaths, he had the lack of remorse and the egocentricity that are key features of the psychopath, as well as the disturbed parenting that we find characterizes the homes of successful psychopaths—perhaps it is in that way that he really differed from you and me.

So Michael Ross's heart was beating fast as he killed people, and it was beating fast again when he was put to death in Connecticut, on May 13, 2005. But unlike John Scripps, it was not just because he was scared. The lethal injection of potassium chloride used on death-row inmates in several states produces death by speeding up ventricular repolarization of the heart and raises the resting electrical potential of the cells of the heart muscle.[89] Ironically, the sluggish cardiovascular functioning that landed these inmates on death row gets speeded up to end their lives. Execution is one way to deal with the problem, but one wonders whether a more effective solution would have been to deal earlier in life with the autonomic factors predisposing some children to adult violence. We'll see in a later chapter how it may well be possible to alter antisocial adolescents' low arousal levels and turn them around in life—without the use of potassium chloride.

FEARLESSNESS OR COURAGE?

There are no simple answers for why people kill, why some are one-off killers, why some are serial killers, or why Theodore Kaczynski went on his campaign of public terror. We have argued that autonomic dysfunction is one component, and that low resting heart rate is a well-replicated risk factor for antisocial and violent behavior. It can predispose some to kill in cold blood. Ted Kaczynski epitomizes the cold-hearted violent offender, as he had a resting heart rate of 54 beats per minute,[90] which would place him in the bottom 3 percent of my temporary-employment-agency sample—a sample already biased toward low heart rates.[91] He had the same sense of fearlessness and low resting heart rate that bomb-disposal experts have. Yet to chalk his violent offending and that of other killers up to one simple bodily process would be wrong.

I once discussed the heart-rate hypothesis with Dan Rather, the anchorman for CBS Evening News and host of 48 Hours. I was working with him on a 60 Minutes interview in New York on the genetics of homicide, and he clearly resonated with the idea of under-arousal and fearlessness. He explained that he too had a low heart rate, and had in his earlier days taken up boxing. He had his own sense of fearlessness and bravado. And perhaps this is illustrated in part by his fierce, relentless, and courageous badgering of American presidents during interviews, for which he has been heavily criticized. Despite possessing a biological risk factor for violence, like Miss Mauritius he was able to find other outlets for his predispositions—by verbal aggression and probing, rather than physical violence.[92]

So the answer to the Unabomber puzzle and others like it has to be more complex and go beyond low physiological arousal. We have clues. Kaczynski had at the minimum multiple features of both schizotypal personality disorder and paranoid personality disorder—features that include odd beliefs, paranoid ideation, no close friends, eccentric behavior, and a blunted affect. Several psychiatrists for his defense, including Raquel Gur, of the University of Pennsylvania, went further, viewing him as having paranoid schizophrenia. Even prosecution psychiatrists admitted he had schizotypal and schizoid personality disorder. We'll see later that these biologically based clinical disorders are themselves risk factors for antisocial and violent behavior.[93] Moreover, according to his mother, Kaczynski was separated from her and the family when he was hospitalized at the age of nine months. This resulted in his afterward being withdrawn, unresponsive, and fearful of separation. Interestingly, separation anxiety disorder can lead to detachment, isolation, and difficulty developing relationships—all of which strongly characterized Kaczynski.[94] We will see later how disruption of bonding due to institutionalization during a critical period of development can affect the brain and, alongside other biological risk factors, trigger violence.[95] There is at least a part solution to the puzzle of this bomber.

The greater puzzle, perhaps, is to understand the fine line between psychopaths and national heroes—why some under-aroused, fearless individuals end up as offenders who take lives, while others with the same predisposition are selfless, courageous, and save lives. Tom

Hanks's character in the movie *Saving Private Ryan* displays enormous bravery and heroism in the line of fire in his rescue of Private Ryan, but as we detect from his shaking hand in the opening scenes in the landing craft just before the storming of Omaha Beach on D-Day, he experiences very significant fear. This is the distinction. He is courageous—despite significant feelings of fear, he performs acts of heroism and selflessness.

In *The Hurt Locker*, Sergeant William James—played by Jeremy Renner—blurs this distinction between courage and fearlessness even further. As the leader of a bomb-disposal unit in Baghdad, is he a bombastic, stimulation-seeking, rule-breaking psychopathic personality? Or is he a superbly professional hero, hell-bent on saving Americans and Iraqis alike at the risk of his own life? Like many psychopaths, he has difficulties connecting at an emotional level with his ex-wife and child. And like many violent offenders, James turns out to be a complex entity who defies any simple classification.[96]

What one sees here—both in these fictional portrayals and in the real-life case of the Unabomber—is a key theme in the anatomy of violence that we will return to. Different biological, psychological, and social risk factors can interact in shaping either violence or self-sacrificing heroism. Violence and terrorism are not just low physiological arousal,[97] yet this is certainly *one* of the active ingredients that, when combined with other influences, can move us toward a more complete understanding of killers like Kaczynski.

The previous chapter outlined how poorly functioning brains can predispose someone to violence. This chapter moved us from the central nervous system into the functioning of the more peripheral autonomic nervous system. In this component of the anatomy of violence we have seen how broken hearts can result in heartbreaking violence. We will now continue our journey back into the brain to look at its physical construction. Lombroso had believed when he peered into the skull of Villella that he had the answer to the cause of crime—a physical, structural abnormality in the brain. Was Lombroso entirely out of his mind? Or might he have been right? Did he have a mind to crime and why the autonomic- and central-nervous-system processes that we have just seen are not working properly? Might violent offenders have broken brains?

5

BROKEN BRAINS

The Neuroanatomy of Violence

Don't you sometimes find that Christmas is just a bit too much? We all get on each other's nerves cooped up at home during the holiday period. That bloated feeling with all the Christmas pudding and turkey. The endless watching of sports on TV. The hangovers and stuffy atmosphere. The unwanted presents that you know you'll have to recycle back for someone else's birthday. The hopelessness of those New Year's resolutions you cannot possibly keep. Yes, we all know how "merry" Christmas can be. In those moments, we can really empathize with Ebenezer Scrooge.

Herbert Weinstein, a sixty-five-year-old advertising executive, was no exception. As soon as the twelve days of Christmas were over, on the evening of January 7, 1991, he and his wife, Barbara, had a major argument in their twelfth-floor Manhattan apartment. This was the second marriage for each, and you may know what that can be like. Disrespectful comments over the other partner's progeny were flying. Herbert's response was to disengage from the arguing and withdraw from the battleground. So far, so good. But disengagement can have an uncanny way of winding up one's partner. After all, everyone likes a good fight once in a while—it lets the steam out. So, not being one to bow out of a fight that easily, Barbara let fly, coming after Herbert and scratching at his face.

Something snapped inside Herbert. He grabbed his wife by the throat and throttled the life out of her. There she was, dead on the floor. That did not look too good, so Herbert opened the window, picked his wife's dead body up, and threw her out. She did a free fall twelve floors down onto East Seventy-second Street, landing on the sidewalk below. Herbert thought it would look like an accident, but on reflection he realized it still didn't look very good. So he crept out of the building, only to be nabbed by the police. They charged him with second-degree murder.

Things were looking bad for Weinstein, but he was a wealthy man and had a good defense team. And his lawyers suspected something unusual in the case. He did not have any prior history of crime or violence. They referred Herbert for a structural brain scan using MRI.[1] They followed this up with a PET scan, which maps brain functioning. If you could see the images you wouldn't have to be the world's leading neurologist to notice that his brain is broken. It was incredibly striking—there was a big chunk missing from the prefrontal cortex. What exactly was happening here? Unknown to anyone—including Weinstein himself—a subarachnoid cyst was growing in his left frontal lobe. This cyst displaced brain tissue in both frontal and temporal cortices.

The neurologist Antonio Damasio was consulted during a pretrial hearing to render his opinion on Weinstein's ability to think rationally and control his emotions. Skin-conductance data were admitted alongside the brain-imaging data to argue that Weinstein had an impaired ability to regulate his emotions and make rational decisions. The defense team went with an insanity defense, and Judge Richard Carruthers was favorably impressed by Damasio's arguments and the testimony of the imaging experts. In a novel pretrial bargain, the prosecution and defense agreed to a plea of manslaughter.[2] This carried a seven-year sentence in contrast to the twenty-five-year sentence Weinstein would have served if he had been convicted of second-degree murder.

It was a monumental decision. No court had ever used PET in this way in a criminal trial.[3] For the first time, brain-imaging data had been used in a capital case prior to the trial itself to bargain down both the crime and the ensuing punishment.[4]

The case of Herbert Weinstein highlights yet again the importance of the brain in predisposing someone to violence. More specifically, the case suggests that a *structural* brain deficit in the left prefrontal cortex results in a *functional* brain abnormality that in turn results in violence.

Cysts such as Weinstein's have an unknown cause and can grow for a long time. They can also be benign, but experts in the case testified that the cyst resulted in brain dysfunction that substantially impaired Weinstein's ability for rational thinking. That bolstered the credibility of his insanity defense.

Recall from chapter 3 that impairment to the frontal cortex is particularly associated with reactive aggression. Revisiting the events from that night we can see that Weinstein's violence was reactive in nature. Arguments had preceded the attack, and his wife had attempted to scratch his face. These are the aggressive verbal and physical stimuli that provoked Weinstein's violent response. Recall our earlier argument that spousal abuse can be caused by a lack of prefrontal regulatory control over the limbic regions of the brain, resulting in reactive aggression in the face of emotionally provocative stimuli. Factor in to the equation that Weinstein had no prior history *in any shape or form* of aggressive or antisocial behavior. In terms of timing, it seems reasonable to suppose that the onset of this medical condition was a direct cause of Weinstein's extreme reactionary violence.

In this chapter we'll build on Weinstein's case in four different ways. We will burrow further into the anatomy of violence by arguing that the brains of some offenders are *physically different* from those of the rest of us.

First, for Herbert Weinstein the structural brain abnormality is so striking that we can all see it. But I'll argue that many violent offenders have structural abnormalities. They may be so subtle that even highly experienced neuroradiologists cannot detect the abnormality, yet they can in practice be detected using brain imaging and state-of-the-art analytic tools.

Second, while Weinstein's brain abnormality likely had its onset in adulthood, I'll suggest that for most other offenders, something has gone wrong with their brain development very early in life. I'll advance a "neurodevelopmental" theory of crime and violence—the idea that the seeds of sin are sown very early on in life.

Third, we'll shift gears a bit in terms of causation. Weinstein's case illustrates how a medical illness late in life can cause brain impairment—but what about younger offenders? We saw in chapters 3 and 4—where we touched on brain imaging and psychophysiology—that violent offenders have *functional* brain impairments. Rather like your car when it misfires or your computer when it runs slowly, there is something just not working right with offenders' brains. So far, we have viewed this as a software problem. Maybe a bad birth messed up the program for normal development, or maybe poor nutrition was the culprit. But now what I'm suggesting is the possibility of *hardware failure*. The idea is that criminals have broken brains—brains anatomically different from those of the rest of us.

Taking a leaf out of Lombroso's nineteenth-century book *Criminal Man*, I'll argue that the world's first criminologist was absolutely correct in espousing *structural* brain abnormalities as a predisposition to violence. He may have been wrong on the precise location in the vermis of the cerebellum, or the ethnic hereditability of these traits, but he was right on the mark in arguing for a structural mark of Cain. This may sound like we're back to the "born criminal" and the destiny of genetics. While I have insisted so far that there is indeed in good part a genetic basis to violence, I'll also highlight here the critical importance of the *environment* in helping to cause the structural brain deformations that we find in offenders.

Fourth, and finally, Weinstein's case deals with severe violence, but are structural brain deformations restricted only to aggressive behavior? I'll argue that they are not, and that their influence runs the gamut of antisocial behaviors and extends into nonviolent crimes—including even deception and white-collar crime. We'll start this part of our journey with a trip back to those temporary-employment agencies in Los Angeles.

BACON-SLICING THE BRAIN

As you'll recall from our earlier discussion of Randy Kraft and Antonio Bustamante, back in 1994 Monte Buchsbaum and I, along with my colleague Lori LaCasse, had shown from our PET functional imaging work that murderers have poor functioning in the prefrontal cortex as well as the amygdala and hippocampus. We had clearly demonstrated for the first time a functional brain abnormality in these homicidal offenders.[5] At that time we were quite ecstatic.

Yet that exhilaration was tempered by a dose of skepticism. For one thing, this was a forensic sample—they were all referred by their defense teams, who suspected that something might be wrong. Would our findings apply to the general population? For another thing, they were all murderers—would our results apply to those who showed a broad range of antisocial behavior? Furthermore, we had shown the presence of functional abnormalities, but we had not really tested Lombroso's hypothesis of physical brain anomalies. How could we overcome these methodological challenges?

The answers all came from temporary-employment agencies. You'll recall from chapter 4 that while prospecting in California I struck gold at temp agencies. There we were able to recruit psychopaths and individuals with antisocial personality disorder. These individuals are free-range violent offenders who are running around right now in the community committing rape, robbery, and murder while you read this book. Robert Schug, one of my gifted PhD students with unusual forensic skills, conducted painstaking in-depth clinical interviews with our participants to assess which ones were psychopaths. We then set to work scanning our sample using anatomical magnetic resonance imaging—aMRI. Unlike functional imaging, aMRI gives a high-resolution image of the anatomy of the brain—just what we need for prying into the structure of the criminal brain.

After just four minutes with a subject we are able to acquire many images of the brain's structure. Then the hard work begins. After brain scanning, we use sophisticated computer software combined with our detailed knowledge of brain anatomy. We identify landmarks in the brain scans that pinpoint exactly where the orbitofrontal cortex and amygdala are. As with a bacon-slicer, we dissect the brain into slices as thin as one millimeter. There are over a hundred of these slices as we move in a coronal direction—from the forehead to the very back of the head. Having a thin slice of a brain results in good spatial resolution—we can visualize tissue as tiny as one cubic millimeter. Just as for your digital camera or TV, the higher the number of pixels within a given area, the better the resolution, and the clearer and sharper the picture.

Then, on each slice, using our neuroanatomical landmarks—the sulci, or grooves, in the brain—we painstakingly trace the area of the brain structure in question. You can see one slice from the prefrontal cortex on the left side of Figure 5.1, in the color-plate section. On the right side you can also see a three-dimensional rendering of a quadrant cut out of the skull to reveal below it the underlying brain tissue in one of our subjects. Just like a slice of bacon that has both red meat and white fat, our brain slices have two tissue types. We first have to trace around the "gray" matter in each slice—the meat, colored green here. This separates the neural tissue from the fat—the white matter—so that we can compute the area of neurons. Add up all these gray neuronal areas across all slices, and we have the number we want—the cortical volume of the brain region of interest.

So what do we find in the prefrontal cortex? Those with a diagnosis of antisocial personality disorder—lifelong persistent antisocial behavior—had an 11 percent reduction in the volume of gray matter in the prefrontal cortex.[6] White matter volume was normal. Antisocial bacon has plenty of fat—just not enough meat, not enough neurons. As we saw in chapter 3, the prefrontal cortex is centrally involved in many cognitive, emotional, and behavioral functions, and when it is impaired, the risk of antisocial and violent behavior increases.

Our antisocial individuals did not differ from controls in wholebrain volume, so the deficit was relatively specific to that critical prefrontal cortical region. But perhaps the brain deficit is not causing antisocial behavior. After all, antisocial individuals often abuse alcohol and drugs, and this could account for the prefrontal gray matter reduction. We therefore created a control group who did not have antisocial personality disorder, but who did abuse drugs and alcohol. We then compared the two groups. The result? The antisocial group had a 14 percent reduction in prefrontal gray volume compared with the drug-abuse control group, a slightly *bigger* group difference than that between normal controls and antisocials.

So drugs are not the cause of the structural brain deficit, but questions still remain. Prefrontal structural deficits have been found in other psychiatric disorders. We also know that those with antisocial personality disorder have higher rates of other mental illnesses, including schizotypal personality, narcissism, and depression.[7] Could the brain impairment have nothing to do with antisocial personality disorder but instead be linked to a different clinical disorder that our antisocials also happened to have?

To deal with this, we created a psychiatric control group that was not antisocial but that was matched with the antisocial group on all the clinical disorders that the antisocial group had. Yet again, we found that the antisocial group had a 14 percent prefrontal volume reduction compared with this psychiatric control group. Our findings cannot be explained away by a psychiatric third factor.

Could the answer instead be family factors? In this case, we think not. We controlled for a whole host of social risk factors for crime, including social class, divorce, and child abuse, but found that the prefrontal cortex–antisocial relationship held firm. And unlike the case of Herbert Weinstein, there were no visible lesions in our antisocial subjects that could account for the volume reduction.

We are left with the possibility that this structural impairment has a subtle early origin. For whatever reason—be it environmental or genetic—the brain is not developing normally throughout infancy, childhood, and adolescence. We'll come back to this "neurodevelopmental" idea later.

The MRI brain scan of Herbert Weinstein showed enormous structural impairment that was very visible. But if you were to compare the MRI scan of an antisocial individual with that of a normal person, you would not see the 11 percent reduction in gray-matter volume. That reduction corresponds to just half a millimeter in thickness of the thin outer cortical ribbon that is colored green in Figure 5.1.[8] The difference is visually imperceptible not just to your eye but also to the eye of the world's best-trained neuroradiologist. Indeed, an expert neuroradiologist would actually judge the brain scan of the antisocial individual to be quite normal. And yet it's not.

We know it's not normal only because we are not making a *clinical judgment* such as medical practitioners make who are looking for visible tumors. We are not taking a brief, global look at this slice to discern outright signs of pathology, as is common neuroradiological practice. We are not looking for a big hole in our slice of bacon. Instead, we are spending hours painstakingly computing the precise volume of gray matter in the prefrontal cortex using brain-imaging software. Doing that, we can identify small differences that have important clinical significance. Herbert Weinstein is just the tallest tree in a forest of brain-impaired offenders. Below such visibly striking cases are a host of violent offenders with more subtle but equally significant prefrontal impairments. Yet in clinical practice such sharks will slip away entirely unnoticed.

Let's face it, findings come and go. Our study was the first to demonstrate a structural brain abnormality in any antisocial group. But perhaps it was just a fluke. We therefore conducted a meta-analysis that pooled together the findings of all anatomical brain-imaging studies conducted on offender populations—twelve in all—and found that this specific area of the brain is indeed structurally impaired in offenders.[9] Since this meta-analysis, yet more studies have observed prefrontal structural abnormalities in offenders.[10] The findings are not a fluke.

To make better sense of what we found, and to understand more fully the implications of this specific structural brain abnormality, we need to take a quick trip to a neurologist's clinic in Iowa. As it happens, it is the clinic of the neurologist who consulted in the pretrial hearing of Herbert Weinstein—Antonio Damasio.

I have briefly mentioned earlier how Damasio, then at the University of Iowa and now at the University of Southern California, made truly groundbreaking contributions to our knowledge of how the brain works. A lot of this knowledge has come from the study of unfortunate individuals who, for one reason or another, have suffered a head injury resulting in brain damage. The silver lining to these clouds, from a scientific standpoint, is that by taking together all the clinical patients with damage to one specific brain region, and by comparing them to patients with lesions in different areas, we can draw conclusions on the critical functions of that brain region. Together with his equally brilliant wife, Hanna Damasio, and other colleagues, Antonio has made fascinating deductions from these patients about the functions of some areas of the prefrontal cortex and related regions, including the amygdala.

One group of patients had lesions localized to the ventral prefrontal cortex, the lower region of frontal cortex. It includes the orbitofrontal cortex, which sits right above your eyes, and the ventromedial prefrontal cortex, which is in line with your nose. The patients showed a striking pattern of cognitive, emotional, and behavioral features that set them apart not just from normal controls, but also from patients with lesions outside of this brain area.[11]

First, at an emotional level, while their electrodermal response system is otherwise intact and responsive, patients with ventral prefrontal damage do not give skin-conductance responses to socially meaningful pictures such as disasters and mutilations. The ventral prefrontal cortex is involved in coding social-emotional events. It connects to the limbic system and other brain areas to generate appropriate emotional responses within a social context, measured here by a sweat response. Without that neural system in place, the individual is emotionally blunted—and we saw earlier that psychopaths and those with antisocial personality disorder are similarly emotionally blunted and lacking in empathy.

Second, at a cognitive level, such neurological patients make bad decisions. In a psychological test called the Iowa gambling task, which was developed by the neurologist Antoine Bechara, subjects have to sort cards into one of four piles. Depending on which pile they place their card in they get monetary rewards or punishments. Unbeknownst to the subject, the decks are loaded. If they pick decks A or B, they might initially get large rewards, but eventually they are hit by even larger losses. Decks C and D give smaller rewards but they also yield much smaller punishments. Over the course of one hundred card plays, normal subjects learn about halfway through to avoid the high-reward/high-loss decks A and B. They instead persist in picking decks C and D, which ultimately give them the best payoff. They show good decision-making in the face of competing rewards and punishments. Patients with ventral prefrontal lesions don't. They instead keep making bad decisions by picking the bad decks.[12]

Even more interesting is what normal individuals show in terms of their sweat responses during the task. About halfway through the task they become cognitively aware of which decks are bad, and which are good. Just prior to that, when they are consciously unaware of the good and bad decks, they contemplate picking from a bad deck. What Antoine Bechara saw on the polygraph was a skin-conductance response (a somatic marker), a bodily alarm bell warning them that they were about to embark on a risky move. Subconsciously, their body knows that bad news is just around the corner, and that they should hold back on their response—but consciously their brain does not. Very soon after this somatic alarm bell rings, normal individuals change their strategy and switch to the good decks—and they become cognitively aware of what's going on. The ventromedial lesion patients? No alarm bell. So they continue to pick cards from the bad decks.

It's not surprising, then, that psychopaths make bad decisions and mess up their own lives as well as those unfortunate enough to be within their social circle. As we saw in chapter 4, the lack of autonomic, emotional responsivity results in an inability to reason and decide advantageously in risky situations. This in turn is very likely to contribute to the impulsivity, rule-breaking, and reckless, irresponsible behavior that make up four of the seven traits of antisocial personality disorder. So we can understand how structural abnormalities to the prefrontal cortex could later result in antisocial personality—they could be the cause of the functional autonomic abnormalities we documented in the last chapter.

The third striking characteristic of these patients, at a behavioral level, is that they exhibit psychopathic-like behavior. A classic example of this, which took place more than 150 years ago and highlights the intricate link between brain and personality, is the case of Phineas Gage. It's an unusual story that has been told before in neuroscience circles, but it is well worth retelling here.

THE CURIOUS CASE OF PHINEAS GAGE

Gage was a well-respected, well-liked, industrious, and responsible foreman working for the Great Western Railway. The fateful day was September 13, 1848. He was organizing the destruction of a large boulder lying in the path of the projected railway track. The work team had chiseled a hole into the boulder for the gunpowder and sand. The gunpowder was then poured into the hole. It was four-thirty in the afternoon.[13]

The next step should have been an apprentice pouring sand on top of the gunpowder. Gage was standing by with a metal tamping rod that was three feet seven inches long and one and a quarter inches in diameter. He was on the verge of using the rod to tamp down and compress the sand on top of the gunpowder to potentiate the explosion. At that critical moment, Gage was distracted by a conversation with his co-workers. After a few seconds he turned back to the boulder, believing that sand had been placed on top of the gunpowder. It had not. He tamped down with the rod right on top of the exposed gunpowder. The metal rod rubbed against the rock and created a spark that ignited the gunpowder. It transformed the tamping rod into a lethal spear that blasted its way right through the head of Phineas Gage.

Gage had been stooped over the hole as he tamped down with his hand. The rod entered his lower left cheek and exited from the top-middle part of his head, creating an open flap of bone on the top of his skull. You can see this flap in Figure 5.2 and the bone-shattering damage the rod created. The deadly missile flew through the air, landing eighty feet away, while Gage was hurled to the ground.

Understandably, all the railway workers thought Gage was as dead as a doornail. But after a couple of minutes he began to twitch and groan, and they realized that he was still alive. They put him into an oxcart and took him to the nearest town. He was carried upstairs into a hotel room and a doctor was summoned. What was the treatment in the nineteenth century when you had a tamping rod blown through your brain? Rhubarb and castor oil.

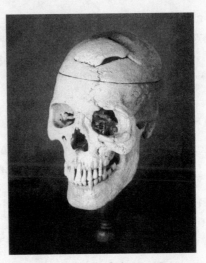

Figure 5.2 Skull of Phineas Gage

You would not think Gage stood a snowball's chance in hell of surviving. But what a miraculous remedy rhubarb and castor oil turned out to be! Gage lost his left eye, but in no less than three weeks, he was out of bed and back on his feet. Within a month Gage was walking around town creating a new life for himself. And it truly was a new life. For in the words of his friends, acquaintances, and employers, he was "no longer Gage":

> He is fitful, irreverent, indulging at times in the grossest profanity (which was not previously his custom), manifesting but little deference for his fellows, impatient of restraint or advice when it conflicts with his desires, at times pertinaciously obstinate, yet capricious and vacillating, devising many plans of future operations, which are no sooner arranged than they are abandoned in turn for others appearing more feasible. A child in his intellectual capacity and manifestations, he has the animal passions of a strong man. Previous to his injury, although untrained in the schools, he possessed a well-balanced mind, and was looked upon by those who knew him as a shrewd, smart businessman, very energetic and persistent in executing all his plans of operation. In this regard his mind was radically changed, so decidedly that his friends and acquaintances said he was "no longer Gage."[14]

We see here, very clearly, that Gage had been transformed from a well-controlled, well-respected railway worker into a pseudo-psychopath—an individual with psychopathic traits. Like many patients with frontal-lobe damage, he was impulsive, irresponsible, and was reputed to have been sexually promiscuous and a drunkard.[15] He was fired by his employer because he was unreliable. He took on a series of jobs and moved around, switching from one job to another. Eventually he went on tour with the tamping rod and appeared in Barnum's American Museum in New York and other public shows (see Figure 5.3). Among his many jobs he worked at an inn in Hanover, New Hampshire, in 1851, looking after horses. A spirited, risk-taking adventurer, he even spent several years in Chile as a stagecoach driver before traveling to California, where he worked on a series of farms until his premature death on May 21, 1860, after a series of epileptic seizures. Despite a most remarkable recovery from what should have been a mortal wound, that tamping rod he carried with him for the remainder of his life eventually got the better of him.

The case was such a remarkable one that medical doctors at the time scoffed at the idea that anyone could survive such an injury, and viewed it as a hoax. It could not possibly be true. While it was indeed a true case, could it nevertheless be unique? Can accidental damage to the prefrontal cortex really transform an otherwise normal, law-

Figure 5.3 Phineas Gage at Barnum's American Museum holding the tamping rod that destroyed his prefrontal cortex

abiding individual into a capricious, psychopathic–like, antisocial individual?

The answer can be found back in Antonio Damasio's and others' laboratories. A large body of evidence has now convincingly shown that adults suffering head injuries that damage the prefrontal cortex—especially the lower, ventral region—do indeed show disinhibited, impulsive, antisocial behavior that does not conform to the norms of society.[16]

But you could counter that adults have brains that are relatively fixed. What about children, whose developing brains show much greater plasticity? Does damage to the prefrontal cortex in youngsters also lead to antisocial behavior? Overwhelmingly, studies of the behavioral changes that follow head injuries in children find that conduct disorder and externalizing behavior problems are common.[17, 18] While some other children develop internalizing behavior problems like anxiety and depression,[19] there is little doubt overall that head injuries in children predispose them to impulsive, dysregulated behavior.[20]

But what if the damage to the prefrontal cortex occurs really early during infancy? Surely there is enormous plasticity of the brain at this developmental stage, allowing it to recover lost functions and to resume normality. Clinical cases of such selective prefrontal damage are rare, but they confirm that prefrontal lesions very early in life can directly lead to antisocial and aggressive behavior. A study from Damasio's laboratory reported on two cases—one female, one male—who suffered selective lesions to the prefrontal cortex in the first sixteen months of life.[21] Both showed early antisocial behavior that progressed into delinquency in adolescence and criminal behavior in adulthood, and included impulsive aggressive and nonaggressive forms of antisocial behavior. Both also had autonomic deficits, poor decision-making skills, and deficits on learning from feedback. Yet again we see that triad of traits that Antoine Bechara and Antonio Damasio clearly documented in those suffering prefrontal damage in adulthood—psychopathic behavior, autonomic impairments, and reduced somatic markers.

I know what you're thinking. The clear limitation is that we are dealing with only two cases—one case more than the one presented by Phineas Gage. However, another laboratory reported on nine cases of children who suffered frontal lesions in the first ten years of life.[22] All nine suffered behavioral problems after the injuries, with seven of the nine developing conduct disorder. Even in the case of the remaining two, they both exhibited either impulsive, labile behavior or uncontrollable behavior.

These cases, when taken together, strongly suggest that damage to the prefrontal cortex can directly lead to antisocial and aggressive behavior. It's an important point. Brain-imaging research showing that murderers and those with antisocial personalities have prefrontal abnormalities demonstrate that a relationship exists. But do prefrontal structural and functional impairments cause crime and violence—or does violence cause the brain impairment? Violent offenders get into fights, and so can acquire "closed" head injuries: the skull is not broken but there is internal damage to the brain. It's certainly possible, yet neurological case studies showing that prefrontal impairments in infancy, adolescence, and adulthood are later *followed* by antisocial, aggressive, and psychopathic-like behavior are telling. They provide striking support for a causal explanation flowing from prefrontal impairments to a disinhibited personality to violence.

DIGGING DEEPER INTO THE PREFRONTAL CORTEX

We have seen from MRI studies that antisocial individuals in the community have structural brain impairments. We have also seen from the clinic that patients with head injuries causing prefrontal structural damage develop antisocial behavior and a loss of somatic markers, resulting in poor decision-making and maladaptive social behavior. We were therefore finding interesting similarities between our antisocial temp workers and the neurological clinical cases of Antonio Damasio and his colleagues. We were excited by these initial findings and wanted to dig deeper into these parallel findings from community to clinic. Two specific issues came to the fore.

First, the autonomic, emotional impairments in the head-injured patients of Damasio and Bechara raised the question of whether or not our antisocial temp workers also showed somatic marker impairments. This was a hypothesis that we tested out. As you may recall from chapter 4 we put our subjects through a stress task in which they had to talk about their worst faults. As pointed out by Damasio, this is a very appropriate task in the context of the somatic-marker hypothesis because it elicits secondary emotions—embarrassment, shame, guilt—that are the province of the ventral prefrontal cortex.[23]

We found that our antisocial, psychopathic subjects not only had a significant volume reduction in prefrontal gray matter, but also showed reduced skin conductance and heart-rate reactivity during the social stressor task. Sure enough, they did lack somatic markers, just as Damasio's prefrontal patients did. Furthermore, when we divided our anti-social group into those with particularly low prefrontal gray volumes and those with near-normal volumes, we found that it was the former group—those with the structural prefrontal impairment—who particularly showed somatic-marker deficits.[24] We were finding an interesting convergence of somatic-marker impairments, prefrontal structural deficits, and antisocial behavior that bore a striking resemblance to the findings on Damasio and Bechara's patients.

The second issue concerned the localization of the structural impairment. Where exactly within the prefrontal cortex did our anti-social, psychopathic individuals have reduced gray-matter volume? Damasio had written an editorial on our original findings, posing the question of whether in future work the deficit may be localized in orbital and medial sectors of the prefrontal cortex.[25] Recall that the tamping rod entered underneath Phineas Gage's eye and traveled straight up his prefrontal cortex. Hanna Damasio had demonstrated from her careful and rigorous reconstruction of the tamping-rod accident that the damage to Gage's brain was localized to the ventral and orbitofrontal part—the lower region—and also the medial or middle part of the prefrontal cortex.[26] What would we find if we made a more detailed analysis of the precise location of the prefrontal volume reduction in our antisocials?

Dividing up the sectors of the prefrontal cortex involved much more complex sulcal landmark identification and tracing of slices; it took us literally years to complete, but eventually we got there. You can see in Figure 5.4, in the color-plate section, what we did. You are looking head-on at one of our antisocial subjects, and we are taking a slice through the frontal cortex. From top to bottom, moving from twelve o'clock to six o'clock, these regions consist of the superior frontal gyrus, middle frontal gyrus, inferior frontal gyrus, orbitofrontal gyrus, and the ventromedial area.[27] In which sector did we see a significant volume reduction in those with antisocial personality disorder?

Three of the five sectors turned up trumps. As Antonio Damasio would have predicted, antisocial individuals showed a 9 percent bilateral reduction in the orbitofrontal gyrus, together with a 16 percent reduction in the volume of the right ventromedial prefrontal cortex.

It is structural impairment to the ventral region of the prefrontal cortex that seems to be particularly implicated in antisocial, psychopathic behavior—the same brain region devastated by the tamping rod on that fateful day for Phineas Gage in 1848.

The third sector provided us with a different but complementary perspective to consider. Our antisocial subjects had a 20 percent volume reduction in the right middle frontal gyrus. In chapter 4 we discussed how neuropsychological research had demonstrated poorer "executive functioning" in antisocial and psychopathic individuals—reduced ability to plan ahead, regulate behavior, and make appropriate decisions. The brain areas classically associated with these executive functions lie in the dorsolateral prefrontal cortex. "Dorso" refers to top and "lateral" refers to side—so "dorsolateral" is the upper, side part of the prefrontal cortex. If you look at Figure 5.4, that's exactly where the middle frontal gyrus is located. And if we look further into the functioning of this brain area, impairment in antisocial offenders makes quite a lot of sense.

Let us consider some of the normal functions of the middle frontal region that have been gleaned from functional-imaging and brain-lesion studies—functions that could well be impaired in offenders. First, the middle frontal gyrus, which makes up Brodmann areas 9, 10, and 46, is part of the neural circuitry that subserves fear conditioning.[28] We saw earlier that criminals and psychopaths have poor fear conditioning. Second, it plays a role in inhibiting behavioral responses,[29] and we know that offenders frequently show disinhibited, impulsive behavior.[30] The middle frontal gyrus is also involved in moral decision-making,[31] and offenders have impaired moral judgment and break moral boundaries.[32] It is further involved in choosing delayed rewards as opposed to immediate rewards,[33] and it is well documented that offenders are less able to delay gratification.[34, 35] It is activated by empathy to pain stimuli,[36] and antisocial individuals lack empathy.[37] This prefrontal subregion is also activated when we look inward and evaluate our own thoughts and feelings.[38] Offenders are characterized by a lack of insight into the harm they perpetrate on people around them.[39]

Clearly the middle frontal gyrus, which is significantly compromised in those with antisocial personality disorder, is heavily involved in cognitive, affective, and behavioral characteristics that antisocial individuals are deficient in. These deficiencies in turn contribute to their antisocial tendencies. We can complete the circle from brain structure to functional deficiencies to antisociality.

In a similar vein, there is more to the ventral region of the prefrontal cortex than effective decision-making. We know that it is involved in
controlling and correcting punishment-related behavior,[40] and in what
neuropsychologists call "response perseveration."[41] And yes, recidivistic
offenders are revolving-door guests in prisons. They seem unable to
learn from their mistakes. They keep on making the same behavioral
responses that resulted before in punishment and prison—what psychologists call perseveration.[42] Fear conditioning is another process gov-
erned by the ventral prefrontal cortex, and we have seen that offenders
have deficits in this area.[43] The ventral area has also been implicated in
compassion and care for others,[44] as well as sensitivity to others' emotional states.[45] Let's face it, we all know that criminals and psychopaths
are not the most caring people in the world.[46] As with the middle frontal gyrus, insight[47] and behavioral disinhibition[48] are also subserved by
this ventromedial region. Offenders are disinhibited and psychopaths
lack self-insight. Interestingly, the ventral prefrontal cortex also helps
to reduce negative emotions during parent-child interactions,[49] and
offenders were likely as children to have thrown temper-tantrums with
their parents. Emotion regulation is another ventral prefrontal function,[50] and emotional dysregulation characterizes impulsively aggressive individuals.[51]

Taking both dorsal and ventral structures together, there are quite
compelling reasons to believe that structural impairments to these
regions can give rise to a constellation of social, cognitive, and emotional risk factors that predispose someone to antisocial behavior and
an antisocial personality. The fact that both ventral and middle frontal brain regions contribute to some of the same functional risk factors for antisocial behavior—poor fear conditioning, lack of insight,
disinhibition—highlights the salience of these well-replicated neurocognitive risk factors. It also tells us that an outcome of antisocial
behavior may be especially likely when both of these regions are structurally compromised.

So far we have dug deeper into the prefrontal cortex and discovered
that the ventral and middle frontal gyrus are the key culprits when it
comes to crime. But these brain areas are guilty not only of their crimes
as charged. Our next level of probing of the prefrontal cortex will implicate these same subregions in a different but equally fundamental societal question—why are men more violent than women?

MALE BRAINS—CRIMINAL MINDS

There's no escaping the fact. Men are meaner than women. But why? Sex differences in crime and violence have traditionally been put down to sex differences in socialization. If you have a little girl, you give her a doll to look after. If you have a little boy, you give him a toy gun to shoot other kids with. We socialize boys and girls differently, and that's why boys bully more than girls. It's seemingly that simple. But have the social scientists really got it right?

An answer can be found by exploring the geography of the prefrontal cortex. What I never told you about in our temp study is that we started off testing women as well as men. But we soon gave up trying to recruit female felons. You women out there are the wonderful angels that make the world go round. It's we men that maketh mayhem. We had recruited just seventeen women in our sample and we were finding that they were not giving us a lot in terms of crime and violence. Plus, our money for the study was tight. So when the going got tough, we dumped the dames and recruited the tough guys. That was a mistake in retrospect, but we still had just enough women in our sample to test a controversial counterhypothesis to differential socialization as a cause of sex differences in crime. Could it be that there are fundamental brain differences between men and women that explain why men commit more crime?

We compared men with women on prefrontal brain volumes. Men had a 12.6 percent volume reduction in the orbitofrontal gray compared with women.[52] That's the underneath part of the prefrontal cortex. Men with reduced ventral gray were more antisocial than men with normal ventral gray volumes. We've seen that already, but what was new in our analyses was that women with reduced ventral gray volumes were more antisocial than women with normal gray volumes. We get the same brain effect in antisocial women that we find in antisocial men. Hold these findings in your prefrontal cortex's working memory for a minute.

Men, of course, were found to be more antisocial and criminal than women, replicating a worldwide finding. No big deal. But what if we look again at this sex difference in crime, this time controlling for the sex difference in ventral gray volume? If we make men and women statistically the same in terms of their ventral volume, we cut the sex difference in crime by 77 percent.[53] So more than half of the reason men and women differ in crime seems to be because their brains are physically different.

I'm not saying that all the difference in crime between men and women can be put down to the brain. And I'm certainly not saying that we should ignore differences in socialization and other social and parenting influences. But what I am arguing is that there are fundamental neurobiological differences between men and women that can help explain the gender difference in crime. It's also striking that we find sex differences in the very same frontal sectors that are linked to antisocial behavior—men and women did not differ in prefrontal sectors that are not related to crime.

These findings do not come out of the blue. Sex differences in prefrontal gray have been documented in several other MRI studies. One imaging study found a 16.7 percent reduction in orbitofrontal volumes in men compared with women.[54] Three other studies have found this same sex difference,[55] including one large study of 465 normal adults.[56] Men have also been reported to show lower activation of the orbitofrontal cortex compared with women when performing a wide variety of cognitive and emotional tasks, including verbal fluency,[57] working memory,[58] processing threat stimuli,[59] and working memory during a negative emotional context.[60] Men simply have different brains from women, and it's pointless to cover up and ignore these fundamental sex differences.

THREE CHORDS OF CAUTION

The position so far looks like this: We've seen that offenders have structural impairment to the prefrontal cortex. We've also seen that they have poor functioning of this same brain region. We documented in a meta-analysis of forty-three brain-imaging studies of offenders involving 1,262 subjects that these structural and functional prefrontal deficits are replicable findings.[61] The structural prefrontal impairment partly explains the sex difference in crime. It's hard to escape from the conclusion that impairment to this brain region—through either environmental or genetic causes or both—predisposes some to an antisocial, disinhibited, impulsive lifestyle.

But before moving on, let's underscore an important fact: no proposed cause of offending—whether it be social or neurobiological—inevitably results in crime and violence. While the dramatic case of

Phineas Gage from Vermont in 1848 originally set up the prefrontal dysfunction theory of psychopathic and antisocial behavior, three more
clinical cases strike a chord of caution lest we take this theory too far.

The first is the remarkable case of an individual known as the Spanish
Phineas Gage—referred to here as SPG—a twenty-one-year-old university student living in Barcelona. It was 1937, the Spanish Civil War was
raging, and nobody was safe. One fateful day he found himself upstairs
in a house being pursued by the opposition in this civil struggle. Almost
cornered, he threw open the window, climbed out onto the windowsill,
and made a bold attempt to escape by shinnying down the drainpipe on
the outside wall.

Unfortunately for SPG, the pipe was old, and it broke away from
the wall. SPG clung on to it for dear life, falling down onto a spiked
metal gate. His head was impaled on the gate, with a spiked point
entering the left side of his forehead, injuring his left eyeball, and coming out through the right side of his forehead. It selectively damaged his
prefrontal cortex, just as the metal tamping rod had blasted a discrete
hole through Phineas Gage's brain.

People came to the rescue. They were able to cut through the bar,
with SPG conscious all the time throughout the ordeal. He even helped
his rescuers to get him off the gate. As with Gage they quickly got
him to medical care, delivering him to the Hospital de la Santa Creu i
Sant Pau in Barcelona.[62] The damage to his prefrontal cortex was quite
extensive, and, just like Gage, he lost vision in his left eye. Again like
Gage he survived the horrific accident, and it was not long before he
was back on his feet, creating a new life for himself. And yet again it
truly was a new life. Just like Gage, he was impatient, restless, impulsive, and would move from one thing to another, unable to properly
finish any single task.

Yet here the striking parallel ends between the American and the
Spanish Phineas Gages. Despite having the usual executive dysfunction
that one expects from such a head injury, and despite his impulsivity,
SPG did not develop the antisocial, psychopathic personality that characterized Gage. Why not?

The answer once again lies at least in part in the environment. At
the time of the accident, he was engaged to his childhood sweetheart.

As they once said in Rome, *amore vincit omnia*—love conquers all. And over in Barcelona love helped conquer the antisocial sequelae that we might normally have expected from this dreadful prefrontal damage. SPG's sweetheart stood by him, and three years after the horrific accident, they were married. Unlike Gage, SPG had spousal support, and his support system did not end there. For the rest of his life he was able to hold down a steady job in one location, unlike Gage, who drifted around for a significant period of his life.

How could this be possible, you may say. You are by now becoming an adroit neuropsychologist, and you know that prefrontal damage invariably leads to the inability to sustain attention, to complete a task, to shift strategies in tackling problems, and to plan ahead. This was indeed true of SPG, who showed significant impairments on frontal-lobe executive tasks. But the environment is again the answer. His parents were wealthy and owned a family firm where SPG was employed for the rest of his life. His poor executive functioning meant that he was never a particularly good worker. He could do only basic manual tasks and always had to be closely supervised and checked. Yet a job it was, and with it came security and occupational functioning.

Lady Luck was not finished with SPG. He not only had a devoted wife and caring, affluent parents to support him, but he also went on to have two loving children who were destined to play a role in his psychosocial rehabilitation. In the words of his daughter:

> As a child, I realized that my father was a "protected" person. When I was young I soon saw what the "problem" was, although I had always suspected it. At 17, I became part of this protection, and I still am.[63]

SPG could hold his broken head high throughout his life. He was always able to bring home the bacon after his hard day's work. He had occupational functioning. He had family functioning. He had love in his life from all quarters. As many of you likely know if you reflect on episodes in your own lives, love truly can overcome enormous adversity. For me this case highlights the critical importance of psychosocial protective factors that can guard against a life of crime in the face of horrendous prefrontal damage.

As with Gage, we see in SPG a man who was not antisocial before the accident that caused the prefrontal damage. Let's now turn to our second chord of caution, but here our case was antisocial *before* the head injury.

This second case study dates from approximately 2000 and concerns a thirteen-year-old-boy from Utah who by all accounts was a bit of a Johnny-gone-rotten.[64] For most of his short life, he had been rotten to the core, with a well-documented history of conduct disorder, risk-taking, hyperactivity, and attention-deficit disorder. Sadly, his parents had long since lost their parental rights, and he lived in a foster home. He was a bad kid, but bear in mind that genes and an early negative home environment likely worked against him to make him what he was.

One day the lonely lad was playing Russian roulette by himself with a .22-caliber pistol. After all, despite the natural beauty of the state, what else is there for a hyperactive, stimulation-seeking, conduct-disordered boy to do in Utah? With the pistol perched underneath his chin and the barrel pointing straight up, he pulled the trigger. The loaded chamber turned, the wheel of fortune spun, and the pistol went off. He succeeded in punching a hole right through his prefrontal cortex.

Again our case was rushed to the hospital. Yet again he miraculously survived his deadly game. The CT scan taken soon after he arrived at the hospital showed that the bullet had punched a neat hole through his brain, selectively damaging the very middle part of his prefrontal cortex in much the same way that Phineas Gage's medial prefrontal cortex was damaged by the tamping rod. If he had wanted to selectively take out this very midline part of the medial prefrontal cortex, frankly, the poor youngster could not have done a better job.

The really unusual aspect of this case is, well, nothing. I mean, nothing really unusual happened afterward. Despite losing at Russian roulette, the boy did not have such a bad ending. His social workers, foster parents, psychologist, and all legal authorities who had been managing his case agreed that he was completely unchanged by the brain damage. He was the same unruly, conduct-disordered urchin that he always had been. But he was not worse. He did not even show any additional cognitive deficits.

As the Americans say, "What gives?" The neuropsychologist Erin

Bigler, who reported this case, reasoned that the young teenager had succeeded in knocking out only the piece of his medial prefrontal cortex that was *already* dysfunctional, the part that had been causing the conduct disorder in the first place. This second case study highlights a truism—that prefrontal damage does not by any means always result in behavioral change in the antisocial direction, particularly if, unlike the American and Spanish Phineas Gage cases, the individual was not normal to begin with.

THE PHILADELPHIA CROSSBOW MAN

Our third case takes this principle to another level. It underscores the point that there can be marked differences in outcome when prefrontal damage strikes. It is yet another Gage-like accident, and, as with the Utah Russian-roulette case, we are dealing with an individual with a deeply entrenched preexisting antisocial condition. But on this occasion there is an astonishing change in behavior after the accident.

This chord of caution deals with a thirty-three-year-old man from Philadelphia who had a history replete with antisocial and aggressive behavior throughout his life—a life-course persistent offender who was pathologically aggressive. He was also depressed. In fact, he was very depressed. He decided to end his life—but in an unusual way. He took a crossbow and—in a manner remarkably reminiscent of the Russian-roulette case—he placed the bow underneath his chin, with the arrow bolt pointing straight up, and he released the trigger.

Like a tamping rod, the bolt shot right up into his prefrontal cortex, and as was the case with the Spanish Phineas Gage, the deadly projectile lodged firmly in his brain. Like the other victims we have witnessed, he was rapidly rushed to medical care. Yet again it's a strange survival story. This unhappy and deeply troubled man was taken to my university hospital—the Hospital of the University of Pennsylvania—to have the bolt extracted from his brain. It selectively damaged the medial prefrontal cortex, just as it had been with Gage and our Russian-roulette case. The missiles in all three cases had essentially the same trajectory, entering from the lower part of the head and exiting from the top of the front part of the skull.

There was a new twist in this case. As with Gage, the Philadelphian Crossbow Man was radically changed by the prefrontal damage—but in the opposite direction. Gage had been transformed from a normal man into a psychopathic-like individual. The Philadelphia Crossbow Man was instead transformed from an aggressive, irritable, emotionally labile antisocial into a quiet, docile, and content man.

The pathological aggression was eradicated overnight. The depression disappeared in a jiffy. It was a miracle cure. Indeed, the only neuropsychiatric symptom that resulted from the damage to a man who had been seriously depressed was that, in the words of his clinician, he became "inappropriately cheerful."[65] He simply cheered up.

This third case study again reveals the complexity of the relationship between brain and behavior, and highlights the striking differences in outcome that can occur as a function of damage to the prefrontal cortex. In the crossbow case, the fact that this disturbed and depressed individual became jolly after the accident is not entirely surprising. Puerile jocularity is one neurological symptom of damage to the prefrontal cortex, and this is what we see here. Indeed, puerile jocularity also characterized the Spanish Phineas Gage. Apparently, he spent a lot of time telling the same old lame jokes and being overly cheerful.[66] So when at your next work party you meet that disinhibited, loquacious extravert who tells bad jokes and laughs at them like there is no tomorrow, make a neurological note to yourself and suspect either a spiked bar, a crossbow bolt—or perhaps just plain old frontal-lobe dysfunction.

Clearly we must be cautious with our prefrontal cortical explanation of crime. Prefrontal damage doesn't always produce antisocial behavior. But let us not forget that overall there is a link between prefrontal structure and violence based on MRI and neurological studies, so we have to be equally cautious not to discount the hypothesis that prefrontal brain damage causes violence.

Let's take this idea a step further from a developmental standpoint. Neurological studies have shown us that brain damage in childhood and adulthood can raise the odds of violence. Now we'll use structural MRI to delineate more precisely that moment in time when something goes badly amiss in brain development—and here we must go back even beyond birth.

BORN TO BOX?

We saw in the Introduction that Cesare Lombroso was fascinated with the idea of a physical brain difference that marked out the born criminal. While no criminal is really "born bad," I believe there is a "neurodevelopmental" brain abnormality in some offenders—a brain that does not grow in quite the way it should.

One indication of brain maldevelopment very early on is a neurological condition called cavum septum pellucidum. Normally everyone has two leaflets of gray and white matter fused together called the "septum pellucidum" that separate the lateral ventricles—fluid-filled spaces in the middle of the brain. You can see that black space in the normal brain in the left image of Figure 5.5, together with the white septum pellucidum line that divides the black ventricles. During fetal development there is in addition a smaller fluid-filled, cave-like gap—or "cavum"—right in between these two leaflets. You can see this black gap separating the two white leaflets of the septum pellucidum in the brain depicted in the right image of Figure 5.5. As the brain rapidly grows during the second trimester of pregnancy, the growth of your limbic and midline structures—the hippocampus, amygdala, septum, and corpus callosum—effectively press the two leaflets together until they fuse. This fusion is completed between three and six months after you are born.[67] But when limbic structures do not develop normally, the cavum between the two leaflets remains—hence the term cavum septum pellucidum.

When we scanned the brains of our subjects from the temp agencies, we found that nineteen of them had cavum septum pellucidum— just like the one shown in the right image of Figure 5.5. We called these the cavum group—those with a visible marker of very early brain maldevelopment. We compared them to individuals with normal brains. Those with cavum septum pellucidum had significantly higher scores on measures of both psychopathy and antisocial personality disorder compared with controls. They also had more charges and convictions for criminal offenses.[68]

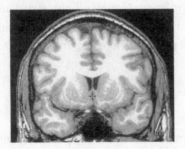
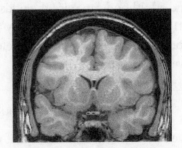

NORMAL BRAIN   CAVUM SEPTUM PELLUCIDUM

Figure 5.5

This research design is the "biological high risk" design. You don't see it too often. We are taking those with the neurobiological abnormality and comparing them with those without the abnormality. But we can also slice this particular pie another way. Let's instead start off by taking those with psychopathy, and compare them to non-antisocial controls on the *degree* to which they have cavum septum pellucidum. Fusion of the septi pellucidi from back to front during fetal development is partly on a continuum. It's a bit like when you zip up your jeans—the zip might not close all the way and there is a gap left. So we can measure the extent to which the septum pellucidum is "zipped up," so to speak.

What we find is that psychopaths have a greater degree of incomplete closure of the septum pellucidum, reflecting some amount of disruption to brain development. But it's not just psychopathy. This is also true of those with antisocial personality disorder as well as those with criminal charges and convictions. It cuts across the whole spectrum of antisocial behaviors.

We see here in the classic clinical design—where we compare those with and without a clinical disorder—a convergence of findings that match those from the biological high-risk design. Different research designs converge on the same conclusion—there is an early neurodevelopmental basis to crime occurring even before the child is born. The evidence for a neurodevelopmental basis to criminal and psychopathic behavior is mounting.[69] As much as traditional criminologists and sociologists would hate to admit it, Lombroso was partly right.

We don't know what specific factors can account for the limbic maldevelopment that gives rise to cavum septum pellucidum. We do know, however, that maternal alcohol abuse during pregnancy plays a role.[70] So while talk of a neurodevelopment abnormality sounds like genetic destiny, environmental influences like maternal alcohol abuse may be just as important.

There is an interesting twist to the link between cavum septum pellucidum and crime. In our study we found that brain maldevelopment was especially linked to features of antisocial personality related to life-long antisocial behavior—things like a reckless disregarded for self and others, lack of remorse, and aggression. Interestingly, boxers are more likely to have cavum septum pellucidum than controls. Is that because the brain damage is caused by being biffed about in the boxing ring rather than the other way around?

Researchers think not, and instead have touted the provocative idea that those with cavum septum pellucidum are "born to box."[71] Their idea is that cavum septum pellucidum nudges the individual into developing an aggressive personality. Those with aggressive tendencies are more likely to take up boxing, making good use of their natural aggression. But could trauma and head injury in our temp workers result in cavum septum pellucidum? We controlled for these factors, as well as many psychiatric confounds, and results remained unchanged. Cavum septum pellucidum by itself predisposes people to antisocial, psychopathic, and aggressive behavior.

For some, therefore, it's an early neurodevelopment disorder that puts their limbic system out of kilter and places them on a path to crime. Add in a degree of frontal-lobe dysfunction, and they lose full control of their basic instincts—whether it's sex or aggression or both.

FEARLESS ALMONDS

It's worth repeating that the complexity of the brain matches the complexity of the causes of crime. When we learn more about our neurobiology in forthcoming decades, we'll see that multiple brain systems are complicit. We have dug down from the surface of the prefrontal cortex into the very deepest chasms of the brain—the cavum septum pellucidum. To mine more knowledge on violence, let's now move away from the very center of the brain into that dysfunctional limbic system that seems not to be developing properly in psychopaths. The key culprit dwelling in this neural neighborhood? We think it's the amygdala.

The amygdala is an almond-shaped structure lying in a deep cortical fold inside the brain—an area called the medial surface of the temporal lobe. There is one in each hemisphere of the brain, about three-quarters of the way down from the top of the brain depicted in Figure 5.6. This part of the brain is critically involved in the generation of emotion. No brain area is more important in the minds of neuroscientists for emotion than the amygdala. Recall that one of the striking features of the psychopath is a lack of affect and emotional depth. Juxtapose this obvious clinical observation with the equally obvious role of the amygdala in the generation of fear, and you come up with a surprisingly simple hypothesis—that the amygdala is structurally abnormal in psychopaths.

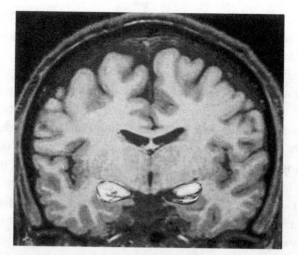

Figure 5.6 Coronal slice of the brain showing the left and right
amygdala toward the base of the brain

Despite its seeming simplicity, nobody had ever tested this hypothesis until my team and I scanned psychopaths and conducted a fine-grained analysis of their left and right amygdalae. In collaboration with our colleagues Art Toga and Katherine Narr at UCLA, we used state-of-the-art mapping techniques to assess the morphology of this brain area in both psychopaths and controls. Art Toga and his laboratory had developed the ability to map group differences on a pixel-by-pixel basis throughout the amygdala. Almost all functional-imaging research findings talk about the amygdala as a unitary structure—largely because the activation patterns seen are quite broad and not localized to any specific subregion. But my astute graduate student from Taiwan, Yaling Yang, reasoned that the amygdala is in reality made up of thirteen different substructures or nuclei, each with different functions. Is the amygdala deformed in psychopaths? And if so, which specific nuclei within the amygdala are compromised?

Yang found that both the right and left amygdalae are impaired in psychopaths—although the deficits are greatest on the right. Overall, there was an 18 percent reduction in the volume of the amygdala in psychopaths.[72] But what specific subareas of the amygdala are structurally compromised? Yang brilliantly mapped out the corresponding amygdala nuclei. Three of the thirteen nuclei were found to be particularly deformed in psychopaths—the central, basolateral, and cortical nuclei.

The specific areas of the amygdala that were deformed in psychopaths are darkly shaded in Figure 5.6. What do these three subregions of the amygdala do?

The central nucleus is strongly involved in the control of autonomic nervous system functions and is also involved in attention and vigilance.[73] Not surprisingly, it plays a particularly important role in classical conditioning, and we saw earlier that fear conditioning is the key to conscience, with psychopaths and criminals having fear-conditioning deficits as well as attentional deficits. The basolateral nucleus is important in avoidance learning—learning not to do things that result in punishment.[74] In this respect, recidivistic offenders just cannot learn when to give up on criminal behaviors that get them punished with imprisonment. The cortical nucleus has been shown to be involved in positive parenting behaviors, and we know what lousy parents psychopaths make. Sum up the functions of the three nuclei of the amygdala that are structurally impaired, and it's not too surprising that psychopaths are functionally compromised in areas important for prosocial behavior.

We think that these structural impairments to the amygdala are likely to be a product of fetal neural maldevelopment. That is, we suspect that something is going very wrong with how this brain structure develops throughout early life in psychopaths. It could be the type of early "health insults" that we will discuss later—like nicotine and alcohol exposure—or some other teratogen that interferes with normal limbic development just as we have seen in cavum septum pellucidum. So it could have an environmental cause.

But it could also be genetic. Unlike the ventral prefrontal cortex and the frontal pole (the very front of the brain), which are quite susceptible to damage resulting from environmental head injuries, the amygdala, with its location deep in the brain, is not generally affected by environmental insults. We simply cannot ignore the possible role of genes in the structural deformations that we observe in psychopaths.

Could the cause of the amygdala deformations be crime and psychopathy itself? Could being cold, callous, and unemotional somehow shrink the amygdala? After all, brain imaging in adults is correlational and does not demonstrate causality. What would help us here are longitudinal brain-imaging studies scanning young children early in life and following them up into adulthood to find out if the amygdala impairment precedes the onset of antisocial behavior in late childhood.

Don't hold your breath. These studies have not been conducted. Young children don't sit still in scanners, and it will be a long time before imaging studies of tiny tots are able to demonstrate whether an abnormal amygdala predicts adult violence and crime. Yet the amygdala analysis in adult psychopaths sets the stage for the idea that amygdala impairments predispose people to later antisocial and psychopathic behaviors—and not the other way around.

Poor fear conditioning is a solid marker for poor amygdala functioning. As we saw in chapter 4, poor fear conditioning as early as age three predisposes someone to crime twenty years later. Yu Gao strikingly demonstrated a link between amygdala functioning in early childhood and adult crime. Causation still cannot be claimed, but the temporal ordering of this relationship has been teased out. Poor conditioning precedes crime by a long chalk. It's about as good as it gets to demonstrating causality, and Yu Gao's results suggested that Yaling Yang's finding of structural amygdala deformations in psychopaths is quite likely a causal predisposition to callous, cold-hearted conduct. Students from China and Taiwan had teamed up to wage war on violence and make new scientific inroads into understanding the brain basis to crime.

PATROLLING SEA HORSES

Moving from the frontal control region of the brain to the deeper limbic emotional areas, we are seeing signs that something is fundamentally wrong with the brain's anatomy in offenders. Their anatomical anomalies are not restricted to these brain regions. If we move just a bit further behind the amygdala, we come to the hippocampus, a critical region shaped like a sea horse that's involved in a variety of functions ranging from memory to spatial ability. Here too we find a structural abnormality in psychopaths, but of an unusual kind.

We saw earlier how hippocampal functioning was impaired in offenders. That functional abnormality is likely caused by structural abnormalities that have been observed in a wide number of studies. In one group of psychopaths that we studied we found that the right hippocampus was significantly bigger than the left.[75] This structural asymmetry is true in normal people too, but it is much stronger in psychopaths. Interestingly, we found this very same asymmetry in our sample of murderers, this time in terms of function.[76]

What causes this abnormality is not known for certain, although there are some interesting clues. If rat pups are moved around early in life into different "homes," they develop an exaggerated hippocampal asymmetry: the right hippocampus grows to be bigger than the left.[77] We found in our interviews with psychopaths that they had been bounced around from home to home much more often than controls in their first eleven years of life—more than seven different homes in psychopaths compared with three in controls.

Another factor is fetal alcohol exposure. When the brains of children suffering from fetal alcohol syndrome are scanned, it is found that the right-greater-than-left hippocampal volume that is found in normal controls is exaggerated by 80 percent.[78] If you have read casebooks on killers, these two clues will be familiar to you. The early lives of violent offenders are invariably characterized by broken homes, substance-abusing and neglectful mothers, and instability. These factors taken together could be the environmental cause of the hippocampal abnormality we see in psychopaths.

Other researchers have similarly observed overall smaller hippocampal volumes in violent alcoholics.[79] In psychopaths, structural depressions have been found in areas of the hippocampus that play a role in autonomic responses and fear conditioning,[80] while we have similarly observed volume reductions in the hippocampus in murderers from China.[81]

What does the hippocampus do apart from helping you remember your boyfriend's birthday and how to get to Walmart from the freeway exit? The hippocampus patrols the dangerous waters of emotion. For one thing, it is critically important in associating a specific place with punishment—something that helps fear conditioning.[82] Just think back to where you were when a bad thing happened—that's your hippocampus helping you remember. So, like the amygdala, it plays a key role in fear conditioning and other forms of learning that partly constitute our conscience—the guardian angel of behavior. Criminals have clear deficits in these areas. The hippocampus is also a key structure in the limbic circuit that regulates emotional behavior.[83] From animal research we know that the hippocampus regulates aggression through projections to the midbrain periaqueductal gray and the perifornical lateral hypothalamus. These are deep subcortical structures that are highly important in regulating both defensive and reactive aggression as well as predatory attack.[84] For example, rats with hippocampal lesions at birth show increased aggressive behavior in adulthood.[85] These hippocampal abnormalities could be linked to the cavum septum pellucidum abnormality we just discussed, because the septum pellucidum forms part of the septo-hippocampal system, a brain circuit that researcher Joe Newman has argued plays a role in psychopathy.[86]

The hippocampus and amygdala are located in the inner side of your temporal cortex. But that's not right in the middle of your brain. What is in the middle is the corpus callosum—a colossal body of over 200 million nerve fibers that connect your two cerebral hemispheres. These fibers—the corona radiata—radiate out from the very center of your brain to the outer areas of your cerebral hemispheres, interconnecting many different brain regions. We measured the volume of the corpus callosum and its corona radiata and found that this volume is much bigger in psychopaths with antisocial personality disorder. It was also longer. And thinner too. A long, thin body of white matter. It's as if there is too much connectivity in the brains of psychopaths—too much cross talk between the two hemispheres.

What do we make of this? Although we often think of psychopaths as antisocial villains with a lot of negative characteristics, they're actually a lot of fun. They have a lot of positive features, especially on the surface. In particular, many psychopaths have the gift of gab. They are very glib, very charming, very good con artists who can convince you of almost anything. Robert Hare—regarded by many as one of the world's leading researchers on psychopathy—has demonstrated, using something called the dichotic listening task,[87] that psychopaths are less "lateralized" for language.[88] We found the same thing in juvenile psychopaths.[89] What does this mean? In many of us, the left hemisphere is largely responsible for language processing—language is strongly lateralized to the left hemisphere. But in psychopaths it's more of a mix of both left and right hemispheres. This might be why they seem to be so adept in their verbal skills. They have two hemispheres—not one—that they can utilize for language processing. This in turn could be due to a larger, better communicating corpus callosum.

We have to remember that psychopaths are a special group of criminal offenders and that we cannot say the same thing about run-of-the-mill violent offenders. But whichever way you look at it, psychopaths appear to be literally "wired" differently from the rest of us.

GETTING THE GOODS

We have moved anatomically from the surface of the brain—the cortex—into the deeper brain regions—the subcortex. Now let's continue our subterranean tour to another deep-brain region—the striatum. In evolutionary terms, this is an old brain structure involved in one basic function common across all species—reward-seeking behavior. For a long time in our laboratory we have felt that psychopathic individuals may be characterized by an oversensitivity to rewards. When there is a chance of getting the goods, they seem to go all out— even at the risk of negative consequences.

The first new study I conducted when I moved from Nottingham to Los Angeles sought to test out this idea.[90] I was an assistant professor. As for all assistant professors when I started out, academic life was not that easy. I was involved in studies in England and Mauritius, but the expectation was that you should also be setting up your own laboratory and conducting work independent from other investigators in order to establish your independence. You have to show that you have what it takes to go it alone.

Easier said than done. I felt lost in L.A. I didn't have a penny for research funds, so whatever research I did would have to be done on the cheap. One piece of luck was that I had two students who wanted to work with me during their summer alongside with Mary O'Brien, a senior professor's graduate student who was interested in child antisocial behavior.

The next bit of luck was that there were a bunch of juvenile delinquents living just down the road from me in the Eagle Rock neighborhood of Los Angeles. I got permission from the Superior Court of California to work with them. They lived in a home as an alternative to being sentenced to a closed institution, and for these teenage boys participating in experiments with young female undergraduates from USC was not unappealing. Forty out of the forty-three kids we approached were keen to be involved in the study.

The third bit of luck was that while I had given up orange juice to save for a down payment on a home, I did have a deck of cards and some plastic poker chips. Taken together, this would be enough for my first study in L.A.

We had the mischief-makers play a game of cards that went something like this. Each card had a number on it. For half of the numbers, selecting them would result in the gain of a poker chip—so this was a reward card. Half of the cards, though, would result in a loss—a punishment card. Touching the card was a response. The subject could touch it to select it, or not touch it to pass. Over the course of sixty-four card plays, the subject had to make as much money as he could—to learn which cards were the winners. We assessed which of the delinquents were psychopaths based on staff ratings of their behavior and personality, and then we compared them to delinquents who were not psychopaths.

The results? My graduate student Angela Scarpa showed that our young psychopaths showed much greater response to the reward cards than the non-psychopaths. They were hooked on rewards, confirming previous studies showing the same in adult psychopaths.[91] Our budding psychopaths actually showed better learning throughout the task too. This suggests that psychopaths can learn—as long as you use rewards to shape their behavior. It was the first time that a reputable journal had published a study on "juvenile psychopaths."[92] Until then, nobody liked the idea that adolescents might actually be psychopaths in the making.

Twenty years went by and we were still mulling over the findings. Could this behavioral difference translate to brain differences in psychopaths? My graduate student Andrea Glenn tested the idea out on our psychopaths from temp agencies.[93] The striatum is a key brain region that is associated with reward-seeking and impulsive behavior. Studies have also showed that it is involved in stimulation-seeking behavior, persistently repeating actions that are related to rewards, and enhanced learning from reward stimuli.[94, 95] Sounds like psychopathic behavior, doesn't it? We found that our psychopathic individuals showed a 10 percent *increase* in the volume of the striatum compared with controls. Results could not be explained by group differences in age, sex, ethnicity, substance or alcohol abuse, whole brain volumes, or even socioeconomic status. They seemed pretty solid.

We reasoned that the increase in striatal size could contribute to the increase in the sensitivity of psychopaths to rewards, and consequently their incessant reward-seeking behavior. To be sure, psychopaths are not alone. We are all driven by rewards. We each want our own stuff. We want masses of money, a decent dwelling, fancy food, wonderful work, fun friends—and let's throw in superb sex for good measure. But the difference between us and psychopaths is that we can say no when tempted by the goodies, whereas psychopaths just want their stuff. And they want it here, and they want it now. For them, reward is a drug that they cannot turn their backs on, and this pushes them along a path of depravity and vice.

Our findings on psychopaths did not stand alone. Increased striatal volumes have also been found in those with antisocial personality disorder,[96] while increased striatal functioning has been observed in violent alcoholics[97] as well as aggressive adolescents and adults.[98] Furthermore, in 2010, just two months after we had published our study touting this neural basis to reward-seeking behavior in psychopaths, a functional brain-imaging study came out from another research group with essentially the same argument.[99] People in the community scoring higher on impulsive, antisocial features of psychopathy were found to be hypersensitive to rewards, this time due to excessive activation of another subcortical brain area when anticipating a reward—the nucleus accumbens. This brain area is strongly involved in the brain's dopamine-reward circuitry, which we discussed in chapter 2. Antisocial individuals really do appear to be turned on more than the rest of us by stuff that takes their fancy.

Rewards are important to offenders, and to them money doesn't just talk—it swears. It's very salient to them. A full 45 percent of psychopaths are motivated by money in the crimes they perpetrate.[100] Studies also show that it takes less money to push psychopaths into violating moral principles than non-psychopaths.[101] But more troublingly, aggressive, conduct-disordered kids show increased activity of the striatum when they view images of other people in pain.[102] Somewhat sickeningly, these aggressive children seem to enjoy seeing people in pain, not unlike a number of serial killers who cruelly torture and maim their victims. Combine this characteristic with frontal-lobe dysfunction and the disinhibited behavior it causes, and you have a cocktail for criminal violence.

However we interpret structural deficits of the amygdala, hippocampus, corpus callosum, and striatum in psychopathic and antisocial offenders, one thing stands out. These structural abnormalities are likely not the result of some discrete disease process or obvious trauma. Such causes would if anything result in overall volume *reductions* to these structures. Our findings are much more complex than that. The right hippocampus is *larger* than the left in psychopaths. The striatum is *larger*. The corpus callosum also has a *bigger* volume. And the corpus callosum is not only *longer* in psychopaths than in controls, it's also *thinner*. So what's the explanation here? It is likely that this shape distortion is neurodevelopmental in nature. The striatum and its associated structures—the caudate and lenticular nuclei—are enlarged, not shrunken. These brain structures are growing abnormally in psychopaths during infancy and childhood. Again we get back to the idea that there is—at least in part—a *neurodevelopmental basis* to psychopathic and antisocial behavior. A born criminal? Not really. But a baby whose brain is compromised in its development? Quite likely.

PINOCCHIO'S NOSE AND THE LYING BRAIN

I want to extend our neurodevelopmental argument by looking at structural abnormalities of the brain that take the form of *advantages*, not disadvantages. We'll combine this theme with a core question. The brains of violent and psychopathic offenders may be deformed, but can this also apply to other offenders? What about me and you when we tell a fib or two? Are there brain bases to less serious forms of offending?

Lying is pervasive. At some level, most of us lie most days of the week. We lie about almost anything. When do we lie most? Community surveys show it's on our first date with a new person. And this gives us a clue as to why we lie so much—it's impression management. If we were brutally honest all the time, we'd likely never get that first kiss. Plus we'd make life really miserable for everyone. Do you really want me to tell you what I honestly think of that dreadful new haircut? That gaudy shirt? Your bad-mannered new boyfriend? No, you don't. So we use white lies to smooth out the rough-and-tumble of everyday social encounters. "That new hairdo suits you!" "That shirt really brings out your personality." "Your new boyfriend is a perfect match for you!" We gain the affection and friendship of others, and at times simply do more good than harm. None of us are saints, but most of us are not serious psychopathic sinners either.

Most of us, that is. For others, lying goes a bit too far. One of the twenty traits of the psychopath is pathological lying and deception. They lie left, right, and center. Sometimes for good reason, and sometimes, perplexingly, for no reason. When I worked with psychopaths, before conducting my induction interview I would review in detail their whole case file. And given that I was working in top-security prisons with long-term prisoners, their files were fairly complete. The information about their life trajectories, behaviors, and personalities gave me a good basis upon which to determine whether the prisoner I was working with was a pathological liar. When someone says something that conflicts with what you know, you have a good opportunity to challenge him. You can check if what he says back to you sounds like sense or seems a sham.

The trouble with psychopaths, though, is that they really are extraordinarily good at lying. Just when you think you've nabbed them telling an enormous whopper, they have the uncanny ability to reel off a seemingly convincing explanation for the discrepancy without batting an eye. Believe me, against your own better professional judgment you could walk out of that interview room believing that you must have gotten your facts wrong—only to read the file again and check in with the senior probation officer and realize he duped you. You really have to experience it to believe it.

It might surprise you to learn that I don't have a clue who is and who is not a psychopath, even after four years of working full time with them in prison and thirty years of academic research. I'm just not that fast on the uptake in this arena. If I met you for the first time and we chatted for an hour, I would be none the wiser as to whether you were a psychopath or not. I'll come back to that later. But it's not just me. Whether you like it or not, you too are completely clueless when it comes to knowing if someone is lying to you or not.

Don't take it personally—we're *all* hopeless, not just you and me. Police officers, customs officers, FBI agents, and parole officers. They are no better than plain old undergraduates in their ability to detect deception.[103, 104] They actually believe they are good at lie detection; they don't even recognize their own mistakes. Doctors don't know when you are lying to them about made-up symptoms in your attempts to get the medications you want.

Why are we so bad at knowing who's a liar? It's because all the things that we *think* are signs of lying are quite unrelated to the ability to detect deception. Think of a time when you did not have any tangible background evidence or context to tell if someone was lying to you, but you judged that person as lying based on how they spoke and behaved. I bet you were basing this on things like their shifty gaze, hesitations in their speech, their fidgeting, or their going off-topic into some detail. In reality, none of these are related to lying.[105] They give us false clues, and we are misled by them.

But how about kids? Surely we are better in judging when a child lies to us? Aren't we?

Well, no, we're not. In one study on this topic, children of different ages were videotaped sitting in a room with the experimenter.[106] Behind them is an interesting toy. The experimenter tells the child he must go out of the room for a few minutes and that the child should not peek at the toy while he's away. The experimenter goes away for a while and comes back. Some kids peek, some do not. The experimenter then asks the child if he or she peeked. Of those who deny peeking, some are telling the truth and some are lying. Experimenters then show the videotapes to a range of individuals to see how good they are at telling when a child is lying. Being correct 50 percent of the time would be the level of chance, because in this scenario, 50 percent of the film clips show a child lying and 50 percent show a child telling the truth.

The tapes are given to undergraduate students. Surely working out if a kid is lying has to be easier than most university exams. But these smart undergraduates are correct 51 percent of the time, not significantly above chance.

So let's see how customs officers fare when viewing the same tapes—they have a boatload of experience in picking out deceptive travelers. They are at 49 percent—below chance levels—though to give them some credit they are not significantly worse than the hapless undergraduates.

Okay, so let's go to cops, as surely they are streetwise about these fledgling psychopathic liars. Nice try, but actually the police are at 44 percent accuracy levels, significantly *lower* than chance, and significantly poorer than undergraduates or customs officers. Next time a cop stops you and accuses you of a traffic violation that you deny and he will not believe your protestations, remind him about this study.

So let's try again. Maybe eleven-year-olds are sophisticated liars, and so we might understand how overall accuracy levels with these kids are at a miserable 39 percent. But can't we tell if a four-year-old is lying? Actually, we cannot. Accuracy levels are at 40 percent at this age, 47 percent at age five, and 43 percent at age six. Parents, you *think* you know what your kids get up to, but actually you don't even have a clue with your own toddler. That's how bad the story is. Sorry, mate, but you really are as hapless as I at figuring out who a psychopathic liar is.

But here's a ray of hope for you. I have two ten-year-old monkeys at home who are always getting into mischief. And yes, Andrew and

Philip are clever and skillful liars—just like most kids. When I want to know who did what, before I pop the question I tell them that it's important to be honest and they should promise to tell the truth. Research indicates that getting young children to talk about moral issues first and then asking them to promise to tell the truth significantly encourages a truthful answer—boosting lie detection accuracy from 40 percent to 60 percent.[107]

This research on children made me and my lab intrigued about what makes a psychopath a good liar. People may be hapless at lie detection, but perhaps machines have a mechanism to better delve inside the minds of Machiavellians. Psychopaths may be able to lie to us face-to-face, but perhaps the signature of a pathological liar may reside below the surface inside their brains. Might pathological liars have a physical advantage over the rest of us when it comes to pulling a fast one?

In our study we assessed whether people had a history of repeatedly lying throughout life.[108] We assessed this in our psychiatric interviews on antisocial personality disorder and psychopathy. We also measured it using questionnaires, and by cross-checking notes between our lab assistants.

For example, on one day our research assistant was struck by the fact that a participant walked on his toes. Upon questioning, our participant told a detailed and convincing story of how he was in a motorbike accident resulting in damage to his heels. The very next day, he was being assessed by a different research assistant on a different floor of our building and he walked perfectly normally. The con only came to light when our research assistants traded notes. A typical pathological lie: deception but without any obvious gain or motivation.

We ended up with a group of twelve who fulfilled criteria for pathological lying and conning by their own admission. But you might reasonably ask how we know if people are telling the truth about their lying. The answer is that—to be honest—we can never be sure that our pathological liars were truthful in admitting that they repeatedly con, manipulate, and lie throughout their lives. But we can be sure that if they are telling the truth, then they are indeed pathological liars. And if they are lying about their lying, then they really have to be pathological liars! So, armed with this logic we went ahead anyway and scanned their brains.

We had two control groups for good measure. One group of twenty-one was not antisocial and did not lie—or at least they claimed

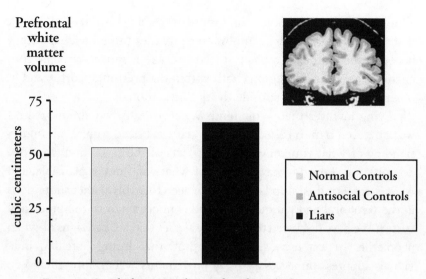

Figure 5.7 Graph showing volume of prefrontal white matter in liars
and controls, together with coronal slice through the prefrontal
cortex illustrating white matter (upper right)

they didn't. These were the "normal" controls. The other group, of six-teen, had committed as many criminal offenses as the pathological liar group—but they were not pathological liars. These individuals made up the "antisocial" control group. These two control groups were then compared with the pathological liar group.

What came out was an unusual finding in the field that must be credited to Yaling Yang, who took the lead on this study. As you can see in Figure 5.7, the volume of white matter in the prefrontal cortex was *greater* in pathological liars than in both control groups. They had a 22 percent volume increase compared with normal controls, and a 26 percent increase compared with criminal controls. The white matter volume increase was particularly true of the more ventral, lower areas of the prefrontal cortex.[109] As you might expect, liars also had significantly higher verbal IQs than the other two groups, but this did not explain away the structural brain differences. As Sean Spence, a leading expert on lying, commented in his editorial on this work, the white matter increase is very unusual, as virtually no other clinical disorder has been associated with this abnormality.[110]

In understanding this finding, we should reflect back on chapter 3, where we discussed how lying is a complex executive function that requires a lot of frontal lobe processing.[111] Telling the truth is easy.

Lying is much harder and requires more processing resources. We think that increased prefrontal white matter provides the individual with a boost in the cognitive capacity to lie because it reflects greater connectivity between subregions both within the prefrontal cortex and in other brain areas. Let's consider lying a little more.

Lying involves theory of mind. When I lie to you about where I was at eleven p.m. on Wednesday, January 7, I need to have an understanding of what you know about the facts of the case—and what you do not know. I need to have a sense of what you think is plausible, and what is not. For this "mind reading" we need to involve other subregions in the temporal and parietal lobes and connect them to the prefrontal cortex. We have discussed the behavioral cues that are bad signs of when people lie. But extensive studies also show that during lie-telling, individuals suppress unnecessary body movements. When I'm telling you the truth about where I was on the night of January 7 and I have nothing to hide, I may gesture with my hands, raise my eyebrows when making a point in the story, and look up into space for a second or two.

Liars tend not to do that. They sit still and suppress motor activity because they are cognitively focusing on their story. All of their processing resources are going into this activity. Suppression requires prefrontal regulation of the motor and somatosensory areas of the brain that control motor and body movements. Greater white-matter connectivity will facilitate that. While liars are busy building the believable façade of their story, they also have to take care not to look too nervous. This involves suppression of limbic emotional regions that include the amygdala. So again, prefrontal–limbic connectivity is important. The more white-matter wiring there is in the prefrontal cortex, the better all these functions can be subserved.

We think that the cause of the greater white-matter volumes in pathological liars is neurodevelopmental. Again, we are talking about an *increase* in volume, rather than a decrease. From a neurodevelopmental perspective, throughout childhood there is massive expansion of brain size. Brain weight reaches adult values between the ages of ten and twelve, with a very significant increase in the absolute volume of white matter by this age.[112] We also know that children become most adept at lying at the same time—by ten years of age.[113] Interestingly, then, the neurodevelopmental increase in white matter parallels developmental changes in the ability of children to lie. This suggests that the increased white matter we find in pathological liars does indeed facilitate their ability to lie. Based on this perspective, we think that the increased prefrontal white matter found in adult psychopathic liars predisposes them to deception and cunning.

The increase in white matter, then, might "cause" pathological lying. But could it be the other way around? You'll likely recall from your childhood the late-nineteenth-century Italian children's story about Pinocchio, the puppet whose nose grew every time he told a lie. Could it be that the act of pathological lying causes the physical increase in white matter in the prefrontal cortex?

This "Pinocchio's nose" hypothesis[114] is not as ridiculous as it may sound. It's the concept of brain plasticity. The more time that musicians spend in practicing the piano, the greater the development of their white matter, especially in childhood.[115] Practicing lying in childhood might particularly enhance prefrontal white matter. But even in adults, extensive practice has been found to correlate with brain structure. London taxi drivers have to undergo three years of extensive training to learn their way around 25,000 convoluted city streets. MRI studies have shown that these taxi drivers have a greater volume of the hippocampus compared with matched controls,[116] and also compared with London bus drivers, who do not undergo such extensive training.[117] Just as working in the gym can build up your muscles, mental effort can flex your brain.

In the case of pathological liars, it's as if a criminal lifestyle makes for a criminal brain. It's a different story from the one Lombroso was telling in Italy in the nineteenth century—the idea that brain impairment causes crime.[118] But we cannot yet discount the alternative environmental explanation that lying causes brain change.

WHITE-COLLAR CRIMINALS WITH BETTER BRAINS

We've seen that common forms of deviance like lying can have a physiological basis. Let's continue our look into less extreme, nonviolent forms of antisocial behavior. What about white-collar criminals who do not get their hands quite as dirty on the streets as blue-collar criminals? Criminologists view white-collar criminals very differently from other offenders. It is accepted that poverty, bad neighborhoods, educational failure, and unemployment are all risk factors for blue-collar crime. But what explains the criminal behavior of bankers, business executives, and politicians? In these cases, the finger is often pointed not at the individual, but at the institution itself for creating a corporate subculture conducive to cultivating crooks who fiddle the books.[119] To traditional criminologists, white-collar criminals are people just like me and you whose better judgment gets swayed by a tempting opportunity at work.[120]

But is the Ponzi-schemer Bernie Madoff essentially an innocent victim of bad judgment in a corrupt corporate setting? Or do offenders like him differ from the rest of us, just as we differ from the "blue-collar" street criminals we've been discussing?

Bernie Madoff made off with a lot of investors' money—an estimated $64.8 billion—bilking thousands of their life savings. He was a seasoned investment advisor, and the con was relatively simple. He got new investors to invest in securities by offering good returns. The good returns were possible because he continuously pulled in new investors, using their money to pay the good returns. He kept this going until someone noticed that there was only one accountant to supposedly vouch for an enormous financial empire. If you are an ex-accountant like me you'll know that's an impossible task.

White-collar crime runs the gamut from extreme examples like this to more common occurrences such as pilfering supplies from work, and other swindles—essentially, any crime that takes place in the work context. Perhaps surprisingly, there has been no biological or psychological theory developed for white-collar crime. There are no "individual difference" theories for this behavior even at the social level—theories that try to explain how such criminals differ from the rest of us. Edwin Sutherland, a renowned criminologist who initially developed the concept of white-collar crime in 1939, viewed these malpractices by the upper crust as a process whereby normal people get indoctrinated by their bosses and co-workers into how to get ahead in business.[121] He felt social and personal factors were of little use in explaining such offending—it was instead essentially a process of learning to seize the opportunity to get ahead.

In essence, this attitude is not too far removed from the normal nature of American business in aggressively competing against rival firms to maximize profit. If you have to push the envelope on business practices, so what? It's not like robbery—you're not hurting or threatening any one individual. And the beauty of it is that you never have to confront the victim, so you don't have to feel too guilty about what you do. It is crime made easy for the person with the smarts to get ahead.

Having read up to this point in the book, you'll understand my perspective on crime in general. White-collar criminals cannot be that spotless, even if their collars are. A macro-social approach that convicts the organization has to be at best a partial explanation because not everyone exposed to a work environment with questionable business ethics commits offenses.

At the University of Pennsylvania, I was fortunate to rub shoulders with William Laufer, a professor of legal studies and business ethics at the Wharton School. Bill brought me up to speed on this neglected area of crime. We had assessed self-reported criminal offending in our community volunteers, and a bunch of them had owned up to white-collar crimes. They had done such things as cheating or conning a business or government agency for financial benefit, using computers illegally to gain money, stealing from work, or telling lies to obtain sickness benefits. It's not as if Bill and I had a nice group of Bernie Madoffs to work with. Clearly, this is pretty run-of-the-mill stuff, but all of these offenses met the criteria of white-collar crime.[122] And for Bill and me it was an initial entry into virgin territory.

We matched twenty-one white-collar criminals with twenty-one individuals who admitted to criminal offending, but who had not perpetrated white-collar crimes. This was important, as our white-collar criminals had also committed offenses outside of the work context, and we needed to control for such offending. This is true of white-collar crime in general.[123] So both groups had the same level of criminal offending. We also matched the groups on age, gender, and ethnicity; the only difference between them was the perpetration of white-collar crime. Working with Yaling Yang, we then compared the two groups on our neurobiological measures, and obtained some interesting group differences.[124]

First—perhaps appropriately for the nature of the white-collar criminal—these offenders had better "executive functioning" as assessed by the Wisconsin card-sorting task. This neurocognitive task measures concentration, planning, organization, flexibility in shifting strategies to achieve a goal, working memory, and the ability to inhibit impulsive responding.[125] Our white-collar offenders did appear to have skills that would normally make for quite a successful business executive.

Second, they gave larger skin-conductance responses to both neutral auditory stimuli and "speech-like" stimuli. They not only gave bigger responses to the initial presentation of these stimuli—indicating

greater attention—but they kept on responding to repeated presenta-
tions of these stimuli. They were able to sustain their attention. This
greater orienting, or "What is it?" response reflects better functioning
of the ventromedial prefrontal cortex, the medial temporal cortex, and
the temporal-parietal junction,[126] areas that we have seen previously to
be dysfunctional in offenders.[127]

Third—and perhaps most interestingly of all—the brains of the
white-collar criminals were physically different from those of the controls. They showed *greater* cortical thickness in several regions of interest. They show greater thickness of gray matter in the ventromedial
prefrontal cortex (BA 11), which is the lower part of the prefrontal
cortex. They also showed increased thickness in a band of cortex that
stretches across the lateral, outer surface of the right hemisphere of the
brain. This includes part of the right prefrontal cortex (the inferior frontal gyrus—BA 44), the right motor cortex (precentral gyrus—BA 6), the
right somatosensory cortex (postcentral gyrus—BAs 1, 2, 3), the right
posterior superior temporal gyrus that forms part of the temporal-
parietal junction (BAs 22, 41, 42), and the inferior parietal region of the
right temporal-parietal junction (BAs 39, 40, 43).

What can we make of these structural brain superiorities in the
white-collar criminals? They are interesting for several reasons. First,
the inferior frontal gyrus is involved in executive functions. This
includes the ability to coordinate thoughts and actions in relation to
internally generated goals, to respond to changes in task demands, the
ability to inhibit a wrong response, to switch from one task to another,
and to decide between conflicting reasoning.[128] This is especially true of
the right hemisphere, where we found the biggest group differences.[129]
Taken together with findings of better executive functioning, increased
cortical thickness of this area is consistent with increased cognitive flexibility and regulatory control in white-collar criminals.

Second, the ventromedial region has been associated with good
decision-making, sensitivity to the future consequences of one's
actions, and the generation of skin-conductance responses.[130] This
structural advantage is again broadly consistent with the better executive functioning, skin-conductance orienting, arousal, and attention
observed in white-collar criminals. But of even greater interest, this
ventromedial region is involved in the monitoring of the reward value
of stimuli, and also learning and remembering what things in life are
rewarding.[131] Intriguingly, we see the anterior, front region of this ven-

tromedial area enhanced in white-collar criminals. Functional imaging studies have shown that this anterior area is specifically associated with abstract rewarding stimuli, particularly money.[132] In contrast, less abstract and more fundamental rewards such as taste are processed in the more posterior region of the ventromedial area, a region that did not differ between the two groups.[133] So, increased thickness specifically in this anterior region of the ventromedial prefrontal cortex suggests that white-collar criminals are particularly driven by abstract monetary rewards like money, as opposed to less abstract rewards.

Third, the premotor area of the precentral gyrus is involved in your ability to monitor your performance, to make decisions, to plan, to program your actions, and to inhibit motor actions depending on the situation.[134] It is also involved in the ability to understand the intentions of others' actions[135] and in social perception.[136] So this structural enhancement is again broadly consistent with adept executive functioning and social cognition in white-collar criminals.

Fourth, enhancement of the somatosensory cortex would be broadly consistent with better somatic marker functioning. Somatic markers are predicated on good functioning of both the somatosensory cortex, where these bodily markers are stored,[137] and the ventral prefrontal cortex, where the somatic markers are processed.[138] We've already seen that this latter area was enhanced in white-collar criminals. Unlike conventional criminals, who have somatic-marker deficits and poor decision-making skills, white-collar criminals may be characterized by relatively *better* decision-making skills.

Fifth, the right temporal-parietal junction is important for social cognition and orienting.[139] Social cognition involves the ability to process social information and to understand others' perspectives.[140] The temporal-parietal junction is also involved in orienting—directing attention to external events—and facilitating responses to these events.[141] Because Brodmann areas 41 and 42 also make up the primary auditory cortex, increased cortical thickness of these areas may help account for the better orienting to auditory stimuli we found in white-collar criminals. This supports the hypothesis that they have better social perspective-taking and the ability to read others, which in turn may place them at an advantage in an occupational context to perpetrate white-collar crimes.

Let's put this all together to grasp the underlying neurobiology of what on the surface was barely considered a crime at all. White-collar criminals have relatively better executive functions and at times are more capable of making good decisions. They are more attentive to what's going on around them and to what people say, as well as being better able to maintain their attention over time. They have a good social sense and know how to read others. They value rewards, particularly abstract rewards like money, and are both motivated and driven by them. They know when to act and when not to act, depending on the social circumstance. They can carefully calculate both the costs and benefits of acting or not acting, depending on the situation. There may indeed be a neurobiological, brain bias to white-collar crime.

In chapter 3 we documented a software failure in the functioning of the brains of violent offenders. Now in this chapter—beginning with Herbert Weinstein—we have seen signs of a fundamental hardware failure in the brains of offenders that could underlie their functional brain impairments, a hard-drive failure that can trip the circuit on violence. This hard-drive defect lies in the frontal cortex and affects behavioral inhibition. It also lies in the amygdala at the level of emotion.

Environmental factors—especially in the form of head injury—play a critical role in causing brain impairments. Yet we have also borne witness to unusual brain abnormalities that implicate greater—not reduced—volume in areas that include the corpus callosum, striatum, and hippocampus. Taken together with the presence of cavum septum pellucidum in offenders, these volume distortions give rise to the hypothesis that offending may be the result of an early *neurodevelopmental* brain abnormality. We have also seen that these brain abnormalities are not specific to serious violence but may characterize nonviolent antisocial behaviors even you may have been committing.

Criminals do have broken brains, brains that are physically different from those of the rest of us. The differences are substantial and can no longer be ignored. This may smack of the "born criminal" and genetics and destiny. Indeed, in many of the prior chapters I have given strong credence to biological and genetic predispositions to violence. Yet this chapter also highlights the critical importance of the *environment* in shaping the structural brain deformations that we find in violent offenders.

But even acknowledging this, our model is still overly simplistic. It's not some neurobiological influence added together with some environmental influence in a simple way that causes violence. As we shall see later, these oppositional processes instead *interact* in complex ways to shape violence. But before reaching that point we need to address the question of what external forces act on the brain to distort its structure and function. And continuing the neurodevelopmental theory of offending I have been outlining here, the next chapter will again focus on very early influences on the brain beyond the individual's control. The seeds of sinful violence are sown early by the grim reaper, and not just at the time of conception. As we are about to see, those seeds are cultivated in utero, at the time of birth, and also in the early postnatal period to give rise to the framework for violence.

6.

NATURAL-BORN KILLERS

Early Health Influences

Peter Sutcliffe had such a difficult birth that doctors didn't think he would survive the night. He arrived at ten p.m. on June 2, 1946, in the Bingley maternity hospital in West Yorkshire. It was just one year after the end of another long war for England and there was a high mortality rate for newborns. But little Peter was a five-pound fighter. In spite of the birth trauma he suffered, the premature baby was released from the hospital after a dramatic ten-day struggle for life.

Following that early biological hit, young Peter grew up in Bingley as a pretty normal kid. He was very much like me. We both were born with birth complications. Both of us were shy lads brought up in the north of England in a typical northern working-class home. We were both small for our age. And both of us were in a big family and brought up Catholic. It seemed that Peter had escaped the clutches of death—but had he? It was when he was a grave digger in Bingley Cemetery in 1967 that he experienced the pivotal moment of his life. He was bent over his spade, digging away at a new grave when he heard it. A vague, echoing voice coming directly from the cross of a nearby Polish grave. Sutcliffe later described the day:

> The mumbling voice had a strange effect. Felt I was privileged to hear it. It had started to rain and I remember looking from

the top of the slope over the valley and feeling I'd just experienced something fantastic. I looked across the valley, and all around, and thought of Heaven and Earth and how insignificant we all were. But I felt so important at that moment. I had been selected.[1]

But selected for what? Slowly, over time, Sutcliffe came to realize that he was the instrument of God's wrath against evil and sexual sin. His mission was to rid the world of the sin of prostitutes.

It was a pivotal psychotic experience. From that point on, despite a happy marriage to a Polish immigrant schoolteacher, Sutcliffe began to dig graves in a very different way. He went from being a broken baby in his mother's womb to becoming one of England's most prolific serial killers, a schizophrenic murderer who ripped open the wombs of thirteen prostitutes in Yorkshire.[2]

In this chapter we will see that for some, the predisposition to a violent life begins even before babies have drawn their first breath. That's right—the birth of the individual may literally mark the birth of the violent offender. As early as the time of conception, health is a strong factor in the equation. And it's health in the public domain that shall be our point of departure in this area of the anatomy of violence.

VIOLENCE AS A PUBLIC-HEALTH PROBLEM

We have seen in the previous chapters that there is substantial evidence for a biological basis to crime and violence. Moving from evolution to genes to central nervous system functioning to autonomic functioning, we have been slowly working our way through the anatomy of violence to argue something that a reasonable social scientist can no longer deny. There is in part a biological basis to violence.

Indeed, the question of *whether* brain deficits in individuals contribute to violence is, frankly speaking, no longer a useful one.[3] Since there is no longer any doubt that brain deficits contribute in some way to antisocial and aggressive behavior, we should instead be asking the more important question, What's happening very early on in life to cause the brain abnormalities that we find in adult violent offenders? Once we can identify these early processes, we are halfway toward new intervention and prevention studies that reshape a child's trajectory away from violent offending. With this knowledge we can begin to reel in the unacceptable level of violence we see not just in the United States, with its high homicide rate, but also everywhere else in the world.

In this and the next chapter I'm going to focus on violence as a public-health issue. While it may seem odd to think of violence in the same way we think about conditions like obesity, AIDS, and flu epidemics, it has become a useful—and increasingly popular—way of approaching the problem. Indeed, the United States' Centers for Disease Control and Prevention (CDC)[4] now views violence as a *serious* public-health problem, and the World Health Organization (WHO), in the first world report on violence, defines this condition as a global public-health problem. Right now we have an epidemic of violence that is the *leading* cause of death across the world for those aged fifteen to forty-four.[5] In the United States, violence is the second-leading cause of death. It's an enormous drain on our health-care system. The CDC puts the cost at $70 *billion* per year,[6] while also acknowledging that this is an incomplete measure of the total cost. It's much more like $105 billion when you add in medical losses, lost earnings, and public program costs related to victim assistance—and that is in 1993 dollars.[7] The actual costs are truly staggering. WHO estimates that gunshot wounds alone currently cost the United States health-care system $126 billion a year, with cutting and stab wounds adding an extra $51 billion to the bill.[8] In England and Wales, the cost of violence is estimated at $63.8 billion every year.[9] Some countries, including Colombia and El Salvador, spend a full 4 percent of their gross domestic product in dealing with *just* the health-related problems associated with violence, let alone legal and judicial costs. Convert that to the GDP of the United States, and it's half a trillion dollars—and imagine how that chunk of change can be better spent.

Clearly violence costs us. But is it really a public-health problem? Do we really need to think of violence in medical terms like this? Yes we do, and that's the change in thinking that is occurring right now. Let me explain. Public health is part of *medicine*. It asks four questions. One, how often and in what situations does violence happen? Two, what are the causes? Three, what are the cures? Four, how can we apply treatments across the board in the general population? It is radically different from sociological perspectives, which view violence as a nonmedical issue. It is different from a clinical perspective that focuses on specific individuals rather than on the broader population. Medical practitioners are right now becoming more and more involved in the treatment and prevention of violence. Even *dentists* are taking this seriously.

Jonathan Shepherd is a professor of oral and maxillofacial surgery in the School of Dentistry at Cardiff University. After moving to Cardiff in 1991 he was shocked not only to see so many victims of violence with facial injuries, but also to find that the vast majority of bar fights that produced these injuries were not reported. Working in unison with law-enforcement agencies, he shared information that allowed the police to get a true picture of where the violent hot spots were in Cardiff. He worked with beer-glass manufacturers, persuading them to replace standard beer glasses with toughened glasses that were much more difficult to break and use as a weapon. The result of these public-health initiatives? A substantial reduction in injuries and a major contribution to making Cardiff not just a much safer city in Wales, but an exciting city to live in.[10] If someone in dentistry can make a difference, surely knowledge from other health fields can also make a contribution to the goal of violence reduction.

For this reason, we will now shift our attention from the dark chambers of our inner biological functioning to the outside, to shed light on how early *environmental* factors contribute to the disruptions we saw in brain and biological processing that were laid out in the previous chapters. What better way to begin this journey than as we began with Peter Sutcliffe, with the birth of the child?

BORN BAD

I found the Rigshospitalet hospital in Copenhagen to be a truly imposing institution on my visit in 1991. Founded on March 30, 1757, and originally named for King Frederick V, it's the national hospital of Denmark. It's a bustling institution with 8,000 personnel and nearly half a million patients to deal with every year. Mary, the crown princess of Denmark, gave birth to her two children, Prince Christian and Princess Isabella, there. Prince Christian's birth, on October 15, 2005, went very smoothly and was marked by a twenty-one-gun salute at noon, with beacons lit all over Denmark in national rejoicing. But for other boys born in the very same Rigshospitalet, birth is not so smooth and regal, and the outcome not quite as glorious.

In 1994 I published our findings on 4,269 live male births occurring at the Rigshospitalet in 1959.[11] Birth complications were assessed by obstetricians assisted by midwives. Examples of delivery complications included things like forceps extraction, breech delivery, umbilical-cord prolapse, preeclampsia,[12] and long birth duration. One year later, social workers went around to all the homes of the mothers and conducted interviews. Had she wanted the pregnancy? Did she ever make an attempt to abort the fetus during pregnancy? Was her child placed in a public institution for any reason for at least four months in the first year of life? These three indicators of maternal rejection of the child were duly noted. When these babies were eighteen years old, we conducted a national search of all court records in Denmark to find out which of the baby boys had been arrested for a violent crime.[13] We then classified them into four groups. Those with neither birth complications nor maternal rejection of the child in the first year of life were the normal controls. Some had birth complications, but had not been rejected by their mothers. Some were rejected, but had a normal birth. And the fourth group had the double whammy—birth complications and rejection by their mothers in the first year of life.

The results were striking. As you can see in the top half of Figure 6.1, the first three groups did not differ significantly from each other, with rates of violence at about 3 percent. It was the fourth biosocial group—the one with both the biological and the social hits—that had the highest rates of violence. This group had three times the average of the other three groups—9 percent of them became violent offenders. Furthermore, although only 4.5 percent of the population had both birth complications and early child rejection, this small group accounted for 18 percent of all violent crimes perpetrated by the entire sample of 4,269—four times higher.[14] It's a classic case of early biological factors interacting with social factors very early in life to shape adult violence.

A lot of violence is committed after the age of eighteen. Would this biosocial interaction also explain this later violence, or is it especially important in explaining early violence? We reassessed the entire birth cohort at age thirty-four for arrests for violent crimes. This resulted in a tripling of the sample size of violent offenders, allowing us to conduct more detailed analyses.[15] The results indicated that the biosocial interaction was specific to violent crime with an early onset. It did not explain violence that started later in life. In addition, we found the interaction to be specific to violent offending—it did not explain nonviolent criminal offending. It seems that a violent birth makes for violent behavior in particular.

Looking back at the three components of "maternal rejection," were there any that were particularly important? Two of them were. First, rearing in a public-care institution in the first year of life was critical. Second, an attempt to abort the fetus also came up trumps. These were the two elements of maternal rejection that interacted with birth complications in producing later violence. In contrast, if the mother simply did not want the pregnancy but took no action, it did not seem to affect long-term outcome.[16] Furthermore, the interaction was found to be specific to more serious forms of violence like robbery, rape, and murder—but not for less serious forms like threats of violence. It seems, then, that birth complications conspire with more severe forms of maternal rejection to launch particularly violent criminal careers.

The problem with our Copenhagen study is that the sample is made up almost exclusively of white babies. More than that, they are white babies from a European country with relatively low levels of homicide. Are these findings some peculiarity of an idiosyncratic Danish culture? What about black babies and other nationalities? These questions were first addressed by two American criminologists who tested the "bad birth and bad mother" hypothesis using a cohort of 867 male and female African-American babies who made up the Philadelphia Collaborative Perinatal Project. Birth complications had been collected on this sample at the times of the babies' births.[17] Criminologists Alex Piquero and Steven Tibbetts followed our original design and broke the larger sample down into the same four groups.[18] Their results are shown in the lower half of Figure 6.1. They're visually striking, showing almost identical findings. Yet again, it was found that those with both birth complications and a disadvantaged family environment were much more likely to become adult violent offenders. The first findings, from Denmark, were not a fluke.

The birth-biosocial interaction is found in Denmark and the United States. What about other countries? So far the interaction appears to be holding up. Pregnancy complications interacted with poor parenting in predicting adult violence in a very large Swedish sample of 7,101 men.[19] In a Canadian sample of 849 boys, an interaction was found between increased serious obstetric complications and family adversity in raising the likelihood of violent offending at age seventeen.[20] In a Finnish

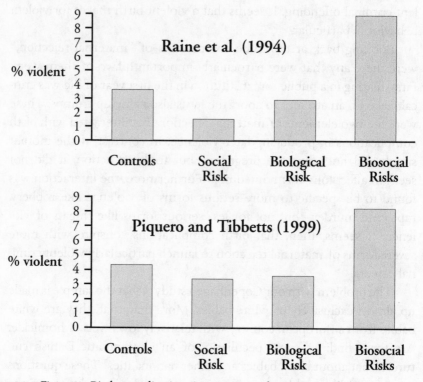

Figure 6.1 Birth complications interacting with negative early home
environments predispose to adult violence

sample, perinatal risk interacted with being an only child[21] in raising
the odds of adult violent offending by a factor of 4.4 in a sample of
5,587 males.[22] Furthermore, interactions between birth complications
and negative home environments in predisposing children to antisocial
behavior have been found in Hawaii[23] and Pittsburgh.[24] Almost wherever you go in the world, you find the same effect.[25] The combination
of birth complications and adverse home environments appears to be
a useful biosocial key that can help open the lock on the causes of violence.

You may nevertheless be asking how exactly birth complications
and negative home environments like maternal rejection combine to
shape adult violence. If we look at birth complications first, the likely
pathway is that they have a negative impact on the brain. Take my birth
as an example. I was born at home as a "blue baby" without intensive-
care treatment. As an adult I have always been hopeless at finding my
way around new places—I have a very poor spatial sense. Some babies

are blue at birth because they suffer from a birth complication termed "hypoxia"—a partial lack of oxygen. Our brains need oxygen to metabolize glucose—a fuel that provides energy for brain cells. Without oxygen, brain cells will start to die in a few minutes. Particularly sensitive to this destructive process is the hippocampus, a part of the brain centrally involved in spatial ability as well as short-term memory, capacities that have been found to be impaired in those who are persistent offenders throughout their lives.[26] Hypoxia at birth was also found in one study to be the best predictor of a lack of self-control,[27] a key behavioral risk factor for crime and especially for explosive, impulsive aggression. As we saw in a previous chapter, the hippocampus is structurally and functionally impaired in violent offenders.[28] Other birth complications such as preeclampsia, maternal bleeding, and maternal infection cause a reduction in blood supply to the placenta, resulting in cell loss not just to the hippocampus but also to other brain areas including the frontal cortex. Consequently, birth complications have multiple neural pathways to a violent outcome.

A more specific pathway by which birth complications can result in behavior problems in children was shown by Jianghong Liu in her analysis of data from the large birth cohort in Mauritius. She demonstrated interconnections between three key processes—birth complications, low IQ, and antisocial behavior. We had assessed prenatal, perinatal, and postnatal birth complications and also measured at age eleven both IQ and externalizing behavior problems—aggression, delinquency, and hyperactivity. Jianghong showed that birth complications were significantly related to increased externalizing behavior problems.[29] She also found that birth complications were associated with lower IQ at age eleven.[30] Low IQ in turn was associated with externalizing behavior problems. The triangulation of relationships was complete. Low IQ *mediated* the relationship between birth complications and later behavior problems—birth complications result in lower IQ, and this in turn results in problem behavior in later childhood—more specifically, aggression, antisocial behavior, and hyperactivity. IQ is predicated on a well-functioning brain, and like other neurocognitive measures it acts as a proxy for brain functioning.

At least five other studies besides Jianghong's have observed direct links between birth complications and behavior problems, delinquency, and adult violence.[31] For example, in Holland two separate studies showed direct relationships between birth complications and exter-

nalizing behavior problems in boys and girls.[32] These and other studies, however, did not test the biosocial hypothesis, and some studies
have not found direct links between birth complications and violence
or have obtained only partial support.[33] At the same time, even more
studies such as the one we conducted in Copenhagen do not find a
direct path between birth complications and problem behavior. Instead,
social processes are critical, seemingly acting as a trigger for the dor-
mant birth-complications risk factor for violence.

In our Copenhagen study we found that a critical component of
"maternal rejection of the child" was being institutionalized for at least
four months in the first year of life. Why was this component of the
social risk factor so important in our study? The life of a young English
boy born during the Edwardian era, in 1907, offers a poignant insight.
John Bowlby was a Londoner who saw his mother for just one hour a
day. She thought that a child could be spoiled by too much attention
and affection. When he was seven years old Bowlby was packed off to
a boarding school, and by his own account had a terrible time there.
In his words, "I wouldn't send a dog away to boarding school at age
seven."[34]

This early experience and poor bonding with his mother proved
to be pivotal for John Bowlby and was to shape his future career. After
graduating in psychology from Cambridge University he worked with
delinquent children before training as a psychoanalyst and psychiatrist,
going on to pioneer a new approach to attachment theory. His classic
book, written at the end of World War II, brings together his own early
experiences and his knowledge of delinquent boys and helps explain
why maternal rejection was so important in our Copenhagen study.

Entitled *Forty-four Juvenile Thieves,* Bowlby's book was an in-depth
analysis of the early home backgrounds of forty-four juveniles who
turned out to be offenders.[35] In those early days of delinquency research
he made the innovative argument that the lack of a continuous and
loving relationship between mother and infant resulted in the inability of the infant to develop a normal personality, and the inability to
form normal interpersonal relationships. His case studies highlighted
in the lives of these forty-four thieves the prolonged separation from
their mothers early in life. This resulted in the absence of a warm, continuous, and intimate relationship between the mother and her infant.
The result? What he termed "affectionless psychopathy." Some of his
illustrations were graphic and dramatic. In two of the affectionless psy-

chopaths, they had each spent nine months in a hospital without *any* visits from either parent.

This social perspective on crime and delinquency was to be fine-tuned a little by other scholars in later years. What transpires is that there is a critical period early in life when being connected with the mother really counts. In humans this starts at about six months and ends after about two years. For this reason breakage of the mother-infant bonding process for at least four months in the first year of life—as experienced by some of our Copenhagen babies—freezes the social-interpersonal development of the infant. That freezing results in the glacial, emotionless psychopath that we see in adulthood. Recall that "Jolly" Jane Toppan was orphaned and institutionalized until the age of five and went on to become a killer nurse—she had exactly this risk factor for psychopathic violence.

As for Bowlby himself, what became of this little boy with his early maternal deprivation and ruthless parental separation? He was likely spared an outcome of "affectionless psychopathy" because right at the get-go, despite the absence of his mother, he had all the attentions of a caring nanny. As others went on to argue, the decisive issue is whether you have someone to bond with—anyone at all.[36] It could be a stand-in mother figure not genetically related to you, like a nanny, or your father, or even an elder sibling who takes on the caregiving role. As long as you have the opportunity to consistently bond with *any* human early in life, you derive the basis for appropriate social relationships.

We've seen here that, as with Peter Sutcliffe, indicators of later violence can emerge by the time we take our first breath of air. For Peter it was not just birth complications, but also schizophrenia, a genetically based mental illness that we'll return to later. But in our developmental search for the origins of violence, is the nine months spent in the womb already too late? Scientists are beginning to trace back the origins of potential evil to points not too long after the moment of conception. The anatomy of violence moves on to events that occur before birth, and our genetic-like perspective here is fittingly found in Genesis and a fable about the origins of humankind.

THE MARKS OF CAIN

Cain, a son of Adam and Eve, has the dubious distinction of being the world's first murderer—and the killer of his own brother. Cain's story is a fitting beginning to the history of homicide. After all, about 20 percent of all homicides take place within the family, and of these about two-thirds can be viewed as reactive aggression[37]—responding aggressively to an upsetting or provoking external stimulus.

Cain was one of these cases. He was absolutely furious with God. God had accepted his brother Abel's sacrificial offering of a sheep—but had rejected Cain's offering of crops. In a fit of rage, Cain displaced his aggression onto Abel and slew him. As punishment, the story goes, God placed a mark on Cain as a curse, and Cain was destined to become a restless wanderer who would walk the earth, never again able to cultivate crops.[38]

The search for a real-life mark of Cain in criminals was a goal of early criminologists, and as we saw earlier, Lombroso, the father of the discipline, was adamant that it could be found. In his thousands of painstaking physical observations of criminals in Italy, Lombroso believed he saw physiological signs of the "born criminal," and witnessed multiple hallmarks of Cain that he called "atavistic stigmata" and that he fervently believed set criminal offenders apart biologically from the rest of us.

Do you have the mark of Cain? Take your right hand, lay it palm up, and relax it. Fold the fingers of this hand a little toward you. Can you see one continuous crease that goes all the way across the top of your palm? Or do you see two main creases that do not join together? If you have a single palmar crease, bad luck. According to Lombroso, you have the atavistic stigmata that makes you an evolutionary throwback to lower species.

Now take off your shoes and socks, stand up, and look down at your feet. Do you see a big gap between the first and second toe? If you do things are not looking good—another strike against you. There are others. If you want to see if you have one that I have, stick your tongue out and look at it in the mirror. Do you see a fissure—a line running down the middle of it? Another mark of Cain.

It sounds completely ridiculous—yet there is some veracity to Lombroso's claims. The "stigmata" outlined above are just three of a number of what are now called "minor physical anomalies." These anomalies have been associated with disorders of pregnancy and are thought to be a marker for fetal neural maldevelopment at about the third or fourth months of pregnancy. For example, during fetal development your ears sit relatively low on your head, but they begin to drift up to their normal positions at about four months of development. If disruption occurs to fetal brain development at this time, there is incomplete embryonic migration of the ear anlage—essentially, the ears will not migrate to their normal position, resulting in low-seated ears.[39] These anomalies are viewed as indirect markers of abnormal brain development. If you want to have a quick look in the mirror at yourself, is the point where your ear connects to your head below your eyes? If so take another gulp.

In case you are getting worried, other minor physical anomalies include adherent (attached) earlobes, electrostatic hair, and curved little fingers. It is believed that they may be caused by environmental tera-togenic influences acting on the fetus, factors such as anoxia, bleeding, infection—or fetal exposure to alcohol.[40] Don't worry if like me you have only one or two minor physical anomalies—it's having a handful that really counts.

Minor physical anomalies—like many other markers for violence—have not been systematically assessed in serial killers. But they have been systematically assessed in research studies on a wide variety of antisocial populations of different ages, ranging from troublesome tod-dlers to violent adults. Beginning with a breakthrough paper in *Science,* minor physical anomalies have even been linked to peer aggression as early as age three.[41] Again, in another study at the preschool level, more minor physical anomalies have been found in aggressive and impulsive boys.[42] Moving on a little into elementary school, boys with problem behaviors have more anomalies.[43] Transferring to secondary school and to the troublesome teens, minor physical anomalies in boys assessed at age fourteen predicted violent delinquency at age seventeen. Interest-ingly, at this age, the relationship was specific to violent offending—it was not observed for nonviolent forms of delinquency.[44] In this study the effects could not be attributed to potential confounds such as family adversity. At about the same age—but from a biosocial perspective—minor physical anomalies in seven-year-olds combined with environ-mental risk factors in predisposing the children to conduct disorder at age seventeen.[45] This highlights again the biosocial key that we saw when we looked into birth complications—the interaction between a biological and a social factor in predisposing someone to antisocial behavior.

Minor physical anomalies assessed by pediatricians at age twelve predicted violent offending at age twenty-one by perpetrators now leaving school and graduating on to violent criminal careers.[46] Yet again, a biosocial interaction was observed with especially high rates of violence found in those with both minor physical anomalies and a history of being raised in unstable home environments. As with birth complications, the presence of a negative psychosocial factor is required to trigger the biological risk factor in adults—and in both cases the effects are specific to violent offending.

It may seem bizarre. Some of Lombroso's ideas may seem very repugnant. Yet over a hundred years after his first theorizing, we can say that Lombroso was at least partially on the mark with his theory for Cain-like atavistic stigmata for criminal offending. We can also say—at least at a superficial level—that the book of Genesis highlights for us external physical indicators of family feuds gone wrong. The key difference is that while the mark on Cain in Genesis was very visible, we never notice anyone's minor physical anomalies. They are imperceptible without a close physical examination.

From a scientific standpoint we get another pointer to the fact that the seeds of violence are sown very early on in life—as early as the prenatal period.

FROM PALM PRINTS TO FINGERS

How often do you look at your fingers? In all likelihood, not very often. But take a look right now at your right hand.[47] With your palm facing you, look at the length of your fingers. Compare the length of the second digit with the length of the fourth digit. The second digit is your index finger, the fourth digit is your ring finger. You'll very likely see that the fourth digit is longer than the second. It is for most people, especially on the right hand. If you can compare yourself with someone of the opposite sex, see who has a longer ring finger relative to the index finger. Males in general have the advantage—they tend to have a longer ring finger compared with their index finger than women do. This gender difference is also true in baboons.

What causes this difference between the genders? Genetics is one explanation, with the same set of genes[48] influencing both genitals and digit length.[49] But in addition, fetal hormone exposure—in particular androgens—plays a critical role. Sometime between ten and eighteen weeks of gestation there is a major surge in testosterone production that among other things produces the primary gender differences we see at birth. It not only masculinizes the nervous system and behavior, but it also influences the ratio of the length of the second to the fourth digit.[50] The higher the testosterone exposure, the longer the size of the ring finger relative to the index finger. Hence men have a relatively longer ring finger than women.[51]

The testosterone explanation of the digit difference seems relatively convincing. Several studies have observed that children with congenital adrenal hyperplasia[52]—a condition caused by high prenatal androgen exposure—shows this male effect of a relatively longer ring finger.[53] Women who have larger waists relative to their hips often have higher testosterone levels, and such women have been found in turn to give birth to children with relatively longer ring fingers.[54] Indeed, because assessing prenatal androgen levels is not easy, this finger difference has been touted as an indirect indicator of the level of androgens during fetal development.[55]

What do we know about people with a more male-like, longer ring finger? For one thing they tend to dominate, show physical advantages, have male-like characteristics, and have personalities linked to aggression. A study in Poland shows that females who have achieved elite status in athletics have relatively longer ring fingers compared with non-elite athletes.[56] And such prowess is not restricted to the track or to Poland. Male British symphony orchestra musicians also have relatively longer ring fingers.[57] On the field, English soccer players who are in the first team have longer ring fingers than those who are in the reserves.[58] Some of you may recall the likes of Paul Gascoigne, Geoff Hurst, Stanley Matthews, Peter Shilton, Glenn Hoddle, Kenny Dalglish, and Ozzie Ardiles—soccer stars who represented their countries in international matches. These twenty-nine stars, as a group, were found to have longer ring fingers than a group of 275 professional footballers who had not played for their country. Furthermore, the more times they had represented their country, the longer their relative ring-finger length.

Another correlate of the long ring finger is sensation-seeking and impulsivity[59]—personality traits that we saw in the previous chapter to be linked to antisocial and violent behavior. People who are relatively lacking in empathy also have longer index fingers,[60] and antisocial, psychopathic offenders certainly lack empathy. Although the evidence is conflicting, men with a longer ring finger tend to have higher attractiveness ratings.[61] Hyperactive children have a longer ring finger,[62] and we know that there is comorbidity between hyperactivity and conduct disorder. Gay men's ring-finger lengths are often in between those of heterosexual men and heterosexual women.[63] It's not true of every study, but in a sense relatively longer ring fingers compared with index fingers go along with male characteristics—high stimulation-seeking, low empathy, and hyperactivity.

Given this, it perhaps comes as no surprise that higher aggression—a very male characteristic—is associated with longer ring fingers. In Canada, male undergraduates who are more physically aggressive have longer ring fingers,[64] with the strength of this relationship being as strong as the relationship between aggression and testosterone. In the United States, male undergraduates with longer ring fingers report being both more aggressive and more likely to engage in male-related play activities.[65] In China we have been finding cross-cultural support for the relationship between high aggression and a longer ring finger in male but not female eleven-year-old schoolchildren.[66]

We often think of aggression in the domestic domain as a bit different from aggression toward strangers. Indeed, the field of domestic violence has been almost completely dominated by scientists with a strong social perspective on intimate partner violence. But relative ring finger length, with its status as a marker for prenatal testosterone levels, sticks up a rude finger gesture to this predominantly social view. Men with long ring fingers are more likely to use threats of aggression against their female romantic partners.[67] They are also more physically violent toward them, and this is especially true for men whose female partners are cheating on them.[68]

By and large, the relative ring finger length relationship with physical aggression seems to be more true for men than for women. So what's going on beneath the Tarzan/Jane stereotypes of aggressive men and nurturing women? I think part of the answer is that women are just less aggressive than men, so aggression scores are more likely to bottom out in women. There is less variability in aggression to explain here. That is something we call a floor effect, and it can suppress relationships. But more likely it's because, as we saw in chapter 1, hardcore physical aggression is costly in an evolutionary sense. Women invest in their offspring more than men, and a woman who initiates violence is likely to be hit in return, which could be a danger to the survival of her offspring—more than would be true for the father of the child. So instead, more "softer" forms of aggression—like gossiping, rumormongering, making others feel guilty, and shutting others out of relationships—are more in the female domain than full-blooded physical aggression. Once we get down to assessing these softer forms of "relational" aggression in women, studies do indeed find relationships between such behaviors in females and longer ring-finger lengths.[69] They also show a relationship between finger ratios and more "reactive aggression"—lashing out at others who have hurt or slighted them.[70]

And what about aggression in the political arena? If you were the leader of a country and in conflict with your neighbor over diamond mines that had just been discovered in disputed territory, how would you react? Suppose you can either negotiate or go to war. Your choice is not entirely as free as you may think. It's partly determined by your relative ring-finger length. Business-school students at Harvard were placed in this game scenario.[71] The interesting parameter was the number of unprovoked attacks the leader would make on the neighboring country. As you might expect, men in general launched more unprovoked attacks than women—32 percent versus 14 percent. Let's remember that by the tender age of one year, boys are already throwing and hitting more than girls.[72] But what's more interesting is that the students with longer ring fingers launched more unprovoked attacks, an effect that was as strong as the gender difference in aggression.[73] If you are a Quaker, check out your political candidates' finger lengths before casting your vote.

Why should this mark of Cain—the longer ring finger than index finger—be a characteristic of aggressive individuals? Of course, the longer ring finger itself is not causing crime. It's more that other factors that go into making a longer ring finger also go into making aggression. We have just seen how higher testosterone in utero is responsible for the digit difference. In chapter 4 we also saw how high testosterone is causally related to aggression. So perhaps we have it here—the longer ring finger is caused by high prenatal testosterone, which in turn fuels aggression. That higher surge in testosterone early in fetal development shapes a more prototypical male brain, which shapes more prototypical male behaviors, including sensation-seeking, interest in sports, low empathy, dominance, and, of course, aggression.

But is there something missing here, a question that begs to be answered? What causes higher testosterone exposure in utero? Smoking cigarettes during pregnancy can result in higher prenatal-testosterone exposure to the fetus that leaves its mark on finger length. We suspect this because mothers who smoke have higher testosterone levels, and this can in turn reduce estrogen exposure to the fetus, resulting in higher fetal testosterone levels. Experimental work in animals has shown a causal connection, with exposure to nicotine in the prenatal period resulting in higher testosterone in the fetuses.[74] Given these links, it's not too surprising that mothers who smoke during pregnancy have male offspring with longer ring fingers than mothers who do not smoke.[75]

There is something elegant in this line of research. Unlike brain-imaging research, where we can observe structural and functional changes to the brain that may be caused by violence and subsequent head injury, the digit difference precedes even the very initial development of antisocial, aggressive, and violent behavior. How do we know that for sure? Ultrasound can give us images of fetuses, but it's not possible to assess finger-length differences from such images. However, researchers in Turkey examined 161 fetuses that had been aborted at different stages of pregnancy and made exact measurements of finger lengths. They established that the gender difference was present by the end of the third month of gestation.[76] There really does seem to be a process in place very early on in life that contributes to aggression many years later.

Relative finger length, then, provides us a window backward in time to view what occurred during fetal development. It suggests not only that Lombroso was partly correct, but also that the pre-birth period is more important than we have previously thought. Of course, mothers cannot control their hormone levels during pregnancy—they are not in any way to blame if their child is exposed to higher testosterone and becomes aggressive later in life. But there are other things she knowingly does that will shape the fate of her child in a negative direction.

SMOKING DURING PREGNANCY

Smoking is not good for your health. But it can do wonders for your violence potential, especially if your mother smoked like a chimney while she was pregnant with you. We now know that if a mother smokes during pregnancy it not only has negative consequences on brain development, but it also leads to increased rates of conduct disorder and aggression in her offspring. A spate of studies has established beyond a reasonable doubt a significant link between smoking during pregnancy and both later conduct disorder in children and violent offending in adults. A number of these studies are impressive in terms of their size, the prospective nature of data collection, long-term outcome, and control for third factors, suggesting that the relationship is causal.

Using the birth cohort from Denmark that included 4,169 males, the psychologist Patty Brennan at Emory University found a twofold increase in adult violent offending in the offspring of mothers who smoked twenty cigarettes a day.[77] She also found a dose-response relationship, with an increase in the number of cigarettes smoked resulting in a linear increase in adult violence. It was an impressive study, and there are many others like it in different countries.

In one birth cohort of 5,966 from Finland, the offspring of mothers who smoked were twice as likely to have a criminal record by age twenty-two.[78] In a follow-up study of this Finnish sample to age twenty-six, a twofold increase in violent crime and repeat offending was found in the offspring of mothers who smoked.[79] In the United States, boys of mothers who smoked ten cigarettes a day during pregnancy were four times more likely to have conduct disorder.[80]

These samples are predominantly Caucasian—are the same effects found for other ethnic groups? They do seem to hold for African-Americans, at least. The same effect of prenatal smoking exposure in increasing both conduct disorder[81] and disruptive behavior problems[82] has been observed in African-American children. One U.S. study showed more than a fourfold increase in conduct disorder in the offspring of mothers who smoked half a pack of cigarettes a day,[83] and another found an increase of six points in behavior problems in three-year-olds exposed to smoking during the third trimester.[84] In New Zealand, a doubling in the rate of conduct disorder was found in the offspring of maternal smokers.[85] You find the same relationship between prenatal smoking and antisocial behavior in Welsh children and adolescents.[86] Wherever you go in the world, you get the same finding.

Of course, you've probably already asked yourself a very good question: Could it be that mothers who smoke during pregnancy are not, on average, the most caring, educated, empathic, and informed parents in the world? Someone willing to subject her unborn child to toxins in the womb may not be providing the best environment after a child is born. To illustrate this point further, in one study a full 72 percent of the offspring of mothers who smoked during pregnancy had experienced either physical or sexual abuse. In addressing this important issue many of the studies have taken pains to control for third factors that could account for the smoking-antisocial relationship. But even then, crime and antisocial personality in the parents, low socio-economic status, low maternal educational level, mother's age at the child's birth, family size, poor child-rearing behaviors, bad parenting, obstetric complications, birth weight, family problems, parental psychiatric diagnoses, attention-deficit/hyperactivity disorder, offspring smoking, and other drug use during pregnancy could not account for the relationship. After that shopping list of confounds, there's not a lot left to control for. Taken together with the dose-response relationship that was also established in several of the studies, these findings appear to be real, and suggest a causal relationship between smoking during pregnancy and later violence.[87]

Every puff counts. Studies repeatedly show that the more cigarettes the mother smokes, the greater the odds of antisocial behavior in her offspring. We'll also see later in the book that many other factors combine together with maternal prenatal smoking to really boost the odds of violence in their offspring.

It's my hope that if you are reading this and you are pregnant you will decide to quit for the good of your little one. But I should warn you that this alone may not be enough. If your husband or co-workers smoke, you are still exposing your baby to the toxic effects of smoking. Lisa Gatzke-Kopp, at Penn State University and a past graduate student of mine, found that *secondhand* exposure to cigarette smoking predicted conduct disorder even after controlling for antisocial behavior in the parents, poor parenting practices, and other biological and social confounds.[88]

How can a few puffs during pregnancy *cause* the fetus to become a fighter later in life? What is the nature of the causal path from fetal nicotine exposure to antisocial behavior? First and foremost, it can partly account for the brain deficits that we saw to be apparent in brain scans of adult offenders. Animal research has clearly demonstrated the neurotoxic effects of two constituents of cigarette smoke—carbon monoxide and nicotine.[89] Nicotine passes across the placenta, directly exposing the fetus. A primary effect is that it reduces uterine blood flow and consequently reduces both nutrients and oxygen to the fetus, producing hypoxia, which can damage the brain. Babies exposed to smoking have been shown to have a reduction in head circumference, indirectly reflecting a reduction in brain development.[90] Studies of brain-scanned adults who were exposed as a fetus to maternal smoking show that they have thinner orbitofrontal and middle frontal gyral thickness—brain areas that we will see in a later chapter are especially implicated in violence.[91]

Because smoking negatively affects the fetal brain, we would expect such exposed infants to show neuropsychological impairments later on in childhood and adolescence—and they do. Studies have documented impairments in selective attention, memory, and speed in processing speech stimuli.[92] A dose-response relationship between increased cigarette smoking and reductions in arithmetic and spelling between ages six and eleven has been reported.[93] We've seen that neurocognitive functions are impaired in offenders, and we also know that such offenders fail in school, where math and spelling abilities are important. Fetal exposure to smoking is a likely contribution to this neurocognitive pathway to antisocial and violent behavior.

Prenatal nicotine exposure, even at relatively low levels, disrupts the development of the noradrenergic neurotransmitter system.[94] This is of particular significance in the context of the autonomic deficits we discussed earlier. Reduction of noradrenergic functioning caused by smoking would be expected to disrupt sympathetic nervous system activity. As we saw earlier, reduced sympathetic arousal as measured by sweat rate has been found in antisocial individuals. Furthermore, when pregnant rats are exposed to nicotine at the levels commonly found in human smokers, the offspring show an enhancement of cardiac M2-muscarinic cholinergic receptors. These receptors *inhibit* autonomic functions,[95] so stimulation of their functioning via smoking would *reduce* autonomic functioning and help explain the well-replicated finding of low resting heart rate in antisocial individuals outlined earlier. It would also help explain the impaired autonomic functions that we have seen in offenders, such as reduced electrodermal fear conditioning. In essence, when the fetus is exposed to smoking, the sympathetic nervous system gets shut down—and the outcome can be an underaroused, stimulation-seeking individual.

One would think that today mothers fully understand and recognize that smoking is bad for their unborn child. Yet the unfortunate reality is that in the United States about a quarter of all pregnant mothers smoke, while in the United Kingdom a quarter of smokers who become pregnant continue to do so during pregnancy.[96] Smoking remains a likely contributor to the violent offending in the offspring of these mothers.

ALCOHOL CONSUMPTION DURING PREGNANCY

In 1992, the double killer Robert Alton Harris was gassed to death in San Quentin prison, the first execution in California in twenty-five years. The terrible nature of his crime, in fact, likely contributed to the new wave of executions in the state. The murders occurred in 1978. Harris and his brother were looking to steal a getaway car for a planned bank holdup when they spotted a couple of teenagers in a green Ford eating Jack in the Box burgers. At gunpoint the boys were forced to drive to a wooded area near a lake under the promise that they would not be harmed. Once there, Harris shot and killed both boys. And it's at this point that, for a jury listening to testimony, it becomes more psychologically gruesome.

Just as Harris was about to execute the terrified boys, one of them—sixteen-year-old Michael Baker—pleaded for his life. According to a witness who shared a cell with Harris after his arrest, Harris boasted that he told the poor boy, "Quit crying and die like a man." In those moments of impending doom, the petrified boy started to pray to God. Harris's response? "God can't help you now, boy; you're going to die."[97] After the executions, and just as callously, Harris calmly finished off the rest of the murdered boys' half-eaten hamburgers, and flicked pieces of the homicidal gore from the barrel of his gun.[98] The heartlessness and clear lack of conscience, combined with the fact that Harris had just been released from prison for another murder, made him a clear candidate for the gas chamber.

What is also true of Harris is that he was born with fetal alcohol syndrome. If smoking is problematic during pregnancy, you can imagine what the negative effects of consuming significant amounts of alcohol may be. Again, the model here is that alcohol consumed by a woman during pregnancy is a significant source of damage to the fetal brain, and this brain impairment predisposes her offspring to violence. There are four features of fetal alcohol syndrome as first established by the pediatrician Kenneth Jones in 1973:[99] exposure to alcohol during pregnancy, craniofacial abnormalities, growth retardation, and central nervous system (CNS) dysfunction as evidenced by learning disabilities or low IQ. The craniofacial abnormalities in fetal alcohol syndrome sufferers can be striking. The middle part of the face is relatively flat, the upper lip is quite thin, and the eyes tend to be widely spaced. The uncanny result of this is that two unrelated babies in a hospital can look alike if they both have fetal alcohol syndrome. The rate of this syndrome is about 3 babies in every 1,000.[100] More common, however, is the condition of "fetal alcohol effects"—in which just some of the symptoms described above are present—with a base rate of approximately 1 percent.

Matching the striking nature of fetal alcohol syndrome is the relationship it bears to crime and delinquency. Perhaps the most comprehensive study conducted to date is that of Ann Streissguth and her colleagues at the University of Washington in Seattle.[101] Although fetal alcohol syndrome is relatively rare, Streissguth was able to obtain an incredible 473 cases of either fetal alcohol syndrome or fetal alcohol effects from the Pacific Northwest, and assessed outcomes for antisocial behavior at age fourteen. A full 61 percent of the sample evidenced juvenile delinquency. Sixty percent were expelled or suspended from school. Forty-five percent showed some form of inappropriate sexual behavior, such as incest, sex with animals, or masturbation in public. More than half of the boys and 33 percent of the girls went on to be arrested or convicted for their offending.

Streissguth's work started off with fetal alcohol syndrome and looked at outcome for antisocial behavior. Another way one could look at it is to start off with a population of antisocial individuals and examine rates of fetal alcohol syndrome and fetal alcohol effects. This is exactly what Diane Fast and her colleagues did, and they found rates of 1 percent for fetal alcohol syndrome—more than three times the expected base rate—and a full 22 percent for fetal alcohol effects.[102] There is little doubt that the mother's intake of alcohol during pregnancy can raise the odds of problematic behavior.

As with smoking during pregnancy one can argue that there is a third factor that underlies the fetal alcohol syndrome–antisocial relationship. Again, a genetically informative adoption study came to the fore in ruling this factor out. Remi Cadoret at the University of Iowa studied the adopted-away offspring of mothers who drank alcohol during pregnancy and found that they too showed higher rates of conduct disorder and adult antisocial behavior compared with adopted children whose biological mothers did not drink during pregnancy. Because the children were adopted away from their biological mothers after birth, their antisocial behavior cannot be attributed to the fact that mothers

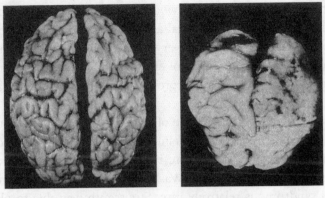

NORMAL   FETAL ALCOHOL SYNDROME

Figure 6.2 Brain of a normal six-week-old baby (left) and brain
of a baby with fetal alcohol syndrome

who drink may be poor caregivers during their child's development. It does look as if the exposure to alcohol during pregnancy is causally related to crime outcome.

The mechanism of action? The brain again has to be the number one suspect. Alcohol exposure ravages the brain during fetal development. The atrophy in brain tissue is striking and is widespread (Figure 6.2). Particularly affected is the corpus callosum, the band of white nerve fibers that connects the two hemispheres and allows for effective communication.[103] Poorer executive functions are also an almost inevitable consequence of fetal alcohol syndrome.[104] Experiments on animals have demonstrated that during the latter half of pregnancy, when the brain is rapidly developing, alcohol exposure results in a loss of neurons. It also affects glutamatergic neurotransmitter functioning, which in turn reduces hippocampal plasticity and the ability to learn.[105] Just as we saw with prenatal smoking exposure, those born with fetal alcohol syndrome show widespread structural and functional impairments when given brain scans in late childhood.[106]

Can pregnant women get away with just one alcoholic drink a week? It seems not. As with smoking, there is a dose-response relationship such that with increasing degrees of alcohol consumption, aggressive behavior and other externalizing behavior problems show a steady increase. A study of African-American mothers demonstrated that having just one alcoholic drink a week during pregnancy was enough to raise the odds of aggression and delinquency in the children.[107] Indeed, this study documented that drinking *any alcohol at all* during pregnancy tripled the odds that the child would have clinically significant delinquency. At a causal level, animal studies yet again show dose-response relationships between increasing amounts of alcohol exposure and increasing degrees of structural brain impairments.[108] It would be unwise for anyone who is pregnant to ignore the potential effects of consuming alcohol during pregnancy.

So is there such a thing as a natural-born killer? If by this question we mean, "Is there an unalterable destiny for violence?," the answer is no. But we have seen here that there are multiple health-related factors that occur right at birth and even before birth that are architects in shaping the landscape of violence. Birth complications, disruption to the developing brain of the fetus, exposure to smoking and alcohol, and testosterone exposure are significant elements in the genesis of violence. Yet these marks of Cain, while biologically based, are essentially environmental processes—not genetic. We get back to the crucial point that biological and social processes are inextricably mixed, and that a true appreciation of the biology of violence needs to take this mix into full consideration.

What is certainly true is that in casting the lots that determine who will become an offender, for some the dice are loaded very early in life. Yet these very early health processes are just the beginning in a mixture of influences that can become toxic. We'll see in the next chapter that health risk factors continue throughout development to create the deadly cocktail. As the example of Peter Sutcliffe shows us, there is more to homicide than birth complications—biologically based mental illness can be a critical factor in shaping a criminal career, as it did with him and with many others.

7.

A RECIPE FOR VIOLENCE

Malnutrition, Metals, and Mental Health

Amsterdam was a bad place to be in the winter of 1944–1945, especially if you were a pregnant mother. It was the beginning of the Dutch Hunger Winter. The Allied invasion of Normandy the previous June eventually gave relief to the Dutch, but in its immediate aftermath it brought misery. The Allies had been blocked at the Rhine and could not free much of the Netherlands from German occupation. In September the exiled Dutch government in London ordered railway workers in the Netherlands to go on strike in order to aid the Allies. They duly followed instructions, but the result was disastrous. German administrators retaliated with a food blockade, cutting off the western Netherlands from its food supplies.[1]

It went from bad to worse. First, winter came very early that year and was unrepentantly harsh. Canals froze over. Food could not be transported. Retreating German troops destroyed bridges and docks, making transportation even more difficult. Second, much of the arable land had been ravaged by warfare and was barren—unable to provide sustenance for the Dutch citizenry. Two more painful blows to aching, empty stomachs.

People began to starve. In November, city residents were rationed only 1,000 kilocalories of food a day. By February 1945 conditions deteriorated further with diets dropping to 580 kilocalories a piece.[2] Ten thousand died of malnutrition, particularly in the cities, which were cut off from the countryside and food. Many thousands more are thought to have died of complications as a result of the famine.[3] And for the remaining millions, life was wretched and depressing. Relief came only with liberation, in May 1945—ending a bitter eight months for the Dutch people.

This seems a peculiar starting point for a window into antisocial personality, but the seeds of violence were being sown in that harsh winter, concealed in out-of-sight little victims—the unborn babies of starved, pregnant women. We know this because in 1963, when the male babies who were in utero during the famine turned eighteen, they underwent compulsory military service, and at that time they were subjected to a psychiatric examination that included formal assessment of antisocial personality disorder.[4] Data collected from these examinations became the foundations of a unique epidemiological study on the effects of prenatal malnutrition on later behavior.

In this breakthrough research study, Richard Neugebauer and colleagues from the New York State Psychiatric Institute conducted detailed analyses on these data. They divided the enormous sample of 100,543 men into those who were exposed to the famine—especially in the large cities in the west, including Amsterdam, Rotterdam, Leiden, Utrecht, and the Hague—and those in the north and south who were not exposed to the famine.[5] The key result? Those exposed to the famine were *two and a half times* more likely to develop antisocial personality disorder in adulthood than those not exposed to the famine. The effects being especially true if the food shortage occurred during the first or second trimester of pregnancy. These findings were the first to demonstrate that poor nutrition during pregnancy predisposes to antisocial behavior in the offspring.

This chapter on nutrition, toxins, and mental health is yet another that highlights the importance of the *environment* in causing the brain impairments that can contribute to crime. From the gut to the teeth to the hair and back to the brain, this particular close-up of the anatomy of violence shows that human and animal studies are building a persuasive picture of how a lack of iron, zinc, protein, riboflavin, and omega-3 in our diets may dump some of us into the violence trash bin. It's a question of how both too little and too much food is bad for us. We'll also see that these dietary deficiencies can be compounded by an overexposure to heavy metals in the environment, including lead and manganese. Finally, we'll round off this physical-health perspective with a mental-health perspective, showing how major mental illness, with its base in biology, also contributes to violence.

My own research into nutritional deficiencies and violence was inspired during a visit with Danny Pine, a brilliant and energetic researcher at Columbia University who had been working on heart rate and cognitive functioning in conduct-disordered children. We were walking to a meeting with Neugebauer, and Danny, with his sparkling glasses and wild beard, which had a life of its own, was talking a mile a minute, as he often does. "And Adrian, you just *have* to meet Richard. *What* a story from Holland—World War II, starvation, crime. It's really something, you're gonna love this." And then he added with a mysterious twinkle and a wry smile, "Don't forget to ask him about the tulip bulbs."

Tulips? What's that all about? That song "When it's spring again, I'll bring again tulips from Amsterdam" flashed through my mind— but how does that fit into an academic meeting on violence? That was as much as Danny left me with before I met Richard Neugebauer and heard firsthand the astonishing story of the Dutch Winter Famine— and the tulip bulbs. Apparently, in the final months of the food blockade the starving Dutch began to eat tulip bulbs. These are toxic, and as we shall see later in this chapter, toxins have been associated with offending. Richard acknowledged that other issues remained unresolved. Only male adults had been studied. What about females? Could this malnutrition story apply to aggressive and antisocial behavior in children? And do social factors like poverty play a hidden role?

These were the issues that percolated through my mind and ultimately stimulated us to look into nutrition in our Mauritius study. When our subjects were three years old, 1,559 of them came to the laboratory with their mothers to be examined by a pediatrician. We looked for five internal and external signs of malnutrition. First, they had their blood analyzed in a lab to assess hemoglobin levels. This gave us a handle on iron deficiency. Second, we had pediatricians conduct a physical examination of each child to look for four other external signs of malnutrition. Do you ever remember as a kid having cracks at the corners of your mouth? I seemed to have them on and off, and I'd poke them with my tongue when they felt hard and dry in order to soften them up.

This is angular stomatitis, caused by a riboflavin deficiency, specifically a deficit in vitamin B_2, but it can also reflect niacin deficiency.[6]

Then the pediatrician would take a good look at the child's hair. What color was it? In Mauritius, almost all of the children have black hair because they are of Indian, African, or Chinese extraction. But some kids had an orange tinge to their hair. It wasn't some kind of funky look that their parents gave them to make them look cute and artsy—it was kwashiorkor. This is African dialect referring to "red hair." It's a sign of zinc, copper, and protein malnutrition that causes dyspigmentation of the hair—essentially a loss of the natural black color.[7] The pediatrician also looked to see if the hair was sparse and thin, a sign of zinc, iron, and protein deficiency.[8] Then, after these two careful looks, the pediatrician would grab a piece of hair and give it a tug. If it came out easily, it was a sign of protein energy malnutrition.[9] There we have the five strands—all clinical indicators of malnutrition.

At this point, Jianghong Liu, who at that time was a research fellow at the University of Southern California, entered the picture. She was the driving force behind the results I'll discuss here. If a child had any one of these significant indicators, she assigned them to the malnourished group. Those who lacked malnutrition were the normal controls. She assessed the kids again at ages eight, eleven, and seventeen, the ages at which we had obtained teacher and parent ratings of their aggressive, antisocial, and hyperactive behavior. The results are shown in Figure 7.1. As you can see, at every single age the malnourished kids had higher scores on all dimensions of what we call "externalizing behavior"—aggression, delinquency, and hyperactivity.[10]

Hold on a second. Aren't kids with poor nutrition more likely to have parents with low levels of education and income? And aren't low levels of education and income social risk factors for childhood behavior problems? Maybe poor nutrition itself makes no active contribution to aggression, but is linked to social deprivation, which causes aggression. Point taken. So Jianghong Liu controlled for poverty and twelve other social factors that could be driving the increase in aggressive behavior in the malnourished kids. The result? The malnutrition-aggression link was obstinate—it just would not budge. And it did not matter whether you were Creole or Indian, a boy or a girl. Poor nutrition does not respect race or gender when it comes to raising the risk of aggression. Furthermore, we also saw a dose-response relationship

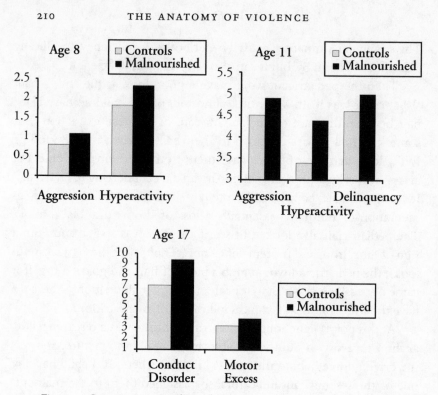

Figure 7.1 Scores on externalizing behavior problems in malnourished
and control groups across three time periods

at age seventeen. If you look at Figure 7.2 you'll see that the more signs of malnutrition the child had, the higher the score for conduct disorder. This result really reinforces the link between malnutrition and conduct disorder.

The type of malnutrition the kids had did matter a bit, though. Iron deficiency was especially important. This ties in with findings from experimental studies on animals showing that iron is involved in DNA synthesis, neurotransmitter production and functioning,[11] and white-matter formation in the brain.[12] If iron benefits the brain, low iron should be a problem. And it is. Experimental studies that have supplemented children's diets with iron show improved cognitive functioning.[13] My angular stomatitis, which reflected a vitamin B$_2$ deficiency, would also play a helping hand in poor cognition, because vitamin B$_2$ enhances the hematological response to iron.[14] Consequently, riboflavin deficiency would reduce iron and further negatively affect cognition. Eat your vitamin-fortified cardboard cornflakes.

It really does seem that poor nutrition, right across the board—

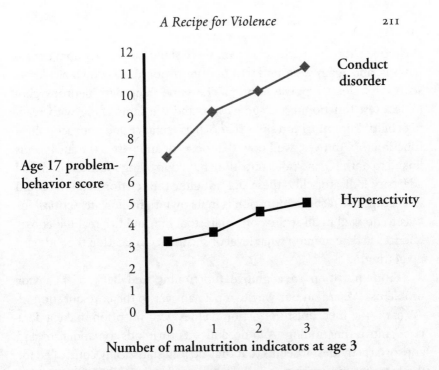

Figure 7.2 Dose-response relationship between signs of malnutrition at age three and behavior problems at age seventeen

across ages and types of problem behaviors—raises the odds of behavior problems in the growing child. But we get back again to a central, fundamental question. What is the mechanism of action, the way in which nutrition—or rather the lack of it—translates itself into aggressive and antisocial behavior? Back to basics. Back to the brain, and back to cognitive functioning.

Jianghong Liu found that the children with poor nutrition at age three also had lower IQs at that age and eight years later at age eleven. She again found a dose-response relationship, with increasing levels of malnutrition resulting in decreasing scores on IQ. If a child had three indicators of malnutrition, her IQ dropped seventeen points. It's a significant tumble: imagine being average in your class and dropping to the bottom 11 percent—not because of who you are, but because of what you don't eat. It did not matter what type of cognitive ability we looked at, malnutrition had an influence on verbal IQ as well as spatial (nonverbal) IQ.

In Mauritius, as in my day at primary school, they take national examinations at age eleven to decide what type of secondary school they will go to. The exams are in English, French, mathematics, and environmental studies. It really decides the rest of these children's lives.

We looked at their performance on these standardized national examinations, and again we found that poor nutrition drives down academic scores in a dose-response fashion. The same thing with neuropsychological test functioning at age eleven, and the same thing with reading ability. Poor nutrition sinks school performance and neurocognitive functioning. And yes, we know that poverty and parental education is linked to both IQ and poor nutrition, but controlling for multiple social adversity indicators like these did not alter the relationship. We could not escape the fact that nutrition is in its own right *absolutely critical* for kids to do well in all realms of intellectual life, and has real-life consequences in determining what level of secondary education the kids end up getting.

From nutrition to cognitive functioning and back to behavior problems. We are on our way to a part-answer to the core question of "What is the mechanism of action?" Does poor nutrition make a dent in cognitive functioning? And do dull wits turn kids to vandalism and antisocial activities? It seems that they do. Liu statistically controlled for the fact that kids with poor nutrition have lower IQ.[15] This technique makes the good and poor nutrition groups equal on intelligence. When that is done, the group difference in antisocial behavior disappears. This vanishing trick identifies poor cognition as a likely mechanism. Poor nutrition leads to low IQ, and this lowering of cognitive ability leads to antisocial behavior.

And it makes sense. You can imagine how low IQ can lead to school failure. You likely did well in school, but imagine what it's like to instead go in every day and get stuck on your reading, get your mind numbed with numbers that don't add up, while all the time most other kids seem to be doing just fine. Day in, day out, you're a failure. A failure for weeks, for months, for years.

It's easy to see how this can result in low self-esteem and a loss of hope. No wonder such kids try to bail out and kick back against the institutional system once they gain the muscle to rebel. Note here that just because poor nutrition acts negatively on the brain to predispose someone to aggression, we are not saying no to social factors altogether. Indeed, poor nutrition is very much an environmental factor. We see here that a negative environment—not getting enough of the right food—results in poor brain and cognitive functioning, which leads some kids down the primrose path to crime and violence. And as we are about to see, it's something of a slippery slope.

OMEGA-3 AND VIOLENCE: A FISHY TALE

Strange stories abound when it comes to trying to explain violence and other devious behavior. Perhaps one of the strangest circulating at the moment is that it's all to do with how much fish we eat. This may sound odd, but if we take a close look at the data, what your grandma always told you may be literally true—that fish food is brain food. And if something affects the brain, it's up for grabs as a causal agent in crime.

We'll begin with a topic in criminology that does not receive as much attention as it should. Why do countries around the world differ so much in violence, and what's the cause of these differences? There are plenty of ideas, old and new. Differences in unemployment rates do not seem to explain international differences in homicide and, perhaps surprisingly, neither does urbanization.[16] A lot of emphasis has been placed on social processes and for good reason, as the correlational data supports it. As we might expect, gross domestic product (GDP) is a strong correlate—the lower the GDP, the higher the violence: a correlation of .68. It really makes sense if we think of poverty as a cause of crime, because a higher GDP goes along with political development, increased democracy, and better education of the people.

A different social mechanism—income inequality—endorses this social perspective. As measured by the Gini index, the higher the income inequality, the higher the homicide rate—a correlation of .57. So the more a country is divided into the haves and the have-nots, the higher the homicide rate. Denmark, Norway, Sweden, and Japan all have relative income equality and low homicide, while countries like Colombia, Botswana, and South Africa have high inequality and high homicide, with the United States in between on both counts.

Interestingly, psychological beliefs also play a role. Some people prefer money, while others prefer love. What would your own pick be? We all differ to some degree, and just like individuals, countries as a whole differ from each other in the relative value they place on love, on the one hand, versus social status, good financial prospects, power, and status on the other. In countries where people believe love is more important than money, there is less violence. Perhaps the Beatles were not far off the mark—all you need is love.

But we need to eat as well as make love. And this is the fishy part. Countries differ an enormous amount in how much fish they eat, just as they differ in their homicide rates. Joe Hibbeln, a leading fish-oil

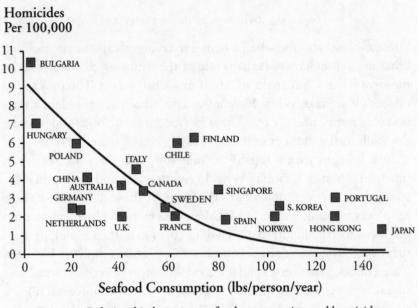

Figure 7.3 Relationship between seafood consumption and homicide rates across the world

expert working at the National Institute on Alcohol Abuse and Alcoholism in the United States put yearly homicide rates and fish consumption together. He found that they were negatively related—at a correlation level of -.63.[17] Take a look at Figure 7.3. It does look as if something may be going on here. Take Japan. They have very low yearly homicide rates—only one homicide per 100,000 people—and they eat well over their own body weight in fish every year. Then you look at eastern European countries like Bulgaria. They eat a measly four pounds of fish a year and rack up homicide rates ten times that of Japan. If you pick out the East Asian countries, they almost follow a straight line, with China at 4.3 homicides/100,000; Singapore at 3.8; South Korea at 3.0; and Japan at 1.2. The greater the fish consumption, the lower the homicide rate.

I showed Joe Hibbeln's provocative data in a talk I gave to the Criminology Department at the University of Pennsylvania in 2005 when I was being interviewed for a job, and one provocative question posed was, "Wait a bit, where's America here?" The United States was not on the graph of twenty-six countries. My colleagues-to-be didn't exactly smell a rat, but felt it was a bit of a slippery story. So they went and looked up the data for the United States for the year in question, and what did they find? Fish consumption right in between the two least-fish-consuming countries, Hungary and Bulgaria, and with homicide rates way up at 9 per 100,000, right next door to the eastern European countries. The correlation of -.63 was large and just as strong as that between GDP and homicide rates.

Explaining differences in violence across countries in the world is one thing, but such explanations may or may not apply to variations in offending *within* a country. Yet even within countries there is evidence that variation in fish consumption is related to antisocial behavior. In a very large sample—11,875 pregnant women from Bristol, England—women who ate more fish during pregnancy had offspring who showed significantly higher levels of *prosocial* behavior at age seven.[18] Put another way, the offspring of mothers who did not eat much fish during pregnancy had more antisocial behavior.

In the United States, a study of 3,581 people from Chicago, Minneapolis, and Birmingham, Alabama, showed that those who hardly ever ate fish had higher levels of hostility than those eating fish at least once a week.[19] There are also more behavior problems and temper tantrums in boys with lower total fatty-acid concentrations as measured from blood.[20] The same is true of aggressive cocaine addicts.[21] Even dogs with low levels of omega-3 have been shown to be more aggressive.[22] Giving your dog omega-3 may do more than give it a sleek, shiny coat.

Let's just suppose for a minute that this is a causal relationship, that bolting down boatloads of sushi and salmon somehow stops you from blowing your fuse. How on earth could this be possible from a scientific standpoint?

There is a reasonable answer based on experimental studies that manipulate the amount of omega-3 that rats have in their diet.[23] Recall from previous chapters that violent offenders have brain structural and functional impairments as well as neurochemical deficiencies. Fish is inevitably rich in fish oil. Fish oil, in turn, is rich in omega-3—a polyunsaturated long-chain fatty acid. Omega-3 has two important components—DHA (docosahexaenoic acid) and EPA (eicosapentaenoic acid). What does DHA do? It is known to play a key role in neuronal structure and function. Making up 6 percent of the dry cerebral cortex, it influences the functioning of the blood-brain barrier that regulates what gets into your brain from your bloodstream. It enhances synaptic functioning, facilitating communication between brain cells. It makes up 30 percent of the membrane of your brain cell and regulates the activity of membrane enzymes. It protects the neuron from cell death. It increases the size of the cell.

DHA also stimulates neurite outgrowth. There is more intricate dendritic branching in the neurons of animals fed a diet rich in omega-3 compared with those fed a normal diet. Dendrites of the cell receive signals from other brain cells, so this dendritic branching translates to more connectedness between cells. The axon that transmits the electrical signal to other cells is longer and has a better sheath to conduct the electrical impulse. DHA regulates serotonin and dopamine neurotransmitters, and we saw in chapter 2 that offenders have abnormalities in those neurotransmitters. We also know that DHA is involved in regulating gene expression,[24] so in theory it can help turn on genes that protect against violence—or turn off genes that increase the probability of violence.

We also saw earlier that cognitive functioning is impaired in offenders. Omega-3 supplementation has been shown to improve learning and memory in animals,[25] and also improves learning in children.[26] So it's not just that omega-3 in theory improves the brain. In practice, it makes a difference in terms of cognitive functioning—and cognitive functioning is critical for performance in school and success in life.

Omega-3 enhances both brain structure *and* function. We saw in earlier chapters that brain structure and function is impaired in offenders. So it's perhaps not all that surprising that we find associations between the amount of fish consumed and the perpetration of violence.

You might still find this all a bit too much to believe. Surely it can't be that simple? And correlation is not necessarily causation, right? You're correct on both counts. But what we will see in a later chapter on treatment is that there is mounting evidence from randomized controlled trials that omega-3 is effective in reducing antisocial behavior—and such trials are as good as it gets in establishing causality and demonstrating a true and meaningful relationship.

But you're likely still not convinced, are you? What use are these malnutrition studies to the United States, or other prosperous nations? Look around, everyone seems pretty healthy and there's plenty to eat. These results must be a problem only in developing countries, like Mauritius.

And you've got a reasonable point here. Visitors to the United States cannot help but be struck by the abundance of food and the big portions that are served up in basic restaurants. And the desserts are veritable mountains of yumminess. You take a look around you and, well, people do look kinda big in America. Rates of obesity are 30.6 percent for the United States and 23.0 percent for the United Kingdom, compared with 12.9 percent for Germany, 10.0 percent for the Netherlands, and 3.2 percent for both South Korea and Japan.[27] There's certainly no lack of sustenance over here in the United States, so what's the deal with all the violence?

There are three complementary perspectives to this issue. First, if you meet or see pictures of adult murderers, it's true that they certainly don't look malnourished. But this belies the fact that as children some of them, like the killers Henry Lee Lucas and Donta Page, were surviving by rummaging around in garbage cans. Page was an unfed, malnourished, scrawny little boy when he was growing up in the ghettos around Washington, D.C. But when as an adult he raped and killed Peyton Tuthill he weighed in at over 300 pounds. The outer appearances of adult offenders can be very misleading, hiding years of malnutrition at a critical early juncture in life when the brain is rapidly developing.

Second, there are two types of nutrients—macronutrients and micronutrients. Kids in America are getting plenty of the macronutrients—carbohydrates, fat, and protein.[28] But the story is different for the second component—the micronutrients that include vitamins and trace minerals, things like iron and zinc. They are "micro" because the amounts we need every day are really small, in the order of micrograms or milligrams. Yet they are critical for the growth and maintenance of body and brain functions. At the level of micronutrients, the World Health Organization argues that up to one half of all the children in the world have iron or zinc deficiency.[29] That's a staggering fact.

Third, we've also got to factor in that there is a wide range in the "bioavailability" of nutrients—the ability of the nutrient in question to get into your bloodstream and act on your brain. Bioavailability is influenced both by a host of genetic factors that determine how well nutrients are absorbed from the gastrointestinal tract, and also by environmental factors such as food inhibitors and enhancers. So, essentially, you can have two people with the same intake of micronutrients, but they may differ radically in terms of the degree to which those micronutrients get into their bloodstreams and act on their brains.

Once again, outside appearances and how well-fed a person appears to be can be very deceiving. Big is not better when it comes to body size and nutrition. Genes and environment, the two big gladiator arenas in which we have been seeing violence played out, can also starve the brain of critical nutrients. Given their potential importance, let's take a brief look at these micronutrients and the roles they may play in violence.

THE MIGHTY MICRONUTRIENTS

What are micronutrients? They include vitamins as well as important trace minerals like iron and zinc. If as a kid you had acne or if you had white spots on your fingernails, as I did, you can suspect zinc deficiency.

Deprive mice of zinc and their aggression increases threefold.[30] Even before birth, zinc deprivation during pregnancy in rats increases their offspring's aggression.[31] Children and adults in the United States with assaultive and aggressive behavior have abnormally low levels of zinc relative to copper.[32] A Turkish study similarly found that violent schizophrenics had lower zinc and copper ratios than nonviolent schizophrenics.[33]

Iron is another important micronutrient. Several studies have found aggressive and conduct-disordered children to be zinc-deficient.[34] One study found iron deficiency in a third of juvenile delinquents.[35] Preschoolers with low iron also show a reduction in positive emotions.[36] This is significant because a lack of positive emotion characterizes conduct-disordered children.[37]

Let's link back to the brain again to understand why these micronutrient deficiencies can predispose someone to violence. Micronutrients like iron and zinc are critical for the production of neurotransmitters and are important for brain and cognitive development. If you reduce dietary levels of zinc and protein in rats during pregnancy, then their offspring show impaired brain development.[38] Adult animals fed a zinc-deficient diet show "passive avoidance learning deficits."[39] This is an inability to learn to inhibit a response that leads to punishment, a cognitive deficit repeatedly found in offenders who have difficulty learning from their mistakes.[40]

We can also link micronutrients to specific brain structures involved in violence. The amygdala and hippocampus, which are impaired in offenders, are packed with zinc-containing neurons. Zinc deficiency in humans during pregnancy can in turn impair DNA, RNA, and protein synthesis during brain development—the building blocks of brain chemistry—and may result in very early brain abnormalities.[41] Zinc also plays a role in building up fatty acids, which, as we have seen, are critical for brain structure and function.[42] The availability of iron in the brain, like zinc, has been shown to affect neurotransmitter production and function.

What results in iron and zinc deficiency? It could be a lack of foods like fish, beans, and vegetables. Bear in mind that micronutrients play an important role in fetal brain development, and up to 30 percent of pregnant mothers with low socioeconomic status are believed to be iron-deficient. Smoking during pregnancy also impairs the transportation of zinc from the mother to her fetus,[43] depriving the fetal brain of a key nutrient. We have already seen that smoking during pregnancy predisposes a woman's offspring to adult violence.

Amino acids are also important because they are what proteins are made out of. Eight of our twenty-two amino acids are essential because our bodies cannot produce them. Animals fed diets reduced in one of these—tryptophan—become aggressive, while high-tryptophan food reduces their aggressive behavior.[44] When tryptophan is experimentally reduced in men and women,[45] they respond more aggressively when provoked.[46] Reversing that scenario, when tryptophan is enhanced, aggressive behavior is reduced.[47]

Low tryptophan likely increases aggression because it impairs the brain's ability to inhibit responses that we should not make. Brain-imaging research has shown that reducing tryptophan reduces functioning in the orbital and inferior regions of the right prefrontal cortex when subjects try to refrain from making a response to a stimulus.[48] We saw earlier that this underneath part of the prefrontal cortex is functionally and structurally impaired in offenders. Because serotonin is synthesized from tryptophan, the amino acid likely predisposes someone to reactive aggression by lowering brain serotonin, the neurotransmitter we saw in chapter 2 to be depleted in impulsive violent offenders.

Where does tryptophan come from? Foods like spinach, fish, and turkey. We see that omega-3 from fish could have a calming effect on aggression. In addition to fish, you might also tell your kids to eat their spinach—even if Popeye is not exactly the best role model for nonaggressive behavior.

TWINKIES, MILK, AND SWEETS

Sugar rush. Many of us have experienced it. We eat a ton of high-carbohydrate foods and drinks and then feel an energy rush that can make us feel able to shoot for the stars. Then we can feel a little agitated, get light-headed and on edge, and make a crash landing. That's what was claimed when Dan White killed the mayor of San Francisco, George Moscone, along with the city supervisor and gay-rights activist Harvey Milk.

Dan White was down in the dumps. Life wasn't working out too well. Having gone from serving in the Vietnam War to working as a police officer and then as a firefighter, he was familiar with high-risk life adventures. But his latest risky venture, a potato restaurant, wasn't working well, and he was out of money. He had resigned his position on the San Francisco Board of Supervisors—a position he had gained with strong union support from both firefighters and the police.

He had also fallen out with Harvey Milk, who was supporting the establishment of a juvenile detention center that had been proposed by the Catholic Church and was located in White's district. Now, while Dan White was a Roman Catholic, he was dead set against having the detention center in his district. He also had a gripe with gays, and Harvey Milk was gay. White had resigned from his political position to focus on his potatoes, but with their failure he went back to Mayor Moscone to regain his position. Moscone was in favor, but Milk was against White's reappointment.

In a fit of reactive aggression, White took a gun and entered San Francisco City Hall through a window to avoid the metal detectors. He went into Mayor Moscone's office and begged him to restore his position. Moscone refused, so White shot him dead. He then went into Harvey Milk's office and shot him dead too.

Enter the Twinkie. At his trial, White's defense team and their psychiatrists argued that he was suffering from depression and had immersed himself in an orgy of junk foods and drinks packed with refined sugar. Bad diet could influence his mood. White was a white working-class heterosexual all-American Catholic who had fought for his country and once saved a woman and her baby from a fire. The jury was made up of predominantly white working-class people who shared White's values. Some openly wept when they heard the pressure he was up against in his life. Instead of first-degree murder and the death penalty, he was found guilty of voluntary manslaughter and received a prison sentence of seven years and eight months.

The San Francisco gay community went nuts. Even Acting Mayor

Dianne Feinstein proclaimed: "Dan White has gotten away with murder. It's as simple as that."[49]

White's defense had been buttressed by $10,000 that the police had raised for him. The result was the "White Night Riots."[50] A crowd of 1,500 quickly gathered that night in the predominantly gay Castro District, where Milk had lived. It grew to an ugly 3,000 who descended onto City Hall and tore the place apart.[51] Police cars were set on fire. After establishing order at City Hall, police retaliated by going into bars in the Castro area and beating up gays. Sixty-one police officers and over a hundred gays were hospitalized for injuries. Dan White eventually committed suicide.

All this because of a little Twinkie?

Not quite, but near enough. Twinkies themselves—sponge cakes with cream filling—were never actually brought up at Dan White's trial, and the term "Twinkie Defense" was a phrase invented by the press. But junk food was brought up at the trial. Could it really contribute to diminished rational thought, as the defense argued? The claim certainly caught on rapidly after the trial. As one protestor put it to reporters as he was setting fire to a police car on that White Night, "Make sure you put in the paper that I ate too many Twinkies."[52]

White's behavior may or may not have been influenced by junk food, and even if it did contribute to the homicides, we are hard-pressed to view this as an excuse—either an excuse for Dan White's outrageous actions or, indeed, an excuse for the reactions of the local community. But if there is a mechanism at play here with respect to aggression, the likely candidate is refined carbohydrates. A number of studies have claimed that dietary changes aimed at reducing sugar consumption reduce institutional antisocial behavior in juvenile offenders. Some of these claims are striking. For example, one early controversial study—a two-year double-blind controlled study of twelve- to eighteen-year-old delinquents—obtained a 48 percent reduction in disciplinary offenses after diets were altered in order to reduce refined carbohydrates.[53] Experimental studies in animals have also demonstrated a causal relationship between low blood sugar and aggression in rats.[54]

Let's move on to Peru and the Quolla Indians for another course in the recipe for violence. The Quolla have a very high rate of homicide and incessantly feud with each other, and have been called, a bit harshly, "perhaps the meanest and most unlikeable people on earth."[55]

One anthropologist who studied them made the keen observation that a significant number of their acts of aggression seemed to be without good cause.[56] He also noticed that the Quolla were often hungry and craved sugar. Could it be that their irrational aggression was due to low blood-sugar levels and reactive hypoglycemia? A glucose-tolerance test, which assesses propensity for low blood-sugar levels, confirmed a relationship between low blood sugar and both physical and verbal aggression in the Quolla.[57] When you next feel irritable and angry for reasons that are not obviously apparent, you might consider a quick *nutritious* nibble to restore your sugar levels—but not a Twinkie.

In Finland, Matti Virkkunen, who is a psychiatrist at Helsinki University, has been repeatedly demonstrating in some important studies very significant metabolic abnormalities in violent offenders that fit the low-blood-glucose idea. In a series of early studies, Matti demonstrated that violent offenders were more prone to hypoglycemia. He demonstrated that aggressive psychopaths had increased insulin secretion, which would explain their low blood-sugar levels.[58] More recently, Matti found low glucose metabolism and low levels of the hormone glucagon in another group of violent Finns.[59] He then found that low glucose and glycogen formation predicted which violent offenders would go on to commit further violence eight years later, with the two measures explaining 27 percent of this future recidivism.[60]

If Matti Virkkunen and others are right, how exactly would the recipe of junk food, hypoglycemia, and low glucose metabolism push a person to violence and aggression? It goes something like this. Diets high in refined carbohydrates can cause extreme fluctuations in blood glucose levels—foods like white bread and white rice. Such foods have the bran, germ, and nutrients stripped from the whole grain, taking away the fiber. Because of the fiber loss, they are rapidly absorbed by the gut, resulting in a large and rapid increase of glucose swishing around in the bloodstream. This in turn triggers an inappropriately large secretion of insulin. Insulin's job is to soak up the excess glucose and convert it into glycogen so that surplus energy can be stored for future use. But too much insulin release results in too much of the available glucose being taken out of circulation. This is bad news for the brain, which requires at least 80 milligrams of glucose a minute to function efficiently. Drop below that mark and you progressively observe symptoms of nervousness and irritability. That combo of increased irritability combined with feeling on edge could be the first step in the development of a full-blown aggressive outburst. It's not too surprising, therefore, that when glucose levels of subjects are experimentally lowered in the laboratory, people report feeling more angry even though there is no provocative stimulus.[61]

But what's really shocking is a recent study by Stephanie van Goozen and her colleagues at Cardiff University in Wales that was conducted on a sample of 17,415 British babies born in 1970.[62] When they were ten years old, the children were asked how often they ate sweets. Van Goozen showed that the kids who ate sweets *every* day were three times more likely to become violent by age thirty-four. They controlled for many factors, and the results remained significant.

If this relationship is causal, what's going on? It could be reactive hypoglycemia. The kids who are helping themselves to candy at age ten are also helping themselves to a lifestyle of unhealthy eating habits— high-energy, highly refined carbohydrates that result in too much sugar too quickly. The resulting rebound of very low blood sugar and symptoms of irritability can predispose a kid to giving someone else a good punch in the face in the school playground. Or, as an adult, a broken bar glass in the face. Keep your kids off the candies.

HEAVY METALS MAKE FOR HEAVY HITTERS

If you think sweets are bad for you, they're nothing compared with other things that can get inside you, mess up your brain, and make you flex your muscles. I'll suggest here that heavy metals can form some of the ingredients in the concoction for crime causation. Let's take a look at a few of the key ingredients.

Lethal Lead

We saw in chapters 3 and 5 that the structure and function of the brains of violent offenders—especially the prefrontal cortex—is compromised. We have also hypothesized that these brain impairments produce secondary effects—emotional, cognitive, and behavioral—which in turn shape violence. Lead is a leading candidate as a source of these structural and functional brain impairments.

First and foremost, lead is neurotoxic, meaning that it kills neurons and damages the central nervous system. The neurotoxic effects of lead have been known for millennia, and efforts to reduce it are not recent. They have a connection to my favorite drink in England—cider. Back in the seventeenth and eighteenth centuries there was a common malady known as Devon colic, a neurological condition that particularly afflicted people in the southwest of England. They grow a lot of apples down in Devon and cider was almost a staple drink there back then. It was thought that the acidic apple juice caused the colic. Yet in the late eighteenth century George Baker, a physician, identified the cause as lead contained in the cider presses. Over the next few decades lead was steadily taken out of the presses. A near-miraculous reduction in Devon colic occurred, proving Baker's hypothesis.

Lead's neurotoxic effects are documented in brain-imaging studies of workers exposed to the metal in their jobs. One study scanned the brains of 532 adult men who had worked in a lead chemical plant.[63] There was a wide range of bone-lead levels in these participants, but an average reading was at the very top of the safety level.[64] Workers with relatively high bone-lead levels had smaller volumes of many brain areas even after controlling for multiple confounds like age and education levels. The fact that the frontal cortex was particularly reduced[65] is very interesting, given that this brain region is involved in violence. This lead effect was equivalent to five years of premature aging of the brain.

So lead workers have brain volume reductions. What about people in the community like you and me who likely have just low to moderate levels of lead in our blood? This question was addressed in a study of 157 individuals from Cincinnati who had had their blood-lead levels measured twenty-three times from the ages of six months to six and a half years.[66] This prospective study again showed that those with high lead levels had low brain volumes. One of the brain regions most affected was the ventrolateral prefrontal cortex, that lower outer region of the front of the brain that is impaired in antisocial and psychopathic individuals. This community sample had an average blood-lead level at age six that was high, but still within the so-called "safe" range as defined by the Centers for Disease Control and Prevention. We can see, then, that those exposed to "safe" levels of lead can suffer from brain impairments. Furthermore, the prospective nature of the study, moving from childhood lead exposure before age six to brain structure at age twenty-three, helps to establish causality.[67]

These studies give clear documentation of the negative impact of lead on the brain, and, intriguingly, they also document that the brain area most frequently found to be compromised in violent populations—

the frontal cortex—is particularly impacted by lead exposure. The next question is whether those with high lead levels are found to be more antisocial.

The landmark study in this area was conducted by Herbert Needleman at the University of Pittsburgh. He found that boys with high lead levels have higher teacher ratings of delinquent and aggressive behavior, and also higher self-reported delinquency scores. It was an impressive and influential study. Similar links have been found in at least six other studies in several different countries.[68] Furthermore, experimental exposure to lead during development increases aggressive behavior in hamsters, thus suggesting a causal link.[69]

Environmental lead exposure, therefore, is a risk factor for antisocial and aggressive behavior in delinquent kids. What about adult crime? And how early in life does this association occur? Answers to these questions were obtained in a methodologically strong study of African-American pregnant women.[70] Both prenatal and postnatal blood-lead levels in their offspring dramatically predicted adult crime in the early twenties and also adult violence. For every 5 microgram increase in prenatal blood-lead levels, there was a 40 percent increase in the risk for arrest.[71] Given that a 5 microgram increase from birth to age five still keeps you well below the limits of what the Centers for Disease Control and Prevention considers safe, this constitutes substantial risk from just a moderate, "safe" amount of lead exposure.

The last study shows that blood lead very early in life is an important predictor of adult crime. We also know that blood-lead levels are maximal at twenty-one months, when children are most exposed to lead.[72] Why is that? You know that toddlers put their fingers in their mouths a lot. And they also get their fingers into every pie they can, including mud pie outside in the garden. Lead lingers well after its release into the environment and stays in the soil for years. Even though gas is now unleaded, the lead residue from the past still lingers in the soil, especially near major roads and freeways.

High blood-lead levels later in childhood can be even more important. One study in Yugoslavia[73] recruited pregnant mothers in 1992, just at the time of the large-scale ethnic conflict between the Serbs and the Croats. The mothers came from two towns near lead smelters. Blood-lead levels in their offspring at age three were more strongly related to destructive behavior than the prenatal measures of blood lead. Similar findings have been obtained in America, with high blood lead at age seven—but not age two—correlating with high antisocial and aggressive behavior at age seven.[74] So lead exposure still matters well after the age of twenty-one months.

Lead research lends itself to an intriguing conceptual point. What has puzzled criminologists is the unpredicted drop since 1993 in violence after a continuous rise, which flew in the face of criminological predictions of further increases. For example, within seven years violent crime in New York had dropped 75 percent. Many sociopolitical explanations were given, but none could account for both the rise and the fall in crime across several decades. Critics of neurocriminology argue that biology cannot, of course, explain differences in violence over time or across regions within a country. Isn't biology fixed and static? Surely it cannot explain secular trends—shifts in violent crime rates across time.

But it can, and dramatically so. In research papers buried in an obscure environmental journal, Rick Nevin documented a strikingly strong relationship between changes in environmental lead levels from 1941 to 1986, and corresponding changes in violent crime twenty-three years later in the United States.[75] So, young children who are most vulnerable to lead absorption go on twenty-three years later to perpetrate adult violence. As lead levels rose throughout the 1950s, 1960s, and 1970s, so too did violence correspondingly rise in the 1970s, 1980s and 1990s. When lead levels fell in the late 1970s and early 1980s, so too did violence fall in the 1990s and the first decade of the twenty-first century. Changes in lead levels explained a full 91 percent of the variance in violent offending—an extremely strong relationship.

Nevin found exactly the same matching of the lead levels and violence curves in Britain, Canada, France, Australia, Finland, Italy, West Germany, and New Zealand.[76] There was cross-cultural replication. Furthermore, in states where lead levels dropped more quickly, later violent crime also dropped more quickly.[77] Variations in lead levels even correlate with variations in crime rates within cities.[78] From international to national to state to city levels, the lead levels and violence curves match up almost exactly.

Kevin Drum, a political blogger and columnist argues that these findings have been completely ignored by criminologists. He contacted criminology experts and none of them showed a scrap of interest.[79] Why? Likely because to recognize that secular trends and both rises and falls in violence can be partly attributed to brain dysfunction—and not to better policing or to gun control or to the end of the crack epidemic — would be to recognize the explanatory power of biology theories. Currently that's something very difficult for many social scientists to accept.

Cruel Cadmium

At a McDonald's next to the post office in the community of San Ysidro, near San Diego, at 3:40 p.m., on July 18, 1984, a middle-aged man walked in with a 9-millimeter semiautomatic Uzi and unloaded 257 rounds of ammunition into the customers. The shooter, James Oliver Huberty, killed twenty-one people and wounded nineteen others.[80] His victims ran the gamut in age from just seven months to seventy-four years.

What on earth made Huberty do it? Cadmium is a very likely culprit. An analysis was made of Hubert's hair after he was shot dead by a SWAT team sniper perched on the roof of the next-door post office. The results were nothing short of astonishing. In the words of William Walsh, the chemical engineer conducting the analysis, "He had the highest cadmium level we have ever seen in a human being."[81] Huberty's lead levels were also high, so he had a double hit. There's no mystery as to why he had multiple metals in his body. Huberty had been a welder for Union Metal for a number of years until he gave it up. The reason he left his welding position? In an exit interview that he gave to his employer upon leaving, he said, "The fumes are making me crazy."[82]

So cadmium can be a killer, not just in people like Huberty, and not just in the United States. Certainly, hair samples from violent offenders in the U.S. show them to have more cadmium than nonviolent offenders.[83, 84, 85, 86] High hair-cadmium levels also characterize U.S. elementary schoolchildren with behavioral problems.[87] The same is true for schoolchildren in China, a leading producer of cadmium. The Dabaoshan mine in the city of Shaoguan in the Guangdong province is a multimetal mine. Water is used to leach the ore, and the waste water is then transported by rivers to local villages, delivering a large dose of heavy metal to the villagers. The result is that the crop region in this countryside has *sixteen times* the recommended level of cadmium. A study of schoolchildren living downstream from the mine showed that hair-cadmium levels explain 13 percent of the variation in their aggressive and delinquent behavior.[88] Cadmium is quite a heavy-metal key on the biological key chain unlocking the etiology of violence.

It's not hard to see how people living near a mine are exposed to cadmium, but what about the rest of us? Not surprisingly, cadmium is a hazardous substance that can cause death and is banned by the European Union for use in electrical equipment. Yet about 75 percent of all cadmium in the United States is used in rechargeable nickel-cadmium batteries rolling around your home right now. Not too harmful there, perhaps, but cadmium does find its way into the environment from municipal waste grounds and fossil fuels because products containing cadmium are rarely recycled.

The people most susceptible to cadmium? Smokers. They inhale about 10 percent of the cadmium content of a cigarette, which gets nicely absorbed into the bloodstream from the lungs.[89] They end up with five times the cadmium levels of nonsmokers.[90] The rest of us get exposed too, because foods like offal (the internal organs of animals) and cereals[91] account for 98 percent of our cadmium intake. In contrast, seafood, which we saw earlier to be associated with *lower* violence, accounts for only 1 percent.[92]

The twist here is that the amount of cadmium acting on your body is a function of other factors. Iron blocks the intestinal absorption of cadmium.[93] Women on vegetarian diets have reduced iron levels and they also have increased cadmium exposure. If they smoke as well they will have an exponential increase in cadmium. This may partly explain why low iron is associated with violence—individuals with low iron levels are more susceptible to the negative effects of cadmium on the brain.

Mad Manganese

Everett "Red" Hodges is one of those larger-than-life characters whose charismatic and witty stories blend with forceful argumentation to make you believe almost everything he has to say. His sons have been both perpetrators and victims of crime. One was a rebel without a cause who got into a load of trouble as a juvenile delinquent. The other was mugged in a parking lot and very badly beaten up, suffering brain damage as a result. "My son was damn-near murdered," Red said in an interview. "I know the anguish and suffering that families go through. And you can't put a price on it."[94]

Red reasoned that if the criminal justice system had done a better job of dealing with the neurobiology of violence, his son and many others would never have been the victim of violence. The anguish of many family members would have been spared.

Red pins the blame on one particular metal—manganese. Having made a good deal of money in a Bakersfield oil field in California, Red Hodges sank a million dollars into funding efforts to investigate his hypothesis. Working with Red, Louis Gottschalk at the University of California, Irvine, demonstrated that three different samples of violent criminals had higher levels of manganese in their hair than controls did.[95] Roger Masters at Dartmouth University similarly showed that areas in the United States with higher levels of manganese in the air have higher violent-crime rates—even after controlling for multiple socioeconomic confounds.[96]

At the same time, the manganese debate is a political hot potato, and it's hard to know who's right and who's wrong. Critics reasonably argue that the evidence is mixed and that we cannot easily untangle cause-and-effect relationships from correlational studies.[97] What helps here are longitudinal studies involving teeth. The cusp tip of the first molar gives a handle on manganese exposure halfway through pregnancy—a time when a fetus's brain is rapidly expanding. Using these teeth, researchers showed that kids with high prenatal manganese levels had disinhibited, antisocial behavior across the board on a host of antisocial-behavior measures.[98]

What causes excessive manganese exposure during pregnancy? A deficiency in iron—the micronutrient that when low is associated with high antisocial behavior—enhances manganese absorption. Women with low iron levels absorb about four times more manganese than women with high iron levels.[99] An early postnatal source of manganese is soy infant formula, which has eighty times the amount of manganese that natural breast milk has. It's possible that the higher IQs found in breast-fed babies may be due to formula-fed babies' being exposed to high manganese, because manganese excretion is controlled by the liver. The livers of babies are underdeveloped, and consequently they are less able to excrete manganese. The excessive manganese could then result in poorer brain functioning and lower IQs.

Put the two together and you begin to build a recipe for violence. Pregnant mothers have a tendency to have low iron. This will result in increased manganese exposure to the fetuses. Then, when the nippers are born, they get soy milk with a hefty dose of manganese that their little livers cannot deal with. The potential result? One more strike on the brain. Higher manganese levels in children can result in impairments in cognitive speed, short-term memory, and manual dexterity.[100] As we noted earlier, this neurocognitive dysfunction predisposes individuals to violence. Furthermore, manganese reduces serotonin, a neurotransmitter that when low causes a predisposition to impulsive violence.

Given this, perhaps it's not too surprising that *fifteen studies* on workers exposed to manganese in all corners of the world—including Chile, Great Britain, Egypt, Poland, Brazil, the United States, Scotland, and Canada—without exception report significant mood disruption, including aggression, hostility, irritability, and emotional disturbances.[101] In Chile the term used is *locura manganica*—meaning "manganese madness." It refers to violence, mood disturbances, and irrational behavior. It's just the type of craziness that James Huberty reported as the reason for leaving his welding job, this time for another mad metal—cadmium.

It has been documented that the aggressive acts of workers exposed to manganese result in "stupid" crimes that are not premeditated and motivated by gain, but more a result of brain impairment resulting in poor emotion regulation and impulsivity.[102] Not surprisingly, low intelligence is an extremely well-replicated risk factor for violent offending, a risk factor that could in part be caused by an excess in manganese.

Mysterious Mercury

Moving from manganese to mercury you might expect the same pattern of results to emerge. But they don't. Mercury is mysterious. Of all the heavy metals, this one may or may not play a role in violence—a fact that is both striking and enlightening. Mercury is toxic to the brain and other body organs, with about half of human-generated mercury coming from coal plants. Dental amalgams are another source, and fish are also argued to be a major dietary contributor.

Despite its toxicity, to my knowledge there are no convincing demonstrations that antisocial and violent individuals have higher mercury levels. It is also surprising that there are so few studies on mercury levels and cognitive ability in community populations. Two major prospective studies that have been done on blood-mercury levels and cognitive-behavioral functioning show conflicting findings.[103] One study, conducted in the Faroe Islands, between Scotland and Iceland, found high mercury to be associated with poorer cognitive functioning.[104] The other study, in the Seychelles, which is just up the road from Mauritius, in the Indian Ocean, found no association between mercury and cognitive-behavioral outcomes.[105] Reviewers are at a loss to explain the discrepancy, putting the difference down to "culture."[106]

Yet if we put together a few seemingly unrelated facts, these geographically contradictory findings can make sense. Where do people get mercury from? Supposedly from eating fish that are high up in the food chain—particularly shark, swordfish, and king mackerel, which are certainly on the no-go list for pregnant mothers. In the Faroe Islands they also eat a whole load of pilot whale, especially outside the capital city. What's the deal with pilot-whale meat? It's not just very high up in the food chain and high in mercury, but it's also low on selenium.

Selenium? This is a mineral that defends the brain against "oxidative stress," a process in which the brain cell takes up too much oxygen, resulting in the production of free radicals that damage DNA and the cell membrane, resulting in cell death. Selenium not only protects against this damage but, more important, it binds with mercury. Like a magnet, selenium latches onto mercury and keeps it from binding with brain tissue, thus preventing brain and cognitive impairment.

If you think about it, fish seem to do okay with all that mercury leaching out of the seabed, and many species are packed with selenium. Going back to the two studies with contrasting results, the high-mercury and low-selenium diet in the Faroe Islands translates into poor cognitive and behavioral functioning. And yet in the Seychelles, pregnant women are also exposed to mercury, eating twelve portions of fish a week. That's a lot, twelve times the consumption of American women. So what is different in the Seychelles and the Faroe Islands? In the Seychelles they do not eat pilot whale, which is low in selenium. Instead they eat fish high in selenium that buffers them from mercury and its cognitive impairments. Their diet thus protects them against any damaging effects of mercury, as well as providing a high dose of the beneficial omega-3. We shall return to omega-3 in a later chapter when we pose prevention strategies to fight violence.

MENTAL ILLNESS MAKES FOR MEANNESS

So we are seeing that biology plays out in the environment and in the physical-health arena when it comes to the makings of malevolence. Some heavy metals take their toll on the brain and predispose people to violence. But health is a multifaceted construct, and it acts in ways other than diet and environmental toxins to shape violence. Let's not forget *mental* health. Biological impairments can also make men mad, and madness can make men mean. Women too, perhaps even more than men. Mental illness has its roots in genes and neurotransmitter abnormalities that mess with our minds. And it's when our minds are mucked up that we are most prone to violence. One prominent and major mental illness that can do this is schizophrenia.

I've long had an interest in schizophrenia because it was, in a way, pivotal in moving me out of accountancy and into criminology. Not that I became psychotic adding up all those numbers at British Airways—although at times I did think I was losing my mind somewhere within those cabin-crew accounts. But this clinical disorder did radically change my life. Haven't we all had those pivotal moments in life when a seemingly chance, inconsequential event changed everything? You pick up some random book, just like you picked up this one, and something clicks. The next thing you know, your life takes a sudden turn—all because of one capricious, unpredictable, and seemingly innocuous experience.

In my case it was a Saturday morning just before lunch in the early summer of 1973, and I was bored to tears working at Heathrow. I knew I'd made a really bad life decision in becoming an accountant, and I was absolutely miserable—had been for months. How had I messed up so badly? I was hungrily hunting for some books at a bookstore in Hounslow where I lived to read over my Saturday lunch treat—an "American" cinnamon apple pie and ice cream—and it leaped out at me. A slim paperback by R. D. Laing and Aaron Esterson entitled *Sanity, Madness, and the Family.*[107] Laing's riveting collection of eleven case studies of schizophrenic patients challenged the prevailing medical model that schizophrenia was a brain-based disorder. Instead this existential psychiatrist argued that schizophrenia had an environmental basis stemming from faulty communication within the family. Schizophrenics have outrageous and bizarre beliefs, but their madness becomes understandable when we consider the context of the family.

I had an epiphany. It was all making sense. So *that's* how I ended up as such an oddball—it was all my nutty parents! It was a revelation that made me determined to understand myself more and to study psychiatry (I ended up studying psychology instead), to challenge the biological model of mental disorder (I eventually did the opposite), and to work in hospitals helping schizophrenic patients (trade that for four years in prison helping psychopaths). Books change our mind-set and sometimes our life—though not always in the way we anticipate, and not necessarily in the right way.

Laing and Esterson weren't exactly right either. Schizophrenia turns out not to be caused by faulty parent-child communication patterns but is instead a debilitating, brain-based, *neurodevelopmental* disorder characterized by delusions, hallucinations, thought disorder, lack of emotion, and disorganized behavior. Affecting about 1 percent of the population around the world, it frequently hits women in their early twenties and men in late adolescence, with about 40 percent of male schizophrenia cases occurring before the age of nineteen—an intriguing fact given that these late adolescent years are also the peak age for violence in men.[108]

What's also intriguing is that when we look at the biological factors that are related to schizophrenia, we find many of the same risk factors that we have seen earlier characterizing violence. Things like frontal-lobe dysfunction, neurocognitive impairment, fetal maldevelopment, birth complications, blunted brain responses to stimuli we should normally pay attention to, and orienting abnormalities. To be sure, crime and schizophrenia are certainly not the same condition. They present very differently to the clinician. And there are risk factors like low resting heart rate that are unique to crime and unrelated to schizophrenia.[109] Yet, at some causal level, there is a degree of common ground.

That common ground expresses itself most strongly when we look at the link between violence and schizophrenia. Large-scale epidemiological studies from many countries around the world now attest to the fact that schizophrenia patients are much more likely than normal controls to have a history of violent and criminal behavior. Turning the issue around, delinquent and criminal populations are more likely to show higher rates of psychotic disorders than the general population. This relationship between violence and schizophrenia is not weak. If you are a schizophrenic male, you are three times more likely to kill than someone of the same social background and marital status who is not schizophrenic. If you are a female schizophrenic, you are twenty-two times more likely to kill than a nonschizophrenic female.[110]

These are striking statistics, and we should be cautious in interpreting them. Many psychiatrists and families of schizophrenic patients do not want to hear this message.[111] It's hard enough for someone with schizophrenia to have to carry the burden of this debilitating illness, let alone to be labeled as violence-prone. It's true that most schizophrenics are not dangerous and neither kill nor perpetrate violence.[112] But the harsh reality is that the neurodevelopmental ravages perpetrated on the brains of schizophrenic patients during childhood and adolescence make them less able to regulate their emotions and hold back their anger as adults.[113]

You might accept that schizophrenia is a neurobiologically based mental disorder. You may even agree that schizophrenics are more likely than others to kill. But you could counter that schizophrenia is a rare mental illness, so surely it cannot account for much violence. And you'd have a point. What we next need to consider, therefore, is that there is a "watered-down" version of schizophrenia with a higher base rate in the general population.

We have exactly that in a clinical condition called schizotypal personality disorder.[114] Instead of hearing voices of nonexistent people, as schizophrenics do, schizotypals mistake an actual noise in the environment for someone speaking. It's not entirely uncommon. I was in my hotel room at a conference in Tuscany washing and shaving in the bathroom sink when I heard a woman very close by, shrilly saying, "Well, hello." Startled, I looked around. I looked in the bedroom. Nobody. How peculiar. I went back to washing, and heard the same thing again. It had to be outside in the corridor. I opened the door to my bedroom, but no one was standing out there. This was seeming more bizarre. Going back to washing and turning on the faucet, I realized that the squeaky female voice was none other than the squeaky tap. About every month or so I hear someone calling my name in the street, and look around to find myself mistaken again. Technically, the symptom is called "unusual perceptual experiences"—you mistake sounds for voices and shadows for objects and people. But I'm all right, I tell myself.

Is it just me who's got a tile loose? Not really. We can measure schizotypal personality quite well using simple self-report questionnaires. I created a measure for it back in 1991 (yes, psychologists really do study the problems they have). It's called the "Schizotypal Personality Questionnaire."[115] It includes questions like this one: "When you look at a person, or yourself in a mirror, have you ever seen the face change right before your eyes"? We found in Los Angeles that 18 percent of supposedly high-functioning undergraduates said yes to this item. "Have you had experiences with astrology, seeing the future, UFOs ESP, or a sixth sense?" Forty-nine percent say they have. "I feel I have to be on my guard, even with friends" has 21 percent endorsement, while 31 percent agreed that "some people think that I am a very bizarre person." When we brought in the students whose total score was in the top 10 percent of the undergraduate population for a clinical interview, 55 percent of this group received a clinical diagnosis of schizotypal personality disorder—equivalent to 5.5 percent of the total undergraduate population, much higher than the 1 percent base rate for schizophrenia.

Now, you could put this all down to the fact that it's L.A. that we're talking about—a safe haven for loonies from other locations to migrate to so they can fit in with all the other nutters and not seem so obviously bananas. And there might just be a smidgen of truth to that West Coast stereotype. But at the same time, the reality is that psychosis has its manifestation at a dimensional level. There are shades of gray here, and there is a surprisingly large minority of people in the population with some characteristics similar to schizophrenia.

Are these individuals more likely to be violent and antisocial? Yes, they are. Whether we look at undergraduates at universities—the privileged offenders—or just individuals in the community, those with higher scores on the "Schizotypal Personality Questionnaire" have higher scores on self-reported measures of crime and violence.[116] They parallel what many others find in clinical populations of schizophrenics. Put together those with schizotypal personality and those with outright schizophrenia and other psychoses, and you really do have a small but significant group at risk for crime and violence.

But why would schizophrenics be more likely to kill than others? One answer can be found at the surface level in the symptoms of schizophrenia. For one thing, one common manifestation of schizophrenia is paranoia. Paranoid schizophrenics are overly suspicious of other people's intentions, and believe others are out to get them. If you believe that, then one reasonable defense is to get them before they get you. Other schizophrenics have delusions of grandeur, which can give them a righteous sense of power and control over others, or a religious grandiosity that may make them feel they have the right to override the sanctity of life. Other schizophrenics have a messianic vision—they are a prophet come to save the world from its debauchery and sins. One way of doing this, of course, is to kill as many prostitutes as you can, just as we saw with Peter Sutcliffe.

There are also features common to schizotypals and psychopaths. These two disorders may seem like chalk and cheese on the surface— the shy, retiring schizotypal versus the brash, confident psychopath. But there is a connection. Schizotypals have constricted affect—meaning that their emotions are blunted and reduced. We similarly see in the research literature on psychopaths repeated evidence of this emotional blunting. They just do not experience emotions in the same way that the rest of us do. Schizotypals also have no close friends outside of their family members, and in a similar fashion psychopaths form only very superficial, fleeting relationships, having an inability to form the deep and meaningful social affiliations that the rest of us do.

These superficial similarities partly explain why schizophrenics are more violent. In the same way that blunted emotions and a lack of social connectedness with other people nudge the psychopath into the perpetration of violence, social disconnection and a lack of feeling can tip the schizophrenic into violence. And if you can't bring yourself to imagine some violent offenders as having schizophrenia-spectrum tendencies, then think again. How many serial killers or murderers have you heard of who at some level were extraordinarily bizarre and acted out strange behaviors? Or had a "had to get them before they got to me" paranoid rationalization for their assaults? Or had really odd beliefs about the world and the people in it? Yes, mad murderers are not uncommon. Recall Ted Kaczynski, the mail bomber, and Peter Sutcliffe, the prostitute killer. Crime connects with schizophrenia—at least part of the time.

Plummeting to a deeper level of analysis, another reason for schizophrenics' being more likely to perpetrate violence lies in the brain. We have known since the 1970s, from the very first brain-imaging studies using CT scans, that schizophrenics have enlarged ventricles—large fluid-filled spaces in the deeper areas of the brain that likely reflect brain atrophy. Since then thousands of brain-imaging studies have documented functional and structural impairments to many brain regions in both schizophrenics and schizotypals, particularly the frontal and temporal lobes.[117] These areas are particularly prevalent in violent offenders.[118] Recall also in our prior discussion of brain imaging that schizophrenics who commit homicide are especially likely to have structural impairments to these brain areas. Consequently, one possible reason schizophrenics are more violent is that they have structural impairments to those brain areas that regulate aggression, as well as disturbances in the limbic system, where emotion is generated.[119]

For some schizophrenics, then, it can boil down to an inability to regulate emotion and acting on the spur of the moment. Things just get a bit out of control sometimes. It's not so much that they meticulously plan an attack or homicide in a cold-hearted fashion. It's more that their disorganized behavior and prefrontal dysfunction results in more reactive forms of aggression—acting impulsively on a provocative stimulus. Indeed, schizophrenics are more likely to kill their own family members than to kill strangers. As many of us know, the home setting can be a tinderbox where what starts as an off-the-cuff comment becomes an out-of-control, blazing argument. Add paranoia and delusions into the mix, and a spark can become a conflagration.

For kids, that spark may come at school. Together with Annis Fung and Bess Lam at City University in Hong Kong, we found that children with high scores on the child version of the Schizotypal Personality Questionnaire had high scores on reactive aggression.[120] In this sample of 3,608 schoolchildren we also found that victimization mediated—or explained—this relationship. Schizotypal kids are picked on because they are odd, shy, and different, and because of that, they reacted by lashing out in anger at others.

The spark igniting the violence tinderbox need not be physical in nature—it might be ideological. Recall from chapter 4 that Ted Kaczynski's bombings were a reaction to industrialization and perceived scientific control over society. In other cases, homicide can be in reaction to social rejection and a sense of hopelessness. That might have been partly true for Kip Kinkel, who was expelled from Thurston High School and on the same day shot his parents before embarking on a mass school killing. Might social isolation have partly triggered Adam Lanza's shooting his mother and then later killing schoolchildren at Sandy Hook Elementary School?

Thus, poor mental health is a risk factor for violence in part because it reflects the type of brain dysfunction that can predispose people to violence. We certainly see a lot of evidence of mental-health disturbances in violent offenders. Not just in disorganized murderers overcome by florid symptoms of psychosis, but also in organized serial killers who can exhibit more muted forms of schizophrenia, as well as overt psychotic symptoms. Here's an example of that muted form and the mix of schizotypal symptoms that include odd beliefs, bizarre behavior, delusional thinking, paranoid ideation, blunted affect, and no close friends.

THE MADNESS OF LEONARD LAKE

I doubt any of you have ever heard of Leonard Lake. Though he killed at least twelve—and as many as twenty-five—men, women, and babies, he is still considered a small fry in the bigger sea of serial killers. People like him slip from public attention, where there are so many other killers basking in an eerie limelight. Yet Lake's case illustrates a mental-health point that is relatively underreported in the literature and needs to be recognized.

Lake had been diagnosed with schizoid personality disorder when he was discharged from the Marine Corps after service in Vietnam. Although he went into psychotherapy, there is no known effective treatment for this personality disorder, one of the schizophrenia-spectrum disorders. Lake was an odd man in many ways. He was fascinated by medieval legends, paganism, and the Vikings. He was once observed to have a large pot on his stove in which he was cooking the head of a goat for soup.[121]

Odd beliefs and behavior like this are characteristic of those with schizotypal personality disorder. One schizotypal I heard being described at a clinical case conference at UCLA wanted to sleep with a goat. Lake's behavior and beliefs were no less bizarre. He had delusions of grandeur and developed a vision of running a survivalist compound in which only the strongest and bravest individuals would survive the apocalypse that was about to come. He believed the world would be destroyed in a nuclear war, but that he would rebuild the human race with his collection of young female sex slaves.[122]

Bizarre beliefs in those with schizophrenia-spectrum disorders don't pop up from random neural misfiring in the brain. Instead, they have some foundation in the social environment. Lake's delusions eerily mimic the main theme in Stanley Kubrick's classic film *Dr. Strangelove*, which was released in 1964. In the movie, the nuclear-arms race is getting out of control and paranoia is running rampant. Brigadier General Jack Ripper initiates a B-52 nuclear attack against the Russians under the belief that a communist conspiracy lies behind the water fluoridation that is sapping his "precious bodily fluids." The Russians have, unbeknownst to the West, developed a doomsday device that is programmed to wipe out the world in the event of an attack on Russia. The U.S. president, under the advice of Dr. Strangelove (a former Nazi weapons expert), develops a plan to occupy deep mine shafts. Selected men—who of course would include the president, Dr. Strangelove, and senior officials—will cohabit with many young women selected for their reproductive fitness and attractiveness so that the men may perform prodigious acts of unselfish reproduction to help repopulate the world.

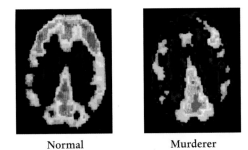

Normal   Murderer

FIGURE 3.1: *Positron-emission tomography (PET) scans showing a bird's-eye view of reduced prefrontal functioning in murderers (top of scan) compared with controls. Red and yellow indicate higher brain functioning.*

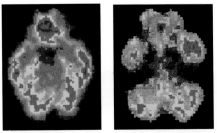

Normal   Bustamante

FIGURE 3.2: *Bird's-eye view of PET scans showing reduced orbitofrontal activation (very top of scan) in the impulsive murderer Antonio Bustamante compared with a normal control*

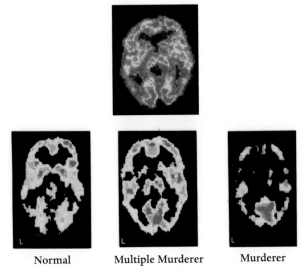

Normal   Multiple Murderer   Murderer

FIGURE 3.3: *Bird's-eye view of functional brain scans (PET scans) of a normal control (bottom left), serial killer Randy Kraft (middle), a onetime impulsive murderer (right), and the author (top)*

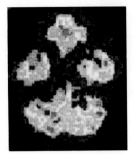 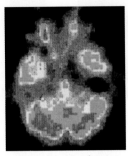 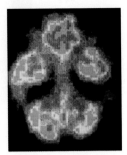

Normal Control   Reactive Murderer   Proactive Murderer

FIGURE 3.4: *Bird's-eye view showing reduced prefrontal functioning (top of PET scan), specifically in a reactive murderer compared with a proactive murderer and a normal control. Red and yellow indicate higher brain functioning.*

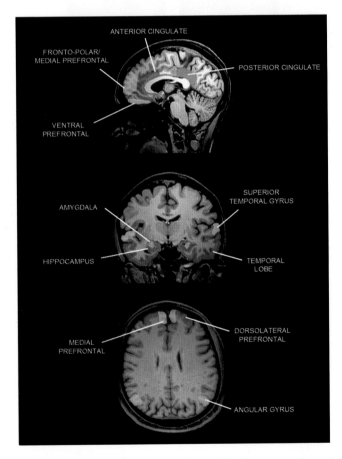

FIGURE 3.5: *Side view (top), head-on view (middle), and bird's-eye view (bottom) of MRI slices showing brain regions associated only with moral decision-making (green), only with violence (red), and areas associated with both violence and moral decision-making (yellow)*

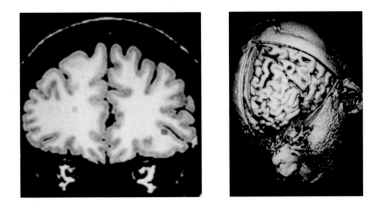

FIGURE 5.1: *Structural MRI scan exposing the prefrontal cortex (right),
and on the left a prefrontal head-on slice showing separation of
neuronal matter (green) from axonal white matter*

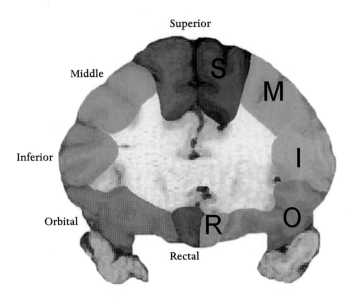

FIGURE 5.4: *Head-on view of the brain showing segmentation of the
prefrontal cortex into gyral sectors to calculate brain volumes in those
with antisocial personality disorder*

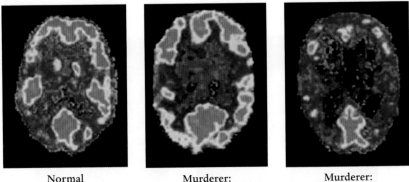

| Normal
Control | Murderer:
Bad Home | Murderer:
Good Home |

FIGURE 8.4: *Bird's-eye view of PET scans showing reduced prefrontal functioning (top of scan) in murderers from good homes. Red and yellow indicate higher brain functioning.*

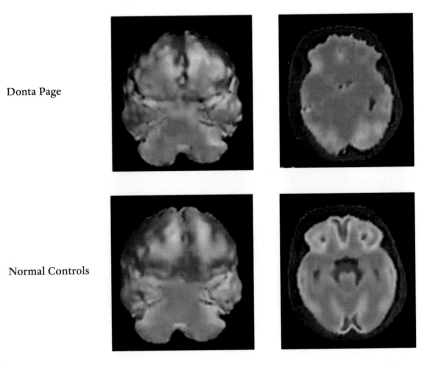

Donta Page

Normal Controls

FIGURE 10.1: *PET scans showing reduced ventral prefrontal functioning in the murderer Donta Page compared with normal controls. The right column shows a bird's-eye view. In the left column, you are looking head-on and slightly up at the brain.*

Did Lake once watch *Dr. Strangelove,* or some similar apocalyptic narrative, and take on board some of these bizarre belief systems? Or were his violent fantasies shaped in part by his tours of duty in the Marine Corps in Vietnam? Or both? Lake certainly had paranoid ideation and believed the wider world was under imminent attack, that it would be wiped out, and that he would need to repopulate the world. He began to act on his beliefs with callous disregard for the suffering of others. He had the cognitive, emotional, and behavioral features of schizotypal personality disorder.

Lake put his vision into operation by setting up a compound in Wilseyville in the rural area of Calaveras County[123] in California. There, in his bunker, he stockpiled arms and food to survive the nuclear fallout, complete with all the necessary shackles, chains, and sexual devices to help him repopulate the post-nuclear world. With a partner, Charles Ng, he lured both men and women using classified ads in which he advertised the sale and exchange of video equipment. Men who replied were immediately killed for their possessions. Women were imprisoned in an underground bunker, where Lake and Ng would make them perform sex-slave rituals in snuff videos, begging for mercy as they were tortured and raped.

Schizotypals score lower on empathy than normal individuals,[124] and Lake's level of empathy was decidedly low. He is recorded telling one of his victims, Kathy Allen, "If you don't do what we tell you, we will tie you to the bed, rape you, shoot you in the head, and take you out and bury you."[125] The reality was to be even worse. Indifferent to the pain he was causing by torturing and raping the women in his bunker, Lake took away the baby of one of his victims, Brenda O'Connor, claiming it was for now in the safe hands of another family. Terrified and hysterical at what might happen to her baby and deluded into believing she could save it, Brenda went along with Lake and Ng's perverted wishes in their snuff videos. The reality was that her baby had already been cut up and buried outside the bunker, and Brenda was to follow after slow torture with sadomasochistic devices.

As was mentioned above, schizotypals have no close genuine friendships outside of their own families, and while they may have superficial associates, these relationships are not deep and meaningful relationships. Lake's social connectedness did not even extend as far as his own family. He killed his brother and also killed one of his few associates for his money and possessions, just as he had killed strangers.

A significant number of schizotypals have obsessive-compulsive personality features.[126] Lake too had his obsessive-compulsive features, taking several showers a day and repeatedly washing his hands—he was compulsively clean as a child. He also made his victims shower before sex.

Another symptom of schizotypal personality is bizarre behavior. Lake would dissect his victims after murdering them, boil the skin off their bones, and place their remains in plastic bags, which he buried in shallow graves around his bunker. Individuals with schizophrenia-spectrum disorders are at risk for suicide,[127] and so it's not entirely surprising that after being captured Lake swallowed a cyanide pill that he had carefully hidden under the lapel of his shirt. He died four days later.

Leonard Lake was not a schizophrenic hearing voices like Peter Sutcliffe or Ron Kray or Henry Lee Lucas. He did not stand out on the street looking disheveled or talking to himself. Instead, he had the kind of symptoms that are not too obvious or noticeable in isolation, but in unison can be clear signs of someone at risk for violent behavior. Clearly not all people with schizotypal personality are killers—far from it—and there were certainly additional factors that made Lake the monster he evolved into. But I suspect that features of schizotypal personality are far more common in violent offenders than today's criminal justice system recognizes, largely because these features are not in and of themselves very striking, pathological, or "abnormal."

After all, did anyone think that Adam Lanza might kill his mother and twenty-six children and adults at Sandy Hook Elementary School in Connecticut in December 2012? At the time of writing—just nineteen days after this tragic event—little definitive is known about his mental condition. Yet to me, he very likely had at the least four of the seven symptoms of schizoid personality disorder: lack of close friends, chooses solitary activities, emotional detachment, and does not desire close relationships or being part of a family. This is the very same diagnosis Leonard Lake was given after his discharge from the Marine Corps. Four out of seven signs are sufficient for a clinical diagnosis. Lanza might also have had the remaining three: takes pleasure in few activities, indifferent to praise or criticism, little interest in sexual experiences. Like Lake, he may also have had additional features of schizotypal personality disorder, including odd appearance/behavior, constricted affect, social anxiety, and odd speech.

I have selected schizophrenia-spectrum disorders from a much wider number of psychological disorders to illustrate that health considerations do not end with physical health. Psychosis and subliminal forms of psychosis—like schizotypal personality—have a strong neurobiological basis and are also clearly related to crime and violence.[128]

There are two very important caveats to repeat, however. First, most schizophrenics neither kill nor are dangerous to others. We should take care not to stigmatize patients with schizophrenia or schizoid personality as both "mad and bad." At the same time, we need to recognize the raised rates of violence in schizophrenics so that they can receive treatment to reduce the likelihood of violence, and thus reduce the stigma.[129] Second, there are many other mental disorders—including depression, bipolar disorder, ADHD, and borderline personality disorder—that are also significant mental-health risk factors for violence. It does not stop with schizophrenia, and of course alcohol and drug use are also major mental-health disorders that increase the risk of violence.

I believe that taken together, the physical- and mental-health risk factors that we have scrutinized in this chapter are convincing components of the anatomy of violence. We'll see later that these constituent pieces are not unalterable. Indeed, we have continued the theme seen in the past two chapters, on broken brains and natural-born killers, that the environment has a role in shaping the biological infrastructure of the violent offender. We'll now move further forward with this recipe for violence to understand how all the different ingredients that we have discussed so far blend together to form a lethal brew.

8.

THE BIOSOCIAL JIGSAW PUZZLE

Putting the Pieces Together

Henry Lee Lucas never really had a chance in life. Right from the beginning he was damaged goods. His father, an alcoholic hobo named Anderson Lucas, who lost both of his legs after falling off a freight train, whiled away his time drinking, selling pencils, and making illegal liquor. Henry himself became addicted to alcohol by the tender age of ten. Drunk most hours of the day, Anderson had no time for Henry—or anyone else, for that matter.

Henry's mother, Viola, was an even worse parent. An alcoholic as well as a prostitute, she gave birth to Henry when she was forty, after she had already abandoned four children to foster homes. Henry; his elder brother, Andrew; his parents; and Viola's pimp all shared the same bedroom in a dirt-floor, ramshackle cabin near Blacksburg, Virginia, without electricity or plumbing. From the time he was a small boy, Henry had to watch his mother having sex with her clients.

Chronically malnourished, Henry was forced to scavenge for food in garbage bins to stay alive. His mother would cook only for her pimp, and the children ate their scavenged food off the floor, as Viola wouldn't wash plates. His first hot meal as a child was when he started attending school and a teacher took pity on him. That same teacher also gave him his first pair of shoes.

His mother psychologically and physically abused him. Once, when he was seven years old, he was too slow to fetch wood for the stove, so his mother hit him hard on the head with a wooden board. Such was the level of neglect that he lay where he had fallen for three full days in a semiconscious state, totally ignored by the rest of his family. Ironically, it was Bernie the pimp who eventually thought something was seriously wrong and took Henry to the hospital, telling doctors he had fallen off a ladder.[1]

This was likely only a fraction of the physical abuse and head trauma Henry endured. For the rest of his life he experienced blackouts, spells of dizziness, and at times felt he was floating on air. Neurological examinations and brain scans later in life revealed evidence of extensive brain pathology, very likely a result of the early maternal abuse and deprivation he had suffered.[2]

Henry was also subjected to sustained psychological cruelty by his mother. When he was seven she pointed out a stranger to him in town, telling him, "He's your natural pa," a fact later confirmed by Anderson, Henry's supposed father.[3] To have such a basic fact of life shattered like that would pull the psychological rug out from under most children's feet, and not surprisingly Henry was devastated and in tears when hearing this news. His sister documents that his mother dressed him as a girl from the time he was a toddler up to his first day at school. His teacher, horrified by his treatment, cut his hair and got him a pair of trousers to wear.

The cruelty of his mother seemed to know no bounds. Viola once saw him enjoying playing with a pet mule. She asked him if he liked his pet mule. He said he did. So she fetched a shotgun and killed the mule in front of his eyes. As if this psychological cruelty was not sufficient to satisfy her, she proceeded to whip and beat the child because it would cost money to have the mule's carcass carted away.[4]

At school Henry was continuously tormented by other children because he was very dirty and smelled terrible. His abject misery was compounded when his brother Andrew accidentally stuck a knife in his face when they were making a swing from a maple tree, puncturing his eye and impairing his peripheral vision. Bad luck morphed into extremely bad luck when a teacher at school swung her hand to hit another child in class, missed, and accidentally caught Henry in the same left eye. The accidental blow reopened the wound, resulting in the loss of his eye.[5]

Henry would go on to become one of the most prolific serial killers in history. He was eventually convicted of eleven homicides committed over a twenty-three-year period, from 1960 to 1983, but he was implicated in a massive 189 altogether. All his victims were female—but we'll return to that issue later. For now his case is particularly salutary in illustrating how a toxic mix of biological and social factors can conspire to create a serial killer.[6]

That mix of biological and social deprivation created a surprisingly efficient killing machine, given the disadvantages Lucas was dealt in life. On the biological side there are three very important risk factors for violence that have been highlighted in previous chapters—head injury, poor nutrition, and genetic heritage from his antisocial parents. These are abetted by a host of social risk factors, including abuse, neglect, humiliation, maternal rejection, abject poverty, overcrowding, being in a bad neighborhood, induction into alcoholism, and complete absence of care and sense of belonging. It was this bitter brew—this very cruel concoction—that turned Lucas into an alcoholic killer.

Lucas's case, while extreme, is not exactly unusual. We'll review in this chapter the scientific evidence showing that when even mild social and biological risk factors coalesce, we can especially expect later trouble. So far we have been identifying the biological factors that go to make up the anatomy of violence. But these are just the bare bones. This chapter aims to flesh out the skeleton by outlining research showing how social factors combine—or interact—with biological risk factors to shape the violent offender.

Criminals like Lucas are a biosocial jigsaw puzzle, consisting of many different and scattered pieces. Even after identification of the biological pieces, it is a challenge to understand how they fit together with the social and psychological processes that decades of prior research have tied to violence.

From this vantage point, we will first turn our attention to understanding how social risk factors come together with biological risk factors to create violence—how they interact in a multiplicative fashion. I'll then show you how the social environment moderates—or changes—the way that biological factors work. I call this the "social-push" hypothesis. We'll see how genes shape the brain to promote violent behavior and yet, at the same time, how the social environment beats up the brain and reshapes gene expression. Finally, we'll piece together the parts of the brain that we have implicated so far and map out more precisely how they collectively give rise to violence.

THE BIOSOCIAL CONSPIRACY: INTERACTION EFFECTS

Henry Lucas was ten when he allegedly became addicted to alcohol. I was eleven when I became addicted to making it. I made wine out of anything I could lay my hands on—potatoes, strawberries, raspberries. Like Lucas I was a scavenger. I even made wine from the blossoms of our goldenrod plant. I bootlegged my brew to visitors and relatives. I used the profits to back horses, running the bet—supposedly from my mother—down to the corner shop, whose owner was a bookie. At fourteen I turned to making lager and I was pretty good at it, except I made the alcohol content too high and people got drunk too quickly, cutting my sales.

When I later began to study adolescent antisocial behavior instead of practicing it, what stayed with me from that extensive experience in brewing was a simple lesson: it takes a complex mix of factors to create the end product. You think of wine and you think of grapes, but of course it is much more. The fermentation of the yeast in the sun with a little bit of sugar. Squishing the fruit to make the must. Adding potassium metabisulfite to kill bacteria and wild yeast. Getting the fermentation process going. Having the acid level just right. Using a hydrometer to measure the specific gravity of the liquid to ensure that there is enough sugar for the yeast to convert it into carbon dioxide and alcohol. There's the racking of the wine by siphoning it off the sedimentary lees at the bottom of the gallon demijohn. Most important, it's not just the mix of ingredients but the right *environment*. You need precisely the right temperature for the yeast and fermentation process.

My rule-breaking behavior had no one specific cause. It had to be a biosocial brew. Like my own hooch, the offender propping up the bar constitutes a merry mix of ingredients. Yet despite enormous knowledge of social factors and some beginning knowledge of the biology of psychopathy by Robert Hare in Vancouver,[7] criminologists and other scientists in the 1970s had not woken up to the idea that these two sets of risk factors *interact*. While I was a neophyte when I started my research career in 1977, and while I felt certain that biology was one component, I was equally convinced that the key chain needed to unlock crime held a lot of different keys—social as well as biological ones.

Unlocking crime would require understanding a complex recipe. Very little in life is simple, and wine, lager, and violence are no exceptions. So the ultimate answer had to be more than the one many sociologists were touting. Add the fact that I have always been a bit contrarian—my first research papers focused on biosocial interactions in explaining antisocial behavior,[8] something radically different from the prevailing perspective in the 1970s, which was dominated by radical criminology espousing Marxist viewpoints.[9]

We saw earlier that birth complications—a biological factor—can predispose someone to later adult violence. The seeds of sin strike early in life with anoxia and preeclampsia damaging the developing brain. But we also discussed how this biological risk factor particularly predisposes someone to adult violence when combined with a social risk factor—maternal rejection of the child.[10] We saw that these findings from Denmark were replicated in the United States, Canada, and Sweden. This was the first convincing scientific demonstration of a biological factor interacting with a social factor early in life to predispose someone to violence in adulthood. But it was not the last.

In 2002 I reviewed all research that had examined biosocial interaction effects in relation to any form of antisocial or criminal behavior. I found no fewer than thirty-nine clear, empirical examples of biosocial interactions.[11] They covered the areas of genetics, psychophysiology, obstetrics, brain imaging, neuropsychology, neurology, hormones, neurotransmitters, and environmental toxins. But before we delve into examples, let me highlight one of two important themes that emerged.

The first theme is that when biological and social factors form the groups in the statistical analysis and when antisocial behavior is the outcome measure, then the presence of *both* risk factors exponentially increases the rates of antisocial behavior. We'll call this the interaction hypothesis. We've just seen an example of this in birth complications and maternal rejection as risk factors raising the rate of violence in adulthood—the outcome measure.

Here's another example, from the work of Sarnoff Mednick, the pioneering and brilliant researcher who was instrumental in bringing me to the United States in 1987. Mednick conducted a study of minor physical anomalies, family stability, and violence. As you may recall from chapter 6, these minor physical anomalies are markers of fetal neural maldevelopment. He found that twelve-year-old boys with more minor physical anomalies committed more violent offending in

adulthood. However, when subjects from unstable, non-intact homes
were compared with those from stable homes, Sarnoff found a biosocial interaction. The combination of minor physical anomalies *and*
being raised in an unstable home environment exponentially increases
the rate of convictions for adult violence at age twenty-one.[12] As you
can see in Figure 8.1, if you were just brought up in an unstable home
environment you have a 20 percent chance of committing violence.
But when minor physical anomalies are added into the mix, that rate
jumps to 70 percent—a threefold increase, just as we witnessed when
birth complications interact with maternal rejection. Danny Pine and
David Shaffer at Columbia University observed a very similar biosocial
interaction, with the combination of social adversity and minor physical anomalies tripling the rate of conduct disorder in seventeen-year-
olds.[13]

Let's put this piece of the jigsaw puzzle into practice in the case of
a significantly violent offender. Carlton Gary, nicknamed "the Stocking Strangler," raped and killed at least seven women aged fifty-five to
ninety. His modus operandi was to break into their homes in Colum-
bus, Georgia, beat them up, rape them, and then strangle them with a
stocking or a scarf. They were all white. What turned him into a killer?

Gary was a series of contradictions. At one level, he was a handsome man who worked as a model on local television. Yet he was also

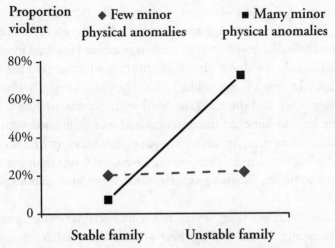

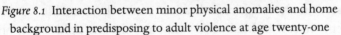

Figure 8.1 Interaction between minor physical anomalies and home
background in predisposing to adult violence at age twenty-one

a pimp and a drug pusher. While he was a caregiver for his elderly aunt by day, he also perplexingly raped and murdered equally elderly white women by night. At the same time as he was committing these murders, he was dating a female deputy sheriff.[14] He was also a bit of a Houdini, a talented escape artist who sawed through the bars of his cell and broke out of a prison in Onondaga County, New York, in August 1977.[15] Even though he broke his ankle in the twenty-foot fall, he made good his escape by jumping on a nearby bicycle. He eventually got a Rochester physician to put a cast on his leg, and for a while was reported to be hopping around like a duck.[16] He also escaped from a South Carolina prison in 1984. He was a persistent offender who had been in trouble since he was a kid—and yet he was a creative man with a reputedly high IQ[17] who often managed to escape the dragnet thrown around him. He successfully talked his way out of an early end to his killing career by accusing another man. All told, he was a bit of a conundrum. Why would a bright, creative, attractive man resort to crime as a way of life? We can discern pieces of that puzzle in his complex biosocial makeup. Here's something of that shuffle.

Gary never really knew his father, having met him only once, when he was twelve. He was all but abandoned by his mother, who could not—or would not—care for him. He was bounced around from relatives to acquaintances *fifteen times* before his first arrest as a juvenile, and we see a clear breakage of the mother-infant bonding process that can predispose a child to become Bowlby's affectionless psychopath.[18] He was also a scrawny young street urchin who, like Henry Lucas, was so malnourished he was forced to rummage around for food in garbage bins. You now know that early malnutrition is an important risk factor for antisocial behavior. Again like Lucas, Gary was allegedly abused by both his mother and the men she lived with. At school during recess one time he was knocked unconscious and was diagnosed with minimal brain dysfunction. Again, we see parallels with Henry Lucas's head injury. Adding to his social deprivation, he had no fewer than five minor physical anomalies, including adherent ear lobes and webbing of his fingers.[19]

We see in Carlton Gary several of the biosocial warning signs we've been discussing. Salient among these are the maternal deprivation we witnessed in the birth-complication study, the unstable home environment we saw in Mednick's study, and the multiple minor physical anomalies that Danny Pine and others have documented.

Head injury and neurological markers of brain dysfunction are further all-too-common risk factors for violence that interact with social risk factors. My postdoctoral student Patty Brennan, now at Emory University, and I documented this in a sample of 397 twenty-three-year-olds, for which early neurological, obstetric, and neuromotor measures had been collected in the first year of life—together with family and social data collected at ages seventeen to nineteen and crime outcome data collected at ages twenty to twenty-two.[20]

Neurological deficits were assessed from an examination conducted in the first five days of life. The pediatrician looked for things like cyanosis (where the skin, gums, and fingernails have a bluish tint to them). When oxygenated, the blood contains a red protein—hemoglobin. When it is blue, it lacks oxygen—and low oxygen impairs brain functioning. At one year of age the babies were also assessed for signs of poor neuromotor development—such as not being able to sit up without support, not reaching for objects until eleven or twelve months, or not holding the head up until after nine months. On the social side, a psychiatric social worker interviewed the mother for measures of family instability, maternal rejection of the child, family conflict, and poverty.

We put all these risk factors into a cluster analysis—a statistical procedure that looks objectively to see if discrete, naturally occurring groups fall out.[21] They did. One group only had poverty. Another only had neuromotor dysfunction and birth complications. The third group had both biological and social risk factors.[22] We also created a normal control group lacking any risk factor. We computed rates of total crime, property offending, and, more important, violent offending.

You can see the results in Figure 8.2. The rate of violence in early adulthood in the poverty-only group was 3.5 percent, compared with 12.5 percent for the biosocial group. As before, we see here more than a threefold increase. The biosocial group also had more than *fourteen times* the rate of total crime of the normal controls. Even though all three groups were of approximately equal size, the biosocial group accounted for 70.2 percent of all the crimes perpetrated by the entire sample.[23] We clearly see here the potency of adding early neurological risk into the equation. These babies were brought into life without sin, and yet they were ushered into the vestibule of violence before they could even sit up on their own.

What we find for adult violence holds for aggressive teenagers.

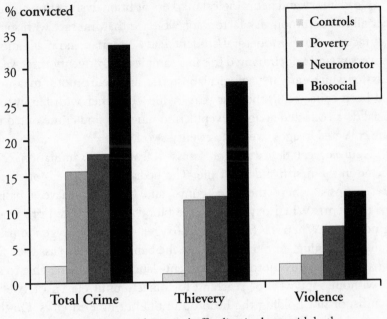

Figure 8.2 Increased criminal offending in those with both
biological and social risk factors

Patty Brennan divided adolescents from Australia into four groups.
One had early social risk factors—poverty, low education, lack of parental warmth, maternal hostility and negative attitude toward the infant,
lack of monitoring, and multiple changes in parents' marital status.
Another group had early biological risk factors—birth complications
and neurocognitive deficits. A third group had both sets of risk factors,
while a fourth group was low on all risk factors. As you can clearly see
in Figure 8.3, 65 percent of the biosocial group who had both sets of risk
factors had serious aggressive outcomes starting early in life compared
to 25 percent of those with just the social risk, 17 percent with just the
biological risk, and 12 percent of the controls.[24] Again in Australia the
combination of birth complications and lack of nurturance is crucial,
as in other countries.

We see the same for another very early risk factor—maternal
smoking during pregnancy. Pirkko Räsänen in Finland found that prenatal smoking doubled the rate of violence in adulthood in an enormous
sample of 5,636 men.[25] Yet if this biological risk factor was combined
with teenage pregnancy, unwanted pregnancy, and slow neuromo-

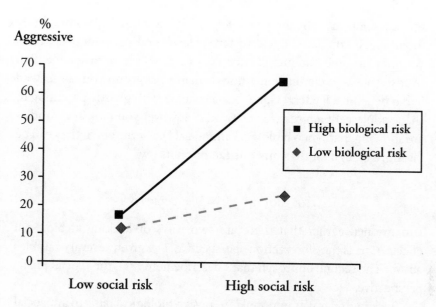

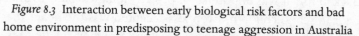

Figure 8.3 Interaction between early biological risk factors and bad
home environment in predisposing to teenage aggression in Australia

tor development, that baby was a staggering *fourteen times* as likely to
become a persistent adult offender.

We again see the seeds of sin conspiring during infancy to create a
deadly weapon in later years. Patty Brennan found a fivefold increase
in adult violence when nicotine exposure was combined with exposure
to delivery complications—but no increase in violence in those who
were nicotine-exposed but lacking delivery complications.[26] Maternal
smoking also interacts with parental absence in predicting early onset
of offending in the United States.[27]

We hear across all these studies a compelling chorus. Social factors
interact with biological factors in predisposing someone to violence. As
we discussed earlier, Caspi and Moffitt amazed the world in 2002 by
establishing that a gene resulting in low levels of MAOA *combined* with
severe early child abuse results in adult antisocial behavior.[28] David Farrington, a world-leading criminologist at Cambridge University, found
that low resting heart rate combined with parental separation before
age ten resulted in voracious violent offending in adulthood.[29] In the
first-ever functional MRI study of any antisocial group, I found that
violent offenders who suffered severe child abuse showed the great-

est reduction in right temporal cortical functioning.[30] Another study found that if you have high testosterone levels and a deviant peer group you may become conduct disordered—yet if you have that same high testosterone and circulate in a non-deviant peer group you are instead led to become a leader.[31] Genes also combine with ghastly parenting to shape adolescent aggressive behavior.[32] However you look at it, studies are showing that when biological and social factors interact, they can be far more malignant than any one factor on its own.

THE "SOCIAL-PUSH" PERSPECTIVE

I mentioned earlier that there are two ways of looking at biosocial effects. One is the "interaction" perspective. I've given several examples above. The second approach that I describe here I call the "social-push" perspective.

Back in 1977 it was unpopular to posit a biological basis to antisocial behavior in schoolchildren. Even less accepted was the belief that biological factors combined with social factors. So when my first research publication as a young student focused on this biosocial perspective, it was a virtual no-go area. Hans Eysenck, Britain's best-known and most controversial psychologist, had already lit a fuse with his controversial book *Crime and Personality,*[33] in which he had the audacity to suggest that crime had a biological basis. Despite the controversy, I believed the book contained a fascinating concept that was related but different—an "antisocialization process." This concept profoundly influenced my work.

The idea was all but lost amid the acerbic criticisms others made of the book. It appears in a section that really resonated with me. Eysenck considered a child whose mother was a prostitute and whose father was a thief—a child in "Fagin's kitchen." He suggested that if that child "conditioned" well or learned quickly from his antisocial home role models he would become a good pickpocket—just like the Artful Dodger in *Oliver Twist*. In contrast, children who do not condition well will paradoxically not be socialized so easily into an antisocial way of life.[34]

I had my chance to examine this idea when I first learned psychophysiological techniques in the laboratory of my PhD supervisor, Peter Venables. That was at York University in 1977. I learned the fundamentals of the eccrine sweat-gland system. I scoured the classical-conditioning literature to design a fear-conditioning experiment. I studied what types of electrodes should be used, and the chemical content required of the gel that helped the silver/silver chloride electrode to make contact with the fingers. I learned to measure bias potentials on the electrodes and to rechloride them when the bias potential was unacceptable. I worked with our technician Don Spaven on generating the auditory stimuli to play over headphones in the conditioning experiment. I tested out the decibel levels with an artificial ear and a really expensive audiometer, snapping the connector between the two and making Don very upset. But soon after that setback, I was ready to get going on recruitment.

I interviewed school headmasters, met with the teachers, and put up recruitment flyers in schools. I went knocking on the doors of parents to get permission and recruited kids into the study. I chased up those who had not responded to my recruitment letter. I then went to the schools and gave the schoolkids questionnaires to fill out to assess their antisocial personality, and to get home background information. The teachers rated their antisocial behavior. I walked to school to pick up the kids, brought them over to the lab, and walked them back again when we had finished. It was a heck of a grind. But it was my first research study, and I was enormously excited—even in the autumn rain and the winter snow. The kids felt pretty good too because they got fifty pence for taking part in the study, about a week's pocket money back in 1978.

We discussed fear conditioning earlier, so you'll recall that it measures anticipatory fear. The task assesses how much a kid sweats when hearing a soft tone that predicts a loud, unpleasant tone. Can they learn—like Pavlov's dogs—to form an association between two events in time? Can they learn that certain events are followed by punishment? Do they have a "conscience"—a set of classically conditioned emotional responses—that makes them feel uncomfortable even at the thought of doing something antisocial?

I found that the environment mattered. If the schoolkids came from a *good* home, then those who conditioned poorly were antisocial.[35] Yet if they came from a bad home, the reverse was true—those who conditioned well were the antisocial ones, Dickens's Artful Dodgers. I was really excited because I got these same findings no matter if it was the teachers rating the antisocial behavior or if it was the child self-reporting on his or her own antisocial personality. Findings were replicating across raters who often disagreed with each other, which suggested that the results were robust. The criminologist and historian Nicole Rafter very generously attributes my first finding as a classic study that got biosocial research in criminology under way,[36] but the reality is that, like many scientists, I was standing on the shoulders of giants.[37]

Where does this lead us? I now want to introduce you to the second biosocial theme that I developed in that review in 2002.[38] So far we've seen that when a biological risk factor interacts with a social risk factor, the outcome is an exponential increase in violence. But "moderation" is another way that social and biological factors can influence each other. A social process can "moderate"—or change—the relationship between biology and violence. That is exactly what the conditioning experiment had demonstrated—that home background moderates the relationship between fear conditioning and antisocial behavior.

Let's take another example, this time from the PET-scan research on murderers that we discussed earlier. I had shown that murderers in general have poor prefrontal glucose metabolism.[39] In another analysis, however, I divided the murderers into those from bad homes and those from relatively normal homes. We assessed eight different forms of home deprivation—factors like child abuse, severe family conflict, and extreme poverty. To get these data we scoured criminal transcript histories, medical reports, newspaper reports, and reports from psychiatrists, psychologists, and social workers. We even interviewed some of the defense attorneys. We then assigned murderers into either a "deprived" home background group or a "non-deprived" group. The question we then asked was, "Which group has the poor prefrontal functioning that predisposes them to violence?"

You can see the answer in Figure 8.4, in the color-plate section. We have here an example of a normal control on the left, who shows good prefrontal functioning—the red and yellow colors at the top. In the middle we have a murderer from a bad home background. And on the right we have a murderer from a good home. You can see that it's the murderer from the relatively *good* home background who shows reduced frontal functioning—the cool colors at the top of the image. And that is the result we observed for the groups as a whole.[40]

The social environment moderates—or alters—the link between poor frontal functioning and murder. The bad brain–bad behavior relationship holds true for murderers from one type of home background—but not for those from a different home.

But how do we explain this? One way to think of it is like this: If you are a murderer, and you come from a bad home, what explains why you are violent? Perhaps here we don't have to look any further than the bad home, which is a well-known social predisposition to violence.

But what if you are a murderer and you come from a *good* home? What causes violence here? It's certainly not the home, because in this case it's pretty good. Instead it has to be something else—a bad brain, perhaps. And that is indeed what we see in Figure 8.4 (in the color-plate section). Murderers from good homes had a 14.2 percent reduction in right orbitofrontal functioning—a brain area of particular relevance to violence. Accidental damage to this brain area in previously well-controlled adults is followed by personality and emotional changes that parallel criminal psychopathic behavior, or what Antonio Damasio has termed "acquired sociopathy."[41]

Let's think back to the case of Jeffrey Landrigan, which we discussed in chapter 2. He had a fabulous home background, with a loving mother, a father who was a geologist, and a sister who was as well educated and straitlaced as her parents. He had all the advantages of life. And yet Jeffrey swiftly spiraled out of control, beginning at age eleven with burglary, and eventually ending in homicide. What was the cause? Here we should suspect genetics and brain dysfunction, given that his biological father—whom he had never seen—was himself on death row for homicide. Great home—yet awful outcome. Gerald Stano was similarly adopted into a loving home six months after birth, but went on to confess to forty-one murders before facing the electric chair. Landrigan and Stano are just two among a number of serial killers reported on by Dr. Michael Stone, a forensic psychiatrist at Columbia University, who were adopted into warm, loving, and supportive home environments.[42] Here we should suspect their genetic heritage, rather than bad homes, as a cause of their violence.

This social perspective on biology-violence relationships is not common in research. As we have seen, the "additive" effect of biological plus environmental risk is the prevailing outlook. And yet the alternative social-push perspective makes some sense, and I feel it can help some parents come to terms with the wayward behavior of their children.

Think about it yourself. Think about people who have a bit of the bad seed about them—a friend, a neighbor, or perhaps a family member who went off the deep end even though his or her siblings stayed on the straight and narrow. Sure, some of them come from classic chaotic homes filled with domestic violence and poverty. But don't some of them have near-normal home backgrounds? Surprisingly loving parents? Two siblings can come from the same family—the same environment, the same upbringing—yet have different outcomes. Here I suggest that you should suspect subtle biological risk factors in nudging your acquaintance into crime, just as we have seen for murderers from good homes.

I often get e-mails from concerned parents desperately trying to help their wayward children. In one such message, a mother described how her seven-year-old son killed a household pet, struck out violently at her, and confessed to his therapist that he enjoyed choking his younger brother. When the mother became pregnant, the child began punching her in the belly and saying that he wanted the baby dead. He showed little remorse and his treatments, including counseling, medication, and hospital stays, did little to help.

Clearly this child is a serious problem, and equally clearly the mother really cares a great deal. Unlike the all-too-common scenario of parental neglect, she is desperately reaching out for help. Yet here it is the son who is callous, uncaring, and lacking remorse. Loving home—unloving child. What can account for such a tragic mismatch?

In this case it might be heritable process. Why? Because what I did not tell you earlier is that this child was adopted.

When children are adopted, it is often because the biological parents do not want their child, or their behavior is such that the child must be taken away from them. We saw earlier how maternal rejection of the child—especially in combination with biological risk factors like birth complications—is a risk factor for later violence. There is a break in the mother-infant bonding process at a critical period in the time before adoption, and it is not easy for a later loving home to mend that break. So here genetic processes may be accounting for the dangerous behavior shown in this child from a good home.

The emergence of genetic and biological factors for antisocial behavior in the midst of a benign home background is something I have termed the "social-push" hypothesis.[43] Where an antisocial child *lacks* social factors that "push," or predispose him to antisocial behavior, then biological factors may be the more likely explanation.[44] In contrast, social causes of criminal behavior may be more important explanations of antisociality in those exposed to adverse early home conditions.[45]

This is not to say that antisocial children from adverse home backgrounds will never have biological risk factors for antisocial and violent behavior—they clearly will. Instead, the argument is that in such situations the link between antisocial behavior and biology is watered down because the social causes of crime can camouflage the biological contribution. Social causation will be more salient in children from adverse homes. In contrast, when the home is normal, but the child is not, then a bad brain may be the culprit. Here the social spotlight on violence is dimmed—and what now shines through is biology.[46]

So far I've illustrated the social-push hypothesis with respect to poorer frontal functioning in murderers from benign home backgrounds, and low fear conditioning in antisocial kids from poor homes. Yet this pattern of results has been found for a whole host of biological risk factors. As a graduate student I observed the social-moderation effect again soon after seeing the conditioning effect, finding that low resting heart rates particularly predispose schoolchildren from *higher* social class homes to antisocial behavior.[47]

More important, a number of other scientists have seen the same thing. Antisocial children from privileged middle-class backgrounds attending private schools in England have low resting heart rates.[48] Antisocial English children from intact but not broken homes have lower heart rates.[49] Low resting heart rate also characterized English criminals without a childhood history broken by parental absence and disharmony.[50] In the Netherlands, Dutch "privileged" offenders—those from high-social-class homes who commit crimes of evasion—show blunted skin-conductance reactivity.[51] In Mauritian children, reduced skin conductance responding to neutral tones at age three—a measure of "orienting," or poor attention—is related to aggressive behavior at age eleven, but only in those from high-social-class backgrounds.[52] Similarly in adults, English prisoners who are emotionally blunted and who come from intact homes—but not broken homes—show reduced skin-conductance orienting.[53] Catherine Tuvblad, in Sweden, found that the environment *moderates* the link between genes and environment. As we might expect from what we learned about genetics in chapter 2, she found a genetic contribution to antisocial behavior in boys, but only those from a *good* home background.[54]

This same moderation effect has been observed at a molecular genetic level where abnormalities in genes related to the neurotransmitter dopamine[55] are associated with early arrests, but only in adolescents from *low-risk* family environments—those who are socially better off. Again, genetic factors shine forth more in explaining antisocial behavior when social risk factors are less in evidence.

My student graduate Yu Gao also documented a moderating effect with the Iowa gambling task—a neurocognitive indicator of orbitofrontal functioning. Our colleagues Antoine Bechara and Antonio Damasio had demonstrated that patients with lesions to the ventromedial prefrontal cortex did poorly on this task and also showed psychopathic behavior.[56] You'll recall from chapter 5 that the orbitofrontal cortex is critical for generating somatic markers that inform good decision-making and that it also facilitates good fear conditioning. This task was given to schoolchildren alongside assessments of psychopathic-like behavior.[57] Gao found that kids who did poorly on the orbitofrontal gambling task were more likely to be psychopaths—but only when they came from normal home backgrounds.[58] Just as I'd previously shown that poor fear conditioning predisposes children from good homes to antisocial behavior, so Gao took a measure of this same orbitofrontal cortex and showed the same moderating result.[59]

Moving from the lab to the real world, we can see the social-push hypothesis in cases of killers. Randy Kraft, the Scorecard Killer, had a very supportive and stable home background. Similarly, Jeffrey Landrigan had the best of home environments, yet went on to become a death-row inmate. Kip Kinkel, a teenager who killed his parents as well as two children at his high school, had a caring home environment in rural Oregon. His parents were devoted professionals, and he had a loving sister. We'll see later the orbitofrontal dysfunction that contributed to his violence. You cannot pin the blame on poverty, bad neighborhoods, or child abuse all the time—certainly not in these cases. Nor is social deprivation so obvious in many more murderers who, while not exactly having heavenly homes as kids, did have homes not much different from yours and mine.

FROM GENES—TO BRAIN—TO VIOLENCE

Social factors interact with biological factors to increase a propensity for violence. They also moderate the relationship between biology and

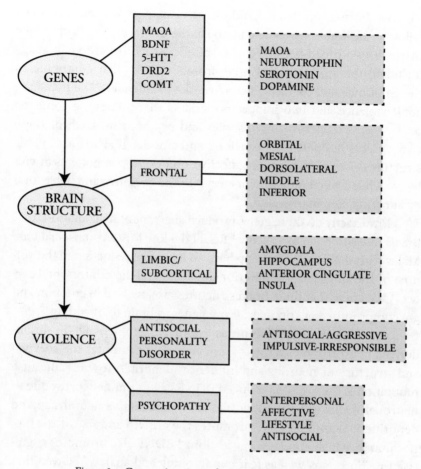

Figure 8.5 Genes give rise to brain abnormalities that in turn predispose to violence

violence. There's a third way to view the influence of the environment on biology, but before we peek into that window on the violent soul we need to step back briefly to genes, the brain, and behavior.

We've already discussed brain mechanisms and the violent mind. We've seen how specific genes link to violence. Now we'll survey the building site where genes provide the scaffolding to structural and functional brain abnormalities supporting the foundations of violent behavior.

You can view my blueprint in Figure 8.5. We start at the top left with genes. They link to both brain structure *and* influence neurotransmitter functioning (such as MAOA). Below that we have brain structure.

The two bottom-up structures thought to support violence are down below in the limbic system and up top in the frontal cortex. Within each of these two broad brain regions, specific structures are identified—including the amygdala and orbitofrontal cortex—that contribute to the emotional and cognitive characteristics of offenders. We then have adult violence and two important variants that predispose someone to it—antisocial personality disorder and psychopathy. Each of these two variants has different behavioral and emotional elements. Limbic structures give rise to the more affective, emotional components of violence, while frontal impairments result in the cognitive and behavioral dysfunction seen in offenders.[60]

How exactly do these genes produce aberrant brain conditions that predispose someone to violence? Recall the low MAOA–antisocial link. Males with this genetic makeup have an 8 percent reduction in the volume of the amygdala, the anterior cingulate, and the orbitofrontal cortex.[61] We know that these brain structures are involved in emotion and are compromised in criminals. From genes to brain to offending.

Let's take the BDNF gene as another example. BDNF—brain-derived neurotrophic factor—is a protein that promotes the survival and structure of neurons and influences dendrite growth.[62] Because mutant mice bred to have reduced BDNF have a thinner cortex due to neuronal shrinkage, we know that BDNF maintains neural size and dendritic structure.[63] BDNF promotes the growth and size of the hippocampus, which regulates aggression.[64] BDNF also promotes cognitive functions,[65] as well as fear conditioning and anxiety.[66] Given that offenders have poor fear conditioning, blunted emotions, and reduced volume of prefrontal gray matter, there is no surprise that the genotype conferring *low* BDNF is associated with increased impulsive aggression in humans.[67] Mice made deficient in BDNF become highly aggressive and prone to risk-taking, just like their human counterparts.[68]

Again, we go from genes to brain to aggressive behavior. While this particular subfield of neurocriminology has a very long way to go, we are starting to connect the dots—beginning with malignant genes, moving into brain impairment, and culminating in crime. Nevertheless, it's going to be more complicated. I'm going to argue that the social environment, far from taking a backseat in this genetic and biological voyage to violence, is driving this Wild West stagecoach.

FROM COMMUNITY TO BRAIN TO VIOLENCE

You now know that the social environment is a causal agent in the brain changes that shape violence. After all, head injury is caused by what happens to you in your social world. You fall down and your head takes a hit. You have a car crash resulting in a whiplash injury. You were shaken as a baby. Whether it is what people deliberately do to you, or life's luckless accidents, your brain gets damaged. And it is that damage that can unleash the devil within you—the unbridled, disinhibited influences that we saw in Henry Lee Lucas, Phineas Gage, and many others.

But the environment is even more powerful in influencing the brain than you might imagine. Let me take you back to your childhood, but perhaps change things around a little. Suppose that now you are living in a neighborhood where violence is more commonplace than normal. You're an eleven-year-old girl or boy, and coming up soon you are going to have a standardized school test on vocabulary and reading. Then, out of the blue, someone living in your immediate neighborhood is shot dead. Compared with other kids in your class who have the same smarts as you but who did not have a dead body dumped on their doorstep, you do more poorly on the test.

This is what Patrick Sharkey, a sociologist at New York University and past student of the leading criminologist Robert Sampson, observed in an innovative data analysis of more than a thousand children in the Chicago Project on Human Development.[69] If a homicide took place in the child's block four days before testing, it reduced reading scores by almost ten points—or two-thirds of a standard deviation. Similarly, it reduced vocabulary scores by half a standard deviation.[70]

How big are these effects? Placing them into context, the relationship between homicide exposure and reading scores is as strong as the relationship between distance above sea level and average daily temperature. It's as strong as the effectiveness of a mammogram in detecting breast cancer.[71] Similarly, the relationship found between homicide exposure and vocabulary scores matches the relationship between IQ scores and job performance.[72] Put still another way, Sharkey estimated that about 15 percent of African-American children spend at least one month a year doing poorly at school purely due to homicides in their neighborhoods.[73] These effects are really not trivial.

We see here that it's not just *direct* social experiences like physical child abuse that can change a child's cognitive functioning. Even in the dark shadow of social experience, something indirect in society can affect your brain. An insidious effect of social experience can profoundly change neurocognitive functioning.

What precisely is going on here in the neighborhoods of Chicago and other cities with a twinning of high homicide rates and poor school performance? Sharkey did not have any neurobiological data on the children he studied, but if he did I would expect to see subtle but meaningful changes in brain functioning in children exposed to neighborhood homicide. We know that excessive release of cortisol in response to stress is neurotoxic to pyramidal cells in the hippocampus—a brain region critical for learning and memory.[74] It kills them off. It seems reasonable to hypothesize that children who hear about a homicide around the corner get scared out of their wits. Is this going to happen to their family? Can they walk to the store safely? Are they going to be next? That fear and stress can translate into temporarily impairing brain functioning and cognitive performance.

If this mechanism is meaningful, you might expect a temporal relationship between the occurrence of the homicide and the reduction in cognitive performance. Suppose you are a child who has heard that someone was killed a few blocks away from you. Would you be more stressed at school if you received that news just a few days ago—or several weeks ago? Likely you would be most affected in the first few days. That's exactly what Pat Sharkey found. The cognitive decline was present when the homicide took place four days before the test, but not when it took place four weeks before.

What about the proximity of the homicide and your level of fear? If it took place in the block you lived in—as opposed to a more distant area of your neighborhood—wouldn't that be a lot scarier? Might it not create a greater cognitive decline? It did. For both reading and vocabulary, homicides in the nearby block had a stronger effect on the child's performance than homicides taking place further away in the neighborhood.

There was a further tantalizing aspect of Sharkey's results. The cognitive decline occurred for African-American children—but not Hispanic children. Why exactly that should be is unclear, but we can hypothesize. It could be that Hispanics feel less threatened by homicides than African-Americans do. Sharkey points out that in communities where

African-Americans lived, 87 percent of the victims of the homicides were African-American, whereas in the murders that affected Hispanics, only 54 percent of the victims were Hispanic.[75] Therefore, a nearby homicide may weigh more heavily on the minds of African-American children, and consequently pull down their test performance more.

I would add another cultural explanation. Because Hispanic homes tend to have a more nuclear family structure and operate under higher levels of social support, there might be a greater social-buffering effect operating in Hispanic homes compared with African-American homes.[76] This would attenuate the effects of the local homicide on cognitive performance. Hispanic families might protect their children from the news of homicide, or may discuss it together more as a family, emphasizing that their children are protected and safe.

Sharkey's results are intriguing because low verbal IQ is an extremely long-standing and well-replicated correlate of crime.[77] It has also been documented that African-Americans have lower verbal IQs than Caucasians,[78] as well as higher homicide rates.[79] Sharkey and Sampson have argued that over time, living in a disadvantaged neighborhood reduces the verbal ability of African-American children by about 4 points.[80] Because a year of schooling is thought to result in IQ improvements of between 2 and 4 points,[81] the 4-point drop resulting from a neighborhood homicide is the equivalent of missing a year or more of schooling. Mess up schooling, and you mess up employment prospects, and we know that after that, adult crime and violence are not far down the road.

Take this even further. If the brains of African-American children are compromised by high rates of homicide that they experience in their neighborhoods, could this result in a vicious circle of increased violence and shootings in African-American neighborhoods, in turn giving rise to further neighborhood stressors and further cognitive decline?

I know this is controversial, but it is also critically important to recognize that the social environment is far more important than many have ever imagined, and complicated in ways we're still trying to understand.[82] Jonathan Kellerman as a clinical psychologist and scientist in Los Angeles was decades ahead of his time when he published a paper in 1977 documenting how environmental manipulations can reduce oppositional and destructive behavior in a seven-year-old boy with XYY syndrome.[83] The environment can overcome genetics. Believe me, this book has changed your brain structure forever. New synaptic connections have been formed throughout your brain in the amygdala, hippocampus, and frontal cortex by what I have just said. Whether you like it or not, those changes will last some time and be hard to eradicate. Social experiences change the brain, likely in all ethnic and gender groups.

THE MOTHER OF ALL EVIL—MATERNAL NEGLECT AND EPIGENETICS

We've seen that there is a substantial genetic component to crime and violence. Despite arguments I've made for a direct causal pathway from genes to brain to antisocial behavior, social processes are also critical. One such process is the lack of motherly love—and the fascinating mechanism of epigenetics.

Epigenetics refers to changes in gene expression—how genes function. We often conceive of genes as fixed and static, but they are much more changeable than commonly believed. True, the underlying structure of the DNA—the nucleotide sequence—remains relatively fixed. But the chromatin proteins that DNA wraps itself around[84] may be altered by the amino acids that make up these proteins. Proteins can be turned on—or turned off—by the environment. That alters how the DNA is transcribed and how the genetic material is activated. Methylation—the chemical addition of a methyl group to cytosine, which is one of the four bases of DNA—can also increase or decrease gene expression.

How does all this occur? Through the environment—and triggered in animals by as little as a mother's lick. The neuroscientist Michael Meaney first demonstrated that rat pups whose mothers licked and groomed them more in their first ten days of life showed changes in gene expression in the hippocampus. They also dealt better with environmental stressors.[85] Indeed, the functioning of more than 900 genes is regulated by maternal licking and grooming in rats.[86] Maternal separation at birth has very similar effects.[87] Gene expression is thought to be especially affected during prenatal and early postnatal periods,[88] and we know that these early periods are critical not just for the brain but for disruptive childhood behavior, which is a prelude to adult violence.[89] Take away maternal care, and there can be profound biological and genetic effects on behavior.

Strikingly, changes in gene expression caused by the early environment appear to transfer to the next generation.[90] Protein malnutrition during pregnancy doesn't just alter gene expression in the offspring; the offspring's offspring—the grandchildren—develop abnormal metabolism even when their own parents were fed quite normally.[91] So the environment not only changes gene expression in the individual—it also has permanent effects that *transmit* to the next generation. The exciting concept here is that although 50 percent of the variation in antisocial behavior is genetic in origin, these genes are not fixed. Social influences result in modifications to DNA that have truly profound influences on future neuronal functioning—and hence on the future of violence.

We can place these alterations in gene expression into a much broader social context of how abuse and deprivation have foundational, long-lasting effects on the brain—over and above any epigenetic effects. Early social, emotional, and nutritional deprivation in humans has been shown to result in reduced functioning of the orbitofrontal cortex, the infralimbic prefrontal cortex, the hippocampus, the amygdala, and the lateral temporal cortex.[92] It also disrupts white-matter connectivity in the brain—particularly the uncinate fasciculus, a fan-like white-matter tract that connects frontal brain regions to the amygdala and temporal brain areas to the limbic areas.[93] Prolonged and chronic stress, including disrupted or poor mothering, disrupts the brain's stress-response system. That results in excessive glucocorticoid release, a reduction in glucocorticoid receptors, an imbalance in the brain's stress-defense mechanisms, and ultimately brain degeneration.[94] Deprivation makes a big dent on the brain.

There are also vulnerable periods when stress can take a greater toll on different parts of the brain. If sexual abuse occurs early, at around ages three to five, for example, hippocampal volumes are reduced. Yet if sex abuse occurs at age fourteen to sixteen, prefrontal cortical volume is reduced instead.[95] This is broadly consistent with the fact that the hippocampus reaches full maturity early in life[96] and is very much affected by excessive release of cortisol in response to stress. In contrast, the prefrontal cortex develops very slowly in childhood, but grows more rapidly during the teenage years.[97] All told, it's not just that stressful rearing environments affect gene expression and neurochemical functioning—they also affect growth and connectivity of the brain.

There is, of course, much more to violence than maternal neglect. Sex abuse is almost always perpetrated by men. As we have discussed earlier, even the best of mothering sometimes cannot override a biological predisposition to violence. Fathers and friends play a role in fostering juvenile delinquency and adult violence as well. Yet it is undeniable that compassionate caregiving is critical for normative child development. When a mother's love is morphed into spiteful hate—as it was with Henry Lucas and others like him—her kids can end up killing. In this context, mothering—and the lack of it—is giving us fascinating insights not just into the pathway to violence, but also into understanding the precise mechanisms by which maternal neglect might operate.

Let's put these pieces into place. We've seen that the lives of violent offenders are replete with maternal deprivation, physical and sexual abuse, other trauma, poverty, and poor nutrition. We've also seen how these social impairments have their hit on specific brain areas—the orbitofrontal cortex, medial prefrontal cortex, amygdala, hippocampus, and temporal cortex—brain areas that are linked to violence. We can conclude that such social deprivation results in long-term wear and tear of the developing brain to produce adolescent angst and aggression—and, ultimately, adult violence. This truly occurs, and it's never too late for the damage to be done. Adults who lived close to the World Trade Center buildings on September 11, 2001—and thus were exposed to very significant environmental stress—showed a reduction in hippocampal gray-matter volumes when brain-scanned three years later.[98] From environment to brain—and, at least in some—to ultimate destructive violence.

BRINGING THE BRAIN BITS TOGETHER

In this chapter we have been piecing together social and biological processes to explain violence. But what about piecing together just the bits of the brain itself? It's an enormously multifaceted, complex organ. We saw earlier, in chapter 5, that multiple brain regions are implicated in white-collar crime, and we know crime and violence come in all shapes and forms. No one discrete brain region or circuit will by itself account for violence.

It is tempting to focus on the prefrontal cortex, given its complexity and the wide empirical support for its involvement in crime. It is even more appealing to invoke a single brain circuit involving two or three regions to help acknowledge this complexity—such as the prefrontal cortex combined with the limbic system, as I outlined above, or the orbitofrontal cortex and its control over the amygdala.[99] Yet a limitation of the approach I have taken so far is that it is overly simplistic. Violence is an enormously complex and multilayered construct. A complete understanding of its neural basis is certainly going to involve multiple distributed brain processes that in turn give rise to broad social and psychological processes that predispose someone to violence. By beginning to recognize and model this neural complexity, I believe we can gain deeper insights into the etiology of antisocial behavior.

In response to the charge of oversimplicity, here's a functional neuroanatomical model of violence.[100] Let's take the anatomy of the brain and first describe the functions of the individual areas concerned—outlining the functional significance of the brain abnormalities we have found so far in antisocial offenders. I'm basing it largely on prior reviews of structural and functional brain-imaging research on offenders.[101]

In Figure 8.6, I group brain processes under three broad headings—cognitive, affective, and motor—alongside the corresponding brain regions. Brain impairments in these areas predispose someone to more complex social and behavioral outcomes that in turn predispose an individual to antisocial behavior in general and violence in particular. No direct relationships are hypothesized from brain dysfunction to antisocial behavior. Instead, the model emphasizes the translation of disrupted brain systems into relatively abstract cognitive (thinking), affective (emotional) and motor (behavioral) processes. These in turn result in more complex social outcomes that represent the more concrete and proximal risk factors for offending in general. So these brain risk factors are not conceptualized as directly causing aggressive behavior, but instead bias thoughts, feelings, and actions in an antisocial direction that then results in violence.

Let's start on the left, with cognitive processes. Here we can see the involvement of the ventromedial prefrontal cortex, the medial-polar prefrontal regions, the angular gyrus,[102] and the anterior and posterior cingulate. Impairment to these regions results in poor planning and organization, impaired attention, the inability to shift response strategies,[103] poor cognitive appraisal of emotion,[104] poor decision-making,[105] impaired self-reflection,[106] and reduced capacity to adequately process rewards and punishments.[107] These cognitive impairments translate into social elements that lead to crime—poor occupational and social functioning,[108] noncompliance with societal rules,[109] insensitivity to

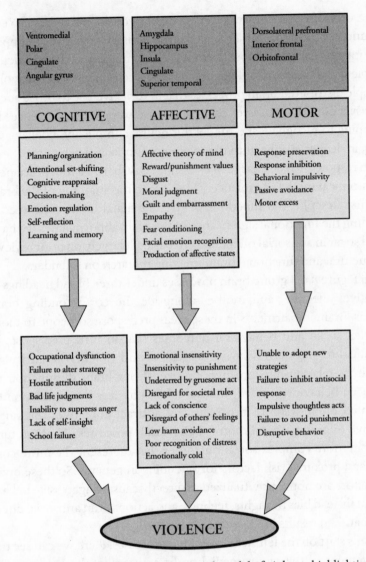

Figure 8.6 Functional neuroanatomical model of violence highlighting cognitive, affective, and motor processes

punishment cues that guide behavior,[110] bad life decisions,[111] poor cognitive control over aggressive thoughts and feelings,[112] overreaction to minor irritations,[113] lack of insight, and school failure.

Turning to the affective processing deficits we see outlined in the top center of Figure 8.6, the neural structures I have highlighted are the amygdala/hippocampal complex, the insula, the anterior cingulate, and the superior temporal gyrus. Impairments to these regions can result in an inability to understand the mental states of others,[114] learning and memory impairments,[115] lack of disgust, impaired moral decision-making,[116] lack of guilt and embarrassment,[117] lack of empathy,[118] poor fear conditioning,[119] poor emotion regulation,[120] and reduction in uncomfortable emotions associated with moral transgressions.[121] These affective impairments can then result in being undeterred from perpetrating gruesome acts on others,[122] callous disregard for others' feelings,[123] poor conscience development,[124] and being unmotivated to avoid social transgressions.[125] It's easy to see how such a set of traits may in turn raise the likelihood of violence.

At the motor level on the right-hand side of the figure, brain areas include the dorsolateral prefrontal cortex, orbitofrontal cortex, and the inferior frontal cortex. Brain impairments here result in response perseveration,[126] motor impairments involving a failure to inhibit inappropriate responses,[127] impulsivity,[128] the failure to shift response sets and passively avoid punishment,[129] and motor excess.[130] In the social context of everyday life this results in the failure to invoke alternative response strategies for conflict resolution,[131] the repetition of maladaptive social behavior,[132] poor impulse control,[133] the failure to avoid punishment, and disruptive behavior.[134]

We see here a flow from basic brain processes to more complex cognitive, emotional, and motor constructs that then translate to real-world practical behaviors that we know characterize violent offenders. It's certainly not a simple model, because violence is not a simple behavior. Yet it conveys the complexity of the problem we are dealing with when we try to put just the brain pieces of the puzzle together. You can imagine the even greater complexity involved when we come to including the macro-social and psychosocial processes that interact with these brain pieces. Furthermore, while I have included multiple frontal, temporal, and parietal regions, imaging research on violence is still a fledgling field. I've certainly simplified it here. There are many more brain regions involved, including the septum,[135] the hypothalamus,[136] and the striatum,[137] among many others.

You may also wonder how violence in particular arises from these cognitive, affective, and motor forces. I view violence at a dimensional, probabilistic level. The greater the number of impaired cognitive, affective, and motor neural systems, the greater the likelihood of violence as an outcome. If, for example, you make poor decisions *and* you don't feel guilt *and* you act impulsively, then that will exponentially increase the likelihood of violence—all other things being equal.

As I hope I have clarified so far, there is not one unique cause of violence. That is why violence is so hard to fathom—and one of the reasons it's fascinating for scientists and the public alike. That is true for the brain too. For some social scientists it's easy to think of the brain as a big blob—and yet in reality it's a mesmerizing mélange of diverse regions, each with intriguing basic functions that contour criminal outcomes. We can see from the brain to basic cognitive-affective-motor processes to social behaviors that raise the risk of riotous behavior that the anatomy of violence is very complex.

Biology by itself is just not sufficient. Instead, we need social risk factors to pull the trigger on an outcome of violence. Although I have here emphasized early social deprivation in the violence jigsaw puzzle, I want to leave the brain uppermost in your mind. That's because the brain goes to the heart of this book's argument—the seeds of sin are brain-based. Despite decades in which scientists emphasized environmental and social processes, the brain is the cardinal transgressor.

This should not be a bitter pill for either social scientists or neuroscientists to swallow. We can get to bad brains through bad genes or bad environments—or, as I have argued in this chapter, through the combination of both. As you read this, greater appreciation for the complexity of violence combined with recent advances in neuroscience are paving the way for a much more sophisticated and integrated journey toward discovering crime causation, a journey that builds on the decades of painstaking sociological and psychological research on crime that social scientists can take credit for. What were two competing perspectives should now be more sensibly viewed as complementary in explaining the causes of crime. For traditional criminologists, what was once an old foe can become a new friend in the fight against violence.

Finally, we should return to our point of departure. Henry Lee Lucas was concocted from a horrendous home brew of head injury, malnutrition, humiliation, abuse, alcoholism, abject poverty, neglect, maternal rejection, overcrowding, a bad neighborhood, a criminal household, and a total lack of love. Violent offenders in general have a history of abuse and early deprivation,[138] and with some exceptions noted earlier, this history particularly characterizes the backgrounds of serial killers.[139] Lucas also had structural and functional brain impairments, as revealed by MRI and EEG examinations, with the frontal poles particularly affected, along with the temporal cortex.[140] Toxicology tests also revealed particularly high cadmium and lead levels, heavy metals that we have seen impair brain structure and function.[141] He can be pieced together and understood from distal structural and functional brain impairments that result in the more proximal cognitive, emotional, and behavioral risk factors for violence—bad decision-making at the cognitive level, callousness at the emotional level, and disinhibition at the behavioral level. These constitute key components of the puzzle making up this multiple murderer.

One unresolved piece remains. Why were all of his victims female? Henry Lucas's first official murder victim was his own mother, whom he killed with a knife when he was drunk. While he believed he was only slapping her with his hand, he later realized that he held a knife when he hit her neck. He was twenty-three years old, and was sentenced to twenty years in prison for second-degree murder, as she ultimately died of a heart attack.[142]

Almost his last victim was Becky Powell, a twelve-year-old juvenile delinquent he had met when he was forty and developed an ambiguous relationship with. At one level he was a loving surrogate father for three years, making sure she was fed, clothed, and cared for—a better parent than he had had. At another level he educated her in stealing and burglary and became her lover. During a tiff when drunk, he stabbed the teenager through the heart, again with a knife. After having sex with her dead body he cut her up into pieces, stuffed her into two pillowcases, and buried her in a shallow grave. He would visit that grave several times, talking to Becky's remains and weeping in remorse.[143] It was the only truly loving relationship he had experienced in his whole life, bringing about a radical change in Lucas, who, surprisingly, confessed to his killings soon after being arrested on a mere weapons charge.

So at opposite ends in Lucas's life we find two love-hate female relationships, with maternal abuse as the core cause of his many killings. Consider the dreadful deprivation of his childhood and the abuse heaped upon him by his alcoholic prostitute mother. The deprivation that she likely experienced herself as a child was passed down to Henry Lee Lucas not just environmentally, not just genetically, but likely epigenetically. We've noted how maternal care is one important ingredient in epigenetics—in gene expression. The complete lack of maternal care likely turned off important genes in Lucas that normally inhibit violence—and turned on genes that promote it. Genetic inheritance passed from one generation to the next. Yet the social environment was truly the factor that turned Lucas into a murderous psychopath.[144] His mother had all the hallmarks of a hateful psychopath, and in killing her, Henry was virtually reliving his intergenerational genetic destiny of psychopathy. As Lucas said himself, "I hated all my life. I hated everybody."[145] He especially hated his mother, and that hatred was in all likelihood turned against other women, even those like Becky Powell whom he came closest to loving.

Recall also the puzzling picture of Carlton Gary, who similarly lacked secure parental bonds and suffered significant early deprivation and malnutrition. Among other perplexing issues in this case is why a handsome African-American man with glamorous girlfriends would resort to raping women over the age of fifty-five. Unusually, all were interracial homicides. All seven of his victims were white women—and yet only 1 in 10 homicides in the United States are interracial. Could this unusual pattern of violence stem from the fact that his mother and his aunt, who also raised him, worked as housekeepers for elderly, prosperous white women? Could complaining, cantankerous white women living at a time when overt racism was more common than it is today have led to hostilities from Gary's caregivers that were passed down to him? Or, alternatively, could Gary's hatred for elderly white women derive from his despising a mother who scarcely existed for him—a subtle redirection of aggression, as we saw for Henry Lee Lucas? And did epigenetics play a supporting role, with deprivation altering gene expression in Lucas for a rebound back to his mother?

What could have been done to save Henry Lee Lucas from a life of serial homicide, and, ultimately, death from heart failure in prison—to say nothing of saving his innocent victims?[146] Are Lucas, Gary, and others like them a lost cause right at the beginning, in early childhood? Genes and brain predispositions to violence are not immutable. As we continue to piece together the different factors, social and biological, that play a role in predisposing individuals to violence, we become better placed to develop appropriate prevention and intervention programs. And that will be the focus of our next chapter—how we might prevent people like Henry Lee Lucas and Carlton Gary from becoming killers.

9.

CURING CRIME

Biological Interventions

Danny seemed to be a hopeless case. In spite of a well-to-do home environment in Los Angeles, complete with the support and care of loving, attentive parents, by the age of three he was stealing constantly. Further into childhood he became a compulsive and adept liar. At the tender age of ten, Danny was not just staying out all night, he was buying and selling drugs. He was known by other neighborhood kids as a nasty piece of work, and because it was a middle-class neighborhood they steered well clear of him. And it wasn't for lack of trying on his parents' part. As his mother recalled, "No matter what the discipline was, or the consequences of his misbehavior, it was never enough. There was no stopping him. We were really at a complete loss for answers."[1]

Danny grew older and stronger, and essentially commandeered his parent's house. He stole cars and appropriated his mother's jewelry for drug dealing. He was getting F's in school. He was a precocious abuser of drugs, graduating from cannabis to speed to cocaine to crystal meth. When he was fifteen he was sentenced to eighteen months in a juvenile detention center. It's a familiar story, with all the early telltale signs of a life of crime, and likely violence too—perhaps another Jeffrey Landrigan in the making.

Out of sheer desperation his parents entered him in a biofeedback treatment clinic after his release from the detention center. These alternative-medicine clinics assess the physiological profiles of individuals with clinical problems to ascertain whether any physiological imbalance can be corrected. How? By helping them become more aware of their biology and teaching them to change their brain. At that point, neither Danny nor his parents actually had any hope that the treatment would do any good. They felt they were just going through the motions—but they turned out to be wrong.

The first clinical evaluation confirmed excessive slow-wave activity in Danny's prefrontal cortex—a classic sign of chronic under-arousal. Then came thirty sessions of biofeedback. Danny sat in front of a computer screen with an electrode cap on his head, which measured his brain activity as he played Pac-Man on the computer. Danny controlled Pac-Man, trapped in a maze, and his task was to move around, gobbling up as many pellets as he could. He could only move Pac-Man by maintaining sustained attention—by transforming his frontal slow-wave theta activity into faster-wave alpha and beta activity. If his attention lapsed, Pac-Man stopped. By maintaining his concentration, Danny was able to retrain his under-aroused, immature cortex, which had constantly craved immediate stimulation, into a more mature and aroused brain capable of focusing on a task.

It was hardly a quick fix. For Danny, the biofeedback training lasted for nearly a year. But a metamorphosis took place over the course of his thirty treatment sessions. He was radically transformed, from an inattentive, F-grade teenager on a downward spiral toward prison into a mature, straight-A, career-oriented student who ended up passing his exams with distinction. It was a complete reversal of fortune.

What accounted for the dramatic change? To begin to answer, we have to look back at what was fueling Danny's antisocial behavior, which started as early as toddlerhood and exploded during adolescence. "I was really bored in school," Danny would say after his treatment was completed, "but all the crimes were really exciting to me. I liked the action, getting away from the cops. I just thought it was so cool."[2]

The thirst for stimulation-seeking is clear. We documented in chapter 4 how children who are chronically under-aroused seek out stimulation to jack their physiological arousal levels back to normal. We know from longitudinal research that schoolchildren with excessive resting slow-wave EEGs are much more likely to become adult criminal offenders.[3] That's exactly what Danny demonstrated in his first clinical evaluation session—excessive delta and theta activity, chronic cortical under-arousal. We also discussed how poor prefrontal functioning predisposes an individual to impulsive homicide. We saw how when the home environment is loving and devoid of deprivation, yet the child is still antisocial, we should expect biology to be the culprit in crime—the social-push hypothesis.

We see in Danny's case an example of how biology is not destiny. The psychophysiological, brain-based predispositions to crime and violence are not immutable. Importantly, Danny himself—albeit with the aid of electronic biofeedback and social support—instituted his own metamorphosis. It's more a case of mind over matter. He had agency in his rehabilitation—and that may have been a critical component in his redemption.

Of course there is no easy solution to crime and violence, and Danny is just a case study. Yet what I want to give you in this chapter is a hopeful message. Rather than giving up when faced with biology-based offending, we can use a set of biosocial keys to unlock the cause of crime—and set free those who are trapped by their biology at an early age.

THE STORY SO FAR

Before embarking on what may work to help kids like Danny, let's summarize what I have been arguing so far, using a theoretical framework to give a context to treatment efforts. You can see it visually in Figure 9.1.

This biosocial model emphasizes the role of genes and the environment in shaping the factors that predispose someone to childhood aggression and adult violence. A key assumption is that *joint* assessment of social and biological risk factors will yield innovative new insights into understanding the development of antisocial behavior.

The right-hand side of the figure outlines the main components of the model. Starting at the top, we have both genes and environment as the causal foundations of later violence. Social risk factors, on the right, have been the understandable focus of social scientists for three-quarters of a century. Biological risk factors, on the left, reflect neuro-criminology, the new and more challenging field of enquiry.

Genes and environment are the building blocks for the biological and social risk factors in the next lower step in the model. Yet you'll also see arrows linking genetics with social factors as well as with biological risk factors. Genes can shape social risk factors for violence such as low

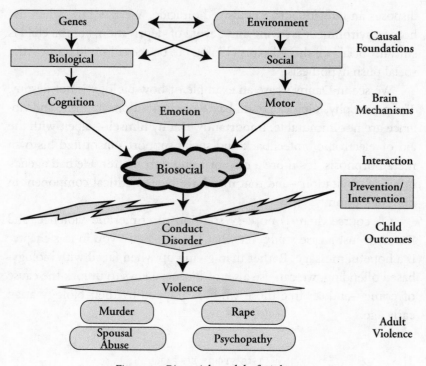

Figure 9.1 Biosocial model of violence

social class and parental divorce.[4] Similarly, social risk factors like environmental stress can impair brain functioning, while living in a risky neighborhood can increase the chance of head injury.

Biological and social risk factors then give rise to brain risk factors that are played out at three levels: cognition (e.g., attention deficits), emotion (e.g., lack of conscience), and motor (e.g., disinhibition) processes. This brain dysregulation can then do one of two things. It can move on to directly give rise to conduct disorder and violence, or it can join forces with social influences to form a biosocial interaction that brings on the teenage thunderstorms of emotion. This biosocial pathway is what I tried to emphasize in the previous chapter, and consequently I place it here as the heart of the model of the anatomy of violence.

Yet there is one piece missing. It is this juncture in our journey— what you see in the dynamic center part of the model—that we will now focus on. The lightning bolts represent striking out the biosocial pathway to adult violence. So what are the biosocial interventions that can block the development of conduct disorder and violence?

One approach to stopping violence—one that we see all too often today—is to wait until the child is already kicking down the doors and becoming unmanageable. Unfortunately, by then it's often too late to effectively correct course. Why not intervene early in life to prevent future violence?

That's what David Olds did in a landmark study that won him the Stockholm Prize—criminology's equivalent to the Nobel Prize. You'll recall that mothers who smoke during pregnancy have offspring who are three times more likely to become adult violent offenders.[5] Birth complications are another risk factor.[6] We also discussed how poor nutrition during pregnancy doubles the rate of antisocial personality disorder in adulthood.[7] We've noted the importance of early maternal care during the critical prenatal and postnatal periods of brain development.[8] Alcohol during pregnancy is also associated with later adult crime and violence.[9] These are the biosocial influences that David tackled.

His sample consisted of 400 low-social-class pregnant women who were entered into a randomized controlled trial. The intervention group had nine home visits from nurse practitioners during pregnancy, with a further twenty-three follow-up visits in the first two years of the child's life—a critical time window in child development. The nurses gave advice and counseling to the mothers on reducing smoking and alcohol use, improving their nutrition, and meeting the social, emotional, and physical needs of their infant. The control group received standard levels of prenatal and postnatal care. Follow-ups were made on the offspring for fifteen years.

The results were dramatic. Compared with controls, the children whose mothers had nurse visitations showed a 52.8 percent reduction in arrests and a 63 percent reduction in convictions. They also showed a 56.2 percent reduction in alcohol use and a 40 percent reduction in smoking. Truancy and destruction of property were reduced by 91.3 percent. These effects were even stronger in mothers who were unmarried and particularly impoverished.[10]

Why was this early intervention so effective? Clues come from other effects of the program. The babies of mothers visited by nurses were less likely to have low birth weight. When the children were age four, the mothers and children were more sensitive and responsive to each other. There was less domestic violence. More of these mothers enrolled their children into preschool programs. The homes became more supportive of early learning. The mothers' executive functioning also improved, and they had better mental health. These improvements were especially true for mothers who were less intelligent and competent.[11] When the children were age twelve, the mothers were less impaired from alcohol and drug use, their partnerships were lasting longer, and they continued to have a greater sense of mastery.[12]

Providing those mothers most at risk for having wayward offspring with health information, education, and support can reverse later adolescent problems that are the harbingers of adult violence. David Olds was tackling not just the social risk factors we see in Figure 9.1, but also the biomedical health factors that join forces with social risk factors to create antisocial behavior. He was tackling the biosocial part of the equation in Figure 9.1, and that's why it worked so well.

The cost of the intervention per mother was $11,511 in 2006—but the government saved $12,300 in food stamps, Medicaid, and other financial aid to the families. The government actually spent less on the intervention group than they spent on the control group.[13] And that's not counting the savings brought about by reducing crime, and the incalculable benefits of improving people's lives.

IT'S NEVER TOO LATE

You'll remember Beauty and the Beast from Mauritius in chapter 4. Joëlle, who became Miss Mauritius, and Raj, the biker who became a career criminal. They were two of the three-year-old children in the study that my PhD supervisor Peter Venables set up—an environmental enrichment from ages three to five that tells us that while it's never too early to start to prevent crime, it's also never too late.

What did our enrichment intervention consist of? It started at age three, had a duration of two years, and consisted of three main elements: nutrition, cognitive stimulation, and physical exercise. The enrichment was conducted in two specially constructed nursery schools. Staff members were brought up to speed on physical health—including nutrition, hygiene, and childhood disorders. They also received training on physical activities, including gymnastics and rhythm activities, outdoor activities, and physiotherapy. They were trained on multimodal cognitive stimulation with the use of toys, art, handicrafts, drama, and music.[14] A structured nutrition program provided milk, fruit juice, a hot meal of fish or chicken or mutton, and a salad, each day. Physical-exercise sessions in the afternoons consisted of gym, structured outdoor games, and free play. The enrichment also included walking field trips, basic hygiene skills, and medical inspections.[15] In fact, there was an average of two and a half hours of physical activity each day. Cognitive skills focused on verbal skills, visuospatial coordination, concept formation, memory, sensation, and perception.

What happened to the control group? These kids underwent the usual Mauritian experience of attendance at *petite écoles* that focused on a traditional ABC curriculum.[16] No lunch, milk, or structured exercise was provided. For lunch, children typically ate rice and bread.

Stratified random sampling was conducted to select which 100 of the 1,795 would enter the environmental enrichment. From the remainder, 355 controls were selected who matched the enrichment group on ten cognitive, psychophysiological, and demographic measures. We then followed up on the children for eighteen years.

What were the results? At age eleven we reassessed the children on a psychophysiological measure of attention—skin-conductance orienting. The bigger the sweat-rate response to the tones played over headphones, the greater the attention that is being paid. The two groups were matched very exactly on this measure at age three—before the intervention began.[17] When they were retested eight years later, at age eleven, the enrichment group showed a 61 percent increase in orienting—a big jump in their ability to focus their attention and be alert to what was going on around them.[18]

We also measured their EEG—brain-wave activity—at age eleven. Brain waves can be grouped into four basic frequency bands. Right now, as you are reading this, fast-wave beta activity predominates because your brain is aroused and activated, scanning this page, absorbing the text, and forming associations. When you are relaxed, alpha predominates. When you are asleep, however, slow-wave delta activity takes over. When you are awake but not very alert, you have more sluggish theta activity. Children in general have relatively more slow-wave theta activity because their brains are immature and still developing. We found that children from the environmentally enriched group showed significantly *less* theta activity than the controls six years after the intervention had finished.[19] Their brains had matured more and become more aroused. In developmental terms their brains were 1.1 years older than those of the controls.[20]

We then followed the children up for another six years, and behavior problems were assessed at age seventeen. The enriched children had significantly lower scores on ratings of conduct disorder and hyperactivity. They were less cruel to others, not so likely to pick fights, not so hot-tempered, and less likely to bully other children. In addition, they were less likely to be bouncing around the place and seeking out stimulation.[21]

We continued to follow them. When they were aged twenty-three we interviewed all the subjects on their perpetration of criminal offending using a structured interview to measure self-reported crime.[22] Those who admitted to committing a criminal offense were categorized as an offender. In addition, we also scoured every single courthouse in Mauritius and searched the records for registrations of offenses that included property damage, drug use, violence, and drunk driving—we excluded petty offenses like parking fines or a lack of vehicle registration. The enriched children showed a 34.6 percent reduction in self-reported offending compared with controls.[23] For court convictions the enriched group had a much-reduced rate of offending, at 3.6 percent compared with 9.9 percent in the control group—but this difference just failed to reach statistical significance.[24] The enrichment really did seem to make a difference—even twenty years later.

That was interesting, but something else piqued our interest even more, which you can see in Figure 9.2. You'll recall that pediatricians had assessed the children for signs of malnutrition at age three—*before* the intervention had begun. On the left-hand side of the figure, kids with normal levels of nutrition at age three who went into the enrichment showed only a small and statistically nonsignificant reduction in conduct disorder. In contrast, when we looked just at those kids who entered the study with poor nutrition, we found that the enrichment showed a 52.6 percent reduction in conduct disorder at age seventeen compared with controls.[25] You can see that on the right-hand side of the figure. Early nutrition status *moderates* the relationship between the prevention program and the antisocial outcome. It works in one group—but not in another. Recall that the prevention program had a lot of ingredients. If nutrition was the active ingredient, you'd expect

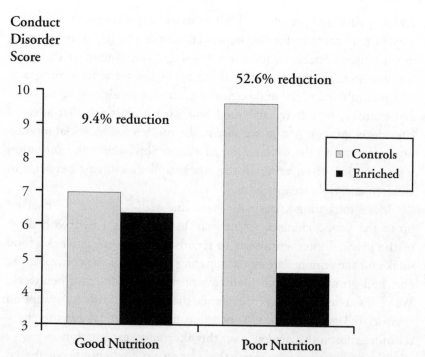

Conduct
Disorder
Score

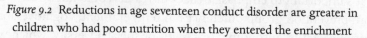

Figure 9.2 Reductions in age seventeen conduct disorder are greater in children who had poor nutrition when they entered the enrichment

the program to work more in kids who had poor nutrition at the get-go—and that's exactly what we found.

It might be that better nutrition makes the difference—but could it be something else? This was the first study to show that early environmental enrichment increases physiological attention and arousal in the long term in humans. That gives us a clue to the mechanism of action—brain change. The prevention program had more physical exercise and outdoor play, and exercise by itself could account for some of the observed effects. Exercise in animals is known to have beneficial effects on brain structure and function.[26] For example, we know that in mice environmental enrichment produces neurogenesis—new brain cells growing in the dentate gyrus of the hippocampus—that is entirely attributable to running.[27] So it could be something as simple as the daily walks and running around in free play that the children in the enrichment group got that improved hippocampal functioning and reduced adult crime.

Another hypothesis is that the increased social interaction with positive, educated preschool teachers in the experimental enrichment may in part account for the beneficial effects. On the other hand, it may be unreasonable to focus on *any* single component of the intervention. Instead, the multimodal nature of the prevention program, which combined social and cognitive components alongside nutrition and exercise, may have facilitated biosocial interactions that affected later development. Just as we saw in the model, the biosocial interaction is central to the explanation of crime. Similarly, with prevention it's a question of covering all the bases to block bullying behavior in children and violence in adults.

More intriguingly, perhaps the crime reduction can be chalked up to the young children eating fish. In Mauritius, I met with three of the original interventionists to reconstruct the typical week's food intake for the enriched group, comparing it to that of the controls. The enriched group had more than two portions of fish extra per week. We've discussed in chapter 7 evidence that increased fish consumption is associated with reduced violent crime, and we'll see later in this chapter more substantive evidence for this alternative explanation.

It's important to emphasize that our results could not be attributed to pre-prevention group differences in temperament, cognitive ability, nutritional status, autonomic reactivity, or social adversity, which were carefully controlled for.[28] The fact that the prevention program reduced crime twenty years later using two different measures of outcome—both self-reporting and objective measures—indicates the robustness of the effects. It's unusual in the field to get results that last. Something in the enrichment is really working to reduce adult crime and violence.

Let's also be careful about the claim. The early enrichment did not eradicate crime. It reduced it by about 35 percent—so that leaves a lot. Obviously, we need more than two years of intensive enrichment to abolish adult crime. And maybe the Mauritius miracle crime cure would not apply to other countries that have a different culture and standard of living. Yet many kids don't get good nutrition, even in the affluent United States, and we think our findings from Mauritius may be particularly relevant to poor rural areas of the United States such as the Mississippi delta region and also to inner cities, where rates of both malnutrition and behavioral problems in children are relatively high.[29]

We were pleased with what was achieved and how early efforts paid off in reducing crime. At the time the study began there were no government preschools on the island at all. One lasting infrastructure contribution made by the research team, which included Peter Venables, Sarnoff Mednick, Cyril Dalais, and staff at the Mauritius Child Health Project, was embodied in the 1984 Pre-School Trust Fund Act, which established government preschools based on the two model nursery schools the group had set up in 1972.[30] Currently, 183 such schools are running in five educational zones in Mauritius—and making a difference in turning Mauritius into a model African country.

OFF WITH THEIR HEADS!

The authoritarian Queen of Hearts in Lewis Carroll's *Alice's Adventures in Wonderland* was a wayward woman with a radical way of dealing with even the smallest of difficulties. "Off with their heads" was her simple solution to every misdemeanor.[31] Although quite heartless, the Queen of Hearts was on the anatomical road to addressing one of the most difficult to treat classes of violent criminals—pedophiles and sex offenders. Surgical castration is the simple, radical, and highly controversial solution some authorities resorted to in order to reduce recidivism rates of sex offenders. Is this a mindless and unethical policy that should be halted? Or does it get to the heart of the matter and provide a workable solution to an intractable problem?

Surgical castration still continues in Germany, ever since a law was passed in 1970 allowing it. It's a voluntary procedure, and only a few are performed every year. Because it sounds barbaric and is so easy to condemn, the German government has put several safeguards in place to regulate it. The offender has to be over twenty-five, and approval is needed from a panel of experts.[32] Nevertheless, it remains a controversial practice in Europe. The Council of Europe's anti-torture committee in Strasbourg, for instance, views it as a degrading treatment that should be halted. But let's reserve judgment until we hear all sides.

It's not only Germany that conducts castration. The Czech Republic has put over ninety inmates under the knife in the past ten years. Pavel is a case in hand. He was imprisoned at the age of eighteen after he gave in to uncontrollable sexual desires for a twelve-year-old that resulted in the boy's death. But even before the crime he knew he had a serious problem. After waking up in the middle of the night in a sweat just two days before the murder, he sought help from his doctor. He was told that the urges would go away. But they didn't, and apparently they became magnified as he watched a Bruce Lee movie, which stimulated his compulsion to use violence to heighten his sexual appetite. He took a knife to the boy and killed him.

After eleven years in prison and psychiatric institutions in the Czech Republic, and just one year before he was due to be released, Pavel asked to be surgically castrated. "I can finally live knowing that I am no harm to anybody," he reported after the procedure. "I am living a productive life. I want to tell people that there is help."[33] Pavel now loves his life in Prague, working as a gardener for a Catholic charity.

For Pavel, removal of his testicles was the price he paid for peace of mind, even if it meant being alone, with neither sex nor romance. It's a tough life, but nevertheless a life that gives him meaning and some degree of dignity. Isn't that better than rotting away in prison, or living every day being torn apart by the wild horses inside that are urging you to desecrate the body of an innocent child?

Debates over the ethics of castration are heated and inevitably revolve around prisoners' rights and the benefits to the individual and society. Let's leave aside the ethics for now, which can be debated at length. Here we'll take a cold, calculated look at the empirical evidence for and against the efficacy of this drastic intervention. Does it work? If it does not make a difference, that would be a compelling argument for eradication of this drastic—and some would say draconian—form of treatment.

We saw earlier how high levels of testosterone are associated with increased aggression, yet these data are correlational, not causal. The etiological assumption behind castration is that lowering testosterone and thus sex drive would lower reconviction rates in sex offenders. But does it?

Good studies of the effects of castration in human prisoners are few and far between. Ethically, you cannot randomly assign one sex offender to castration and one to an alternative treatment. The study that comes the closest to the impossibly ideal experiment was conducted by the medical researchers Reinhard Wille and Klaus M. Beier in Germany in the 1980s.[34] Wille and Beier followed up ninety-nine castrated sex offenders and thirty-five non-castrated sex offenders for, on average, eleven years after release from prison. Such a sample covers about 25 percent of all castrations in the period from 1970 to 1980, and is therefore reasonably representative of this population. Subjects could not be randomly assigned to experimental and control conditions as would be demanded by a rigorous randomized controlled trial. Nevertheless, the thirty-five controls had all requested castration—but ended up changing their minds. As such they constitute as close a control group as can be ethically achieved.

Recidivism rates for sexual offenses over the eleven-year post-release period were 3 percent in castrated offenders compared with 46 percent in the non-castrated offenders—a dramatic fifteenfold difference. The 3 percent reconviction rate in castrated sex offenders is consistent with rates found in other studies that have not been as rigorous as that of Wille and Beier. Rates of reconviction in castrated sex offenders from these ten other castration studies range from 0 percent to 11 percent, with a median of 3.5 percent. These data provide further support for considerably lower reconviction rates in castrated sex offenders. Bear in mind that 70 percent of castrates in Wille and Beier's study were satisfied with their treatment. It's certainly not a panacea for pedophilia and other sexual offences, but should it be entirely ruled out if appropriate safeguards can be guaranteed?

What about the wider literature? One review of 2,055 castrated European sex offenders showed recidivism rates ranging from 0 percent to 7.4 percent over a period of twenty years,[35] results very similar to those in the Wille and Beier study. Yet another review, by Linda Weinberger, a professor of clinical psychiatry at USC, documents the low incidence of sexual recidivism following physical castration in many different countries, commenting that "the studies of bilateral orchiectomy are compelling in the very low rates of sexual recidivism demonstrated among released sex offenders."[36] At the same time she cautions that it is hard to generalize to present-day high-risk offenders, and recognizes the ethical difficulties. However, a commentary on this review cautions that it is important not to *underestimate* the potential importance of castration when considering the release of an offender.[37]

It sounds grotesque, doesn't it? The holier-than-thou among you will be wringing your hands in horror at the barbarity of this surgical intervention. But *you* don't have to live your life as a pedophile in a top-security prison, do you? *You* don't have to face the daily taunts—and danger of being raped—that these men face. *You* don't have your mug shot on the Web for all to see after your release so that people know exactly where you live. *You* don't have to be responsible for controlling sexual urges that are very difficult to contain. Shouldn't people like Pavel at least be given the *option* of castration under conditions guaranteed to have no external coercion?

Fortunately—or perhaps unfortunately, depending on your perspective—there are less drastic methods of dealing with sexual offenders: chemical castration. Here anti-androgen medication is given to reduce testosterone—and hence lower both sexual interest and performance. In the United States medroxyprogesterone—or Depo-Provera—is used to increase circulating progesterone. In the United Kingdom and Europe, cyproterone acetate is used, which competes with testosterone at androgen receptors in the brain. Other medications include leuprolide, goserelin, and tryptorelin. In all cases, they reduce testosterone to prepubertal levels.

Nobody actually doubts that these medications significantly reduce sexual interest and performance. Yet again, the methodological and scientific issue is whether they reduce re-offending. Friedrich Lösel at the Institute of Criminology at Cambridge University conducted a meta-analysis and concluded that the effects of chemical castration are actually stronger than with other treatment approaches, a very telling result.[38]

Because it is somewhat less controversial than physical castration, chemical castration is offered in Britain, Denmark, and Sweden on a voluntary basis to sex offenders. Nevertheless, policy became tougher in Poland since 2009, when offenders who rape either a child under the age of fifteen or a close relative *have* to undergo chemical castration after release from prison.[39] This came about in part after a man was accused of having two children with his young daughter—akin to the case of Josef Fritzl in Austria. Eighty-four percent of the Polish population supported the policy.[40] In South Korea, a new law was put into effect in July 2011 that allows judges to sentence offenders who have committed crimes against children under sixteen to receive chemical castration. In Russia, chemical castration can be recommended by a court-appointed forensic psychiatrist for those who have attacked children under the age of fourteen.[41]

In the United States, at least eight states have had laws on chemical castration ever since it was introduced into the Penal Code of California in 1996. In both California and Florida, treatment with Depo-Provera is mandatory for repeat sex offenders and may also be used in some cases with first-time offenders, such as those who have committed a sex crime against children under the age of thirteen. In California, treatment is administered by the Department of Corrections and must begin a week before the parolee is released. It must be continued until the Department of Corrections deems that the offender no longer needs treatment.[42] In Wisconsin, the Department of Corrections can prevent the release of a child sex offender if he refuses to undergo chemical castration.[43] Texas, like Germany, allows surgical castration on a voluntary basis, and, as with Germany, safety procedures are put in place. Offenders must be older than twenty-one, have at least two prior sex-offense convictions, have undergone at least eighteen months of other treatment, and also understand the side effects of the surgery.

The debate is heated. The American Civil Liberties Union argues that chemical castration violates sex offenders' constitutional rights that pertain to privacy, due process, and equal protection, and the Eighth Amendment's ban on cruel and unusual punishment. Others argue that with appropriate controls the treatment is in the best interest of both the individual and society. One editorial in the *British Medical Journal* argued that doctors should avoid becoming agents of social control, and documented the potential side effects of castration, including osteoporosis, weight gain, and cardiovascular disease. At the same time, this editorial argued that when the individual has sexual urges that are hard to control, biological treatment makes sense. The editorial argued that anti-androgen drugs are effective and that offenders are capable of making an informed choice on whether or not to take the drugs. Furthermore, it went on, while some argue that freedom of choice may be lost when the prisoner has to choose between long-term detention and drugs, prisoners *should* be given a choice, and preventing this choice borders on the ethically questionable.[44]

You can answer the question yourself. Imagine you are a sex offender in prison with murderers, rapists, and psychopaths. Would you like to be allowed a choice—a choice between long-term detention, or chemical castration and release?

Nobody is coercing you right now in pondering the answer. You are free to decide. I know what I'd want. I think if you had spent four years in top-security prisons as I have you'd want to be allowed to make a decision and you would want chemical castration. Or perhaps you think that sex offenders are beneath contempt and should rot in hell.

One of the problems with chemical castration is that it affects our right to reproduce. It goes against our evolutionary makeup and mind-set. There is another alternative. What if we stay with the medical model for treatment of offenders but we do not compromise their ability to have children? We'll take a journey to explore this further.

FLIGHT 714: THE ADVENTURES OF TINTIN

You never know what might happen when you get on a plane. Every time I walk down that gangway, images of unforeseen disasters flit through my mind. But before I go on, let me introduce you to my boyhood hero, Tintin. This sixteen-year-old newspaper reporter was the invention of Hergé, a Belgian writer and cartoonist who was an important influence not just on myself but also on Andy Warhol.[45] Tintin's life revolves around writing crime stories and traveling internationally to solve mysteries. He is a boyish, avant-garde swashbuckler who pushes the envelope on puzzles to stop crime—and has fun at the same time. I was brought up with Tintin as a boy. I bought all the Tintin books. I tracked down Hergé in person and got him to sign several of my books, including *Flight 714*. And here I am today, a boy trapped in an adult's body, writing about the causes of crime and traveling the world to stop it.

Now to *Flight 714*—the penultimate story in the twenty-three-volume Tintin series. Tintin is in the tropics in Jakarta and catching a plane to Sydney with an eccentric millionaire. A fight breaks out and the plane gets hijacked by terrorists. The criminals want the millionaire to spill his bank account number. He's injected with a truth drug by the dastardly Dr. Krollspell—a sort of Josef Mengele parody—under the orders of his evil boss, Roberto Rastapopoulos. That's where the medication comes in.

Now cut to my story. As with Tintin it begins benignly enough. I got on board United flight 895 bound to another tropical country— Hong Kong—on Thursday, July 17, 2007. I settled into my bulkhead aisle seat, had my dinner, and then sank into reading Jonathan Kellerman's *Rage*. An academic detective crime story, of course. And that's when it happened.

There was an urgently ominous announcement: "We have a situation. If there is a doctor or—[pause]—a psychologist on board, could you please make yourself known to the flight attendant."

I got a queasy feeling and it wasn't my dinner. Usually they want a doctor, but they also said "psychologist." Well, I might be a psychologist, but I'm also a wimp. The truth is that just before that announcement I heard a racket coming from the section ahead of me past the toilets on the port side. It had distracted me from my book. Then two flight attendants zipped past me. Even more yelling. Maybe Rastapopoulos was on board.

I took a look back up the long aisle behind me to check the passenger seat lights. Come on. Surely there had to be a doctor on board. Was anyone coming to the rescue? But the aisle lights were as dark as a graveyard. I turned back in my seat and began to feel a bit desperate. It seemed like the sort of thing Jonathan Kellerman could solve. He's a psychologist as well as a best-selling crime writer. Maybe he was on board? Maybe there was a Dr. Krollspell lurking in the wings. I looked behind me again. All I could see was a sea of faces looking up the aisle at me to see what was going on beyond me.

Think, Raine, think, you idiot. I thought it through carefully. I decided on the only sensible, professional, and responsible course of action for a professor in criminology. I kept reading Kellerman.

You know what, though? You look up from your book, gaze into space, and say to yourself, "You cowardy custard." Emotional quicksand was quickly covering me in a suffocating swathe of guilt. I looked around again. Not a blinking sausage, no cavalry. I wasn't the only wimp. Okay, to hell with it, here goes. I rang my bell.

There certainly had been a brawl. I walked with the flight attendant up to the front, where a scuffle was still taking place between a male flight attendant and a passenger. My escort gave me a comprehensive, articulate, and professional appraisal of the situation: "He just went nuts and whacked the woman next to him!" I was behind the desperado, so I pinned his arms behind his back while the attendant whipped his tie off. In a jiffy we'd tied the assailant's arms behind his back. The woman was still yelling, but we ignored her, as it was all Chinese. We shoved our prisoner into a window seat. I jammed myself beside him, with the steward next to me to block up the aisle seat. We had secured the situation.

The next thing I know the steward had me switch seats because the pilot wanted to talk to me. They brought me to the cockpit. And that's where the cool bit began and I got to feel like I was Tintin. They wanted to patch me through on the intercom to an MD on the ground. The pilot got out of his seat, I jumped into it, and he instructed me on how to work the communications system. Have you seen Steven Spielberg's movie *The Adventures of Tintin*? Remember when Tintin is in the pilot's seat? The confined space. The instrument panel dazzling your eyes at one level. Then you gaze out of the cockpit window and you're floating on those fluffy white clouds. So very much at peace, so Tintinesque, gliding up there in heaven above the hot struggles of the poor. It's what a British Airways ex-accountant always wanted.

The duty doctor down below snapped me out of it. He knew I was an expert on violence and a psychologist. What's my professional evaluation of the level of danger to other passengers? How can we secure the situation? I say we can deal with it by giving the guy a hefty dose of my Temazepam—a short-acting benzodiazepine. I always bring it on board international flights because I have trouble sleeping on planes due to the noise at night—you know—owing to having my throat cut that night in Turkey. I suggest a good dose, 30 milligrams. The doctor thinks 15 milligrams. We ended up with the 15 milligrams—and that did help calm the villain down. Then we made an emergency landing in Anchorage so that security forces could board the plane and offload the blighter.

I have to say I felt pretty good about it. As the other passengers lined up to land in Hong Kong, I was slapped on the back by the copilot and applauded. United flew me business class for the rest of my round-the-world journey. All in a day's work as a criminologist, I said. Yup, my boyhood Tintin dreams had finally come true.

But back to medication. It really does arrest aggression. Unlike in the case of flight 895 I'm not talking about sedation to quell violence. We've witnessed major advances in psychopharmacology together with substantive evidence that some medications are surprisingly effective in reducing aggressive and violent behavior.

Let's start with children. What's the most common cause of children under the age of nine being referred for psychiatric services? It's none other than behavior problems.[46] The majority of these hospitalized children are receiving medication to treat aggression.[47] Clinical practice is backed up by surprisingly strong empirical support for the effectiveness of drugs as an intervention for childhood aggression. A meta-analysis of forty-five randomized, placebo-controlled trials conducted in children by Elizabeth Pappadopulos[48]—not to be confused with Rastapopoulos—has shown that medications are surprisingly effective in treating aggression, with an overall effect size of 0.56—which is of medium size in terms of the strength of the relationship.[49]

A wide variety of medications have been found to be effective in reducing aggression. The most effective are the newer generation of antipsychotics,[50] which show a large effect size of .90.[51] Stimulants like methylphenidate are also very effective, with an effect size of .78.[52] Mood stabilizers have a medium effect size of .40, while antidepressants have a small-to-medium effect size of .30. The same story that we see in children holds true in adolescents.[53] Two meta-analyses of drug treatments of aggression in juveniles, together with other reviews and meta-analyses of drug efficacy with aggression and antisocial behavior in child and adolescent populations all show the same story.[54, 55, 56] What's clear is that drug treatment is effective in reducing aggression across a wide range of psychiatric conditions in childhood and adolescence—including ADHD, autism, bipolar disorder, mental retardation, and schizophrenia.[57]

How does medication compare to nonmedical treatment of aggressive and violent behavior? My colleague Tim Beck, in my department at the University of Pennsylvania, originally developed cognitive-behavior therapy, which is widely effective in treating a whole range of clinical disorders. It is the most effective and well-accepted treatment for aggression. The overall effect size? Conservatively it is .30.[58] So the overall effect sizes obtained for medications compare very well to the best psychosocial interventions.[59] Indeed, effect sizes for atypical antipsychotics and stimulants if anything exceed the best non-pharmacological treatments.

Skeptics will scrutinize this claim assiduously and can come up with a reasonable retort. Perhaps the medications are treating other conditions like depression, ADHD, and psychosis that aggressive children also have, and that accounts for the reduction in aggression. For example, children with psychosis get crazy ideas into their heads about other kids picking on them, so they strike out in a defensive, aggressive rage. So yes, risperidone works well in reducing the aggression because it's cutting out the craziness—one cause of the aggression. However, many studies have clearly demonstrated the effectiveness of medications with children coming to the clinic *primarily* with antisocial/aggressive behavior rather than psychosis.[60] Meta-analyses of the literature also show that stimulants reduce aggression independent of their effect in reducing

ADHD symptoms.[61] There is even evidence that stimulants and atypical antipsychotics are effective in reducing aggression in preschoolers.[62] It's a bitter pill for many criminologists and psychologists to swallow, but medications do work in controlling and regulating aggression in children and adolescents.

Can medications work too in quelling outbursts in adults? Surprisingly, there is much less research here, probably because once you become an adult and are violent, you are viewed as evil and we lock you up. We don't want to help you anymore. A double-blind, placebo-controlled, randomized trial allocated impulsive aggressive male community volunteers to one of three anticonvulsants.[63] All three medications significantly reduced aggressive behavior.[64] The same result has been found in several randomized controlled trials for treating impulsive forms of aggression in prisoners.[65]

Why *on earth* would anticonvulsants, normally used to stop epileptic seizures, work with reducing aggression? We know these medications have a calming effect on the limbic regions of the brain—particularly the amygdala and hippocampus, where epileptic seizures begin. We saw earlier that impulsive, emotional murderers have excessive activation of these limbic subcortical regions, so anticonvulsants may help reduce their impulsive emotional rage attacks by calming their emotional limbic system.

LET THEM EAT CAKE

Let's continue our travels to find a different cure to crime. La Pirogue is the jewel in the crown of the beautiful island of Mauritius. With its golden sands and tropical gardens surrounding traditional-style thatched rooms, it is a haven of peace and tranquillity. It is my favorite hotel in the world.

Utoeya is also a utopian picturesque island, this one located in the Tyrifjorden fjord outside of Oslo in Norway. With its pretty little beaches it is similarly a summertime resort for young people. And it was here on the evening of July 22, 2011, that eighty-four people lost their lives while I unknowingly sat on the beach at La Pirogue in Mauritius, watching the sun set slowly over the coral reef.

I had flown in just the day before on flight MK 647 from Singapore, where I had been working with my colleagues on our fish-oil study on conduct-disordered children. The biotech company Smartfish, whose headquarters are in the Oslo Innovation Center, supplied the Joint Child Health Project in Mauritius with an omega-3 drink. I had a connection with its cofounder Janne Sande Mathisen, as she had gone to Darlington Technical College, which was just a few blocks from 69 Abbey Road, the house I was brought up in as a child. I had an unexpected e-mail from her on that fateful day:

> Just 20 minutes ago there was an enormous explosion in central Oslo—affecting the governmental buildings. We could hear the explosion even though we live 20 minutes (by car) from the center. It is most likely a bomb and a terror attack. This has never happened before, and it will have strong impact here.

What Janne had heard was a massive blast from a 2,000-pound fertilizer car bomb placed in the center of Oslo, which exploded at 3:17 p.m., damaging ministry buildings including the prime minister's offices and killing eight people.

A short while after, at about five p.m., an armed "policeman" took a ferry across the Tyrifjorden fjord just outside of Oslo to the island of Utoeya to "investigate" the bombing. Landing on the island, which was filled with teenagers taking part in a youth camp for the Labor Party, he called the students toward him. They dutifully came, whereupon he promptly shot them. Anders Behring Breivik continued his shooting spree for an hour, during which time he killed sixty-nine individuals, mostly teenagers, fifty-six of them shot in the head. Thirty-three more were shot but survived. It was the worst peacetime massacre in Norway's modern history.

Those victims had been drawn to that island for its charm and peace, to relax in the countryside and beaches just as I was doing in Mauritius. Yet as I sat in my paradise watching the sun setting over the Indian Ocean, their paradise was being invaded by a sandy-haired, blue-eyed devil. As I heard the crashing of the waves on the coral barrier reef outside my room at La Pirogue, there was the crushing of their young souls outside Oslo. Yet in the sea in both Norway and Mauritius there might just be a part-solution to this kind of mindless violence—fish.

I first got the idea on a visit to Mauritius a decade ago. It was November 2002, and I had just revised our findings from our earlier study showing how early environmental enrichment particularly reduced conduct disorder in kids with poor nutrition—an enrichment that included more fish. I was in the airport in Mauritius, wanting to buy something to read on the plane going to Hong Kong. There is one, and only one, small book shop there, and it largely sells books in French. There were literally two short shelves with books in English. And there I saw it, Andrew Stoll's *The Omega-3 Connection*,[66] which had come out the previous year.

Going through it on the flight, I read his summary on the early studies suggesting that omega-3 might help with depression, ADHD, and learning difficulties. There were no studies of aggression or antisocial behavior, but he speculated:

> We await the results of future studies in our nation's schools and prisons, and hope that at least part of the answer may be as simple as an omega-3 fatty acid.[67]

Perhaps he was right. The staff at the Joint Child Health Study in Mauritius tested the idea in a randomized, double-blind, placebo-controlled trial of omega-3 supplementation in children and adolescents. Participants were drawn from the Mauritius Child Health Project. One hundred children drank one pack of the Norwegian Smartfish Recharge juice per day. It's only a 200-millileter drink (less than a cup), but packed into it is a whole gram of omega-3. They took that for six months. One hundred other children were randomized into the placebo control group and received the same juice drink, but lacking the omega-3. Parents then rated their children's behavior problems at the beginning of the study, six months later (at the very end of the treatment), and for a third time six months after the treatment had ended.

The results were intriguing. As you can see in Figure 9.3, both groups showed a reduction in aggression after six months of taking the drinks. That shows there was a placebo effect—that the fruit-juice drink without the omega-3 was doing just as good a job as the omega-3 drink. However, six months after the end of the treatment, the control group had returned almost to its pretreatment levels of aggression, whereas the omega-3 group continued to show even further reductions in aggression, delinquency, and attention problems. It was a significant interaction between treatment group and time, with the groups really

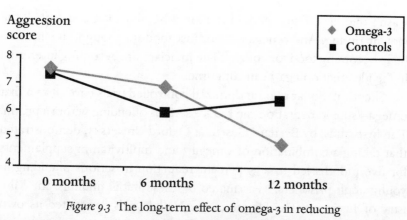

Figure 9.3 The long-term effect of omega-3 in reducing
aggression in children

diverging in outcome a full year after the study had begun.[68] These results provide some initial support for the idea that omega-3 can help in the long term in reducing behavior problems in children, a significant precursor of adult crime and violence.

Why would we expect omega-3 to reduce aggression? In a way it's surprisingly simple. We've seen throughout this book that there is a brain basis to violence. We discussed earlier how omega-3 enhances brain structure and function by increasing dendritic branching, enhancing synaptic functioning, boosting cell size, protecting the neuron from cell death, and regulating both neurotransmitter functioning and gene expression. So omega-3 might partly reverse the brain dysfunction that predisposes one to aggression.

I was initially surprised that there would be a long-term change. Wouldn't any initial results wash out after the Smartfish drink was discontinued? But Joe Hibbeln, a leading figure in the field, explained to me that the half-life of omega-3 in the body could be about two years—it stays in the body ready for re-uptake and it can make a lasting change in the brain.[69] So it stands to reason, at least in theory, that by improving brain structure and function omega-3 could help reduce violence in the long term.

The idea that nutrition could help is not new. In 1789, when the revolting French peasants in Versailles were baying for the blood of their queen, Marie Antoinette is reputed to have said, "If they have no bread, then let them eat cake." Brioche—a rich form of bread that she was supposedly referring to, may not have helped much, but she wasn't that far off the mark in thinking that nutrition could quell the violent rioting. And omega-3 is not just food for thought, it's increasingly becoming food for court.[70] The judiciary are becoming interested in the idea that omega-3 can cut crime.

Skeptical? So far two randomized controlled trials have shown that omega-3 supplementation can reduce serious offending within a prison. The first study, by Bernard Gesch, at Oxford University, demonstrated that taking a combination of omega-3 and multivitamin supplements for five months led to a 35 percent reduction in serious offending in young adult prisoners.[71] Fascinated by these initial findings, the Ministry of Justice in The Hague in the Netherlands conducted its own study on young-adult offenders and found that omega-3 and multivitamins for eleven weeks reduced serious offending within the prison by 34 percent—results almost identical to the British study.[72]

Wherever you go around the world, it seems that omega-3 may make a difference. In Australia, six weeks of omega-3 supplementation reduced externalizing behavior problems in juveniles with bipolar disorder.[73] In Italy, normal adults taking omega-3 for five weeks showed a significant reduction in aggression compared to controls.[74] In Japan, a randomized controlled trial of omega-3 in adults reduced aggression.[75] In Sweden, a randomized controlled trial found that ADHD children with oppositional defiant disorder showed a 36 percent reduction in their oppositional behavior after fifteen weeks of omega-3.[76] In Thailand, a randomized, double-blind trial of the omega-3 fatty acid DHA resulted in a significant reduction in aggression in adult university workers.[77] In the United States, women with borderline personality disorder randomized into supplementation of the fatty acid EPA for two months showed a significant reduction in aggression.[78] Another American study, this time a four-month randomized, double-blind, placebo-controlled trial of fatty-acid supplementation in fifty children, showed a significant 42.7 percent reduction in conduct-disorder problems.[79]

It's all too simple, you say. And strictly speaking you are right. Violence is complex. In omega-3 we are looking at only one ingredient of a much bigger nutritional package that can feed violence-intervention efforts. We saw earlier how eating candy is correlated with crime. Blood-sugar lows can blow the lid off containing aggression. Not eating enough can make one tough. Micronutrient supplementation of both zinc and iron helps accelerate recovery of hippocampal functioning following iron deficiency in rats.[80] We also know that a lack of protein results in EFA (essential fatty acids) deficiency, while micronutrient deficiencies contribute to impaired EFA bioavailability and metabolism.[81] You're right, it's not simple.

Omega-3 is certainly not the sole solution on the nutrition front—there are many more nutritional factors to consider. And nutrition itself is just one piece of the much bigger jigsaw puzzle. Not all omega-3 studies have come up trumps.[82] Nevertheless, these international findings are initial appetizers that should tempt us to consider further how nutrition can nix crime and violence. A body of knowledge is being built up that gives us an alternative perspective to drugs as a solution. Societal distaste for any "Prozac for prisoners" proposition could be tempered by the more palatable alternative medical approach of "fish for felons." It could potentially prevent future disasters.

Anders Behring Breivik was initially argued to have a psychotic disorder—paranoid schizophrenia—that resulted in the Norwegian tragedy. We discussed earlier how schizophrenia is related to violence. Is it entirely a coincidence that the very first study to prevent the development of psychosis in adolescents and young adults was based on omega-3?[83] Is it a coincidence that the early environmental enrichment in Mauritius that included an extra two and a half portions per week reduced not just adult crime, but also adult schizophrenia-spectrum personality traits, especially in those who had poor levels of nutrition before the enrichment?[84] Future studies following the Norwegian Smartfish study on the island of Mauritius may ultimately provide prevention of slayings like the ones that took place on Utoeya island in Norway.

MIND OVER BRAIN MATTER MATTERS

Changing the brain to change violence may not necessarily require drugs or any invasive form of therapy—or even more benign biological interventions such as nutritional change. Let's turn back to biofeedback and Danny. By feeding back to him his brain activity, he was able to learn how to increase activation of the prefrontal cortex. That gave him agency and the ability to better regulate his behavior. But can biofeedback like this really stop violence?

Research on individuals with antisocial personality disorder claims to show that intensive EEG biofeedback involving from 80 to 120 sessions does improve their behavior.[85] That is promising, but the clear limitation is that to date much of the evidence is based on case studies. Randomized controlled trials are needed to more conclusively demonstrate efficacy. We still have a long way to go with this particular biological intervention.

But Buddha may help put us on the path to permanent brain change without drugs or invasive treatment. Mind over matter. Maybe meditation can change the brain for the better.

The technique itself is fairly simple. You would have one training session for eight weeks, each one lasting about two hours. You would practice the technique one hour a day at home, six days a week.[86] You would be taught to become more aware—or more mindful—of your internal mental and bodily state. Attention might, for example, be focused on breathing, becoming more aware of your present-moment experiences, and mindfully going through your whole body's sensations and feelings. You are taught to take a compassionate, nonjudgmental stance to yourself—to not, for example, beat yourself up during training if your mind wanders from the task. Later on you would be taught to become aware of yourself in the here and now.[87]

Doing all that will change your brain—permanently. In 2003, a leading neuroscientist, Richie Davidson, from the University of Wisconsin, Madison, performed a breakthrough meditation study. People were randomized into either a mindfulness training group or a control group that was put on a waiting list for training. Richie demonstrated that just eight weekly sessions of mindfulness training enhanced left frontal EEG activity.[88] Manipulate the brain through mindfulness, and better mood and psychological functioning can result.

One study from Davidson's group showed how focusing on a mental state of compassion and loving kindness for others enhanced brain regions involved in empathy and mind-reading. Participants' ability to process emotional stimuli was enhanced, bringing on line the amygdala and the temporal-parietal junction of the brain.[89] Functional imaging research has also shown that expert meditators have greater activation in brain regions involved in attention and inhibition.[90]

It's not just that meditation changes the brain during the time of meditation. People who have practiced meditation over a long period later show that at rest—in a non-meditation state—their brain has shifted toward increased attention and alertness as measured by gamma activity—a form of high-frequency EEG activity involved in consciousness, attention, and learning.[91] The more hours of practice, the greater the brain change taking place. Meditation is producing long-lasting positive effects on the brain.

Mindfulness practice changes not just brain function but also brain structure. One study scanned subjects before and after an eight-week mindfulness course, with controls again being put on a waiting list. The mindfulness group showed a significant increase in the density of cortical gray matter after treatment—a tangible physical change.[92] Enhanced areas included the posterior cingulate and the temporal-parietal junction, areas involved in moral decision-making. The hippocampus was also enhanced, an area critical for learning, memory, conditioning, and aggression regulation[93] and that is impaired by extreme stress.[94] So even though the hippocampus reaches full maturity early in life,[95] its structure can still be enhanced through later environmental change. Another brain-imaging study documented that extensive meditators have increased cortical thickness in the prefrontal cortex compared to controls.[96] Mindfulness remodels the brain—physically.

Hold in your mind for a while the evidence that meditation can change your brain. Now let's ask whether it changes crime and violence. Perhaps surprisingly, meditation training with prisoners has been going on for quite some time. Transcendental Meditation (TM) was made popular during the swinging '60s by its founder, Maharishi Mahesh Yogi, a charismatic figure who was a guru to the Beatles. By the beginning of the 1970s it was already practiced in California prisons.[97] Since then meditation studies have spread to Texas,[98] Massachusetts,[99] and India.[100] Scientific reviews have argued that meditation in prisoners reduces their anxiety and stress levels, increases their psychological well-being, and reduces their anger and hostility. More important, one literature review on meditation in offenders has argued for not just a reduction in post-release drug and alcohol use, but also reduced recidivism.[101] Even women arrested for domestic violence have shown reduced aggression, alcohol use, and drug use after twelve sessions of mindfulness training.[102]

One large-scale study gave mindfulness training to 1,350 inmates and showed significant reductions in their hostility, aggression, and other negative moods. Interestingly, the improvements were stronger in women than in men. Among the men, improvements were stronger for minimum-security prisoners than for maximum-security prisoners— although all groups did improve. It seems that meditation most helps offenders who are not so severely criminal. One recent randomized controlled trial in normal adults, most of whom were female, showed that mindfulness significantly reduces anger expression and improves the ability to regulate emotions.[103] It might be therefore that this intervention could particularly help female offenders.

What are we to make of this? The claims are intriguing, but the reality is that we sorely need a randomized controlled trial to demonstrate that mindfulness training really can reduce violence. Unlike the studies by Davidson and others on brain change, no such study appears to have been conducted on offenders. Granted, Transcendental Meditation has a funky past, with its prior claims of levitation abilities and other supernormal powers. Mindfulness meditation, with its origins in Buddhism, might also seem quirky by association with the TM movement. Yet there is now unquestionably a strong body of scientific support—based on randomized controlled trials—documenting its efficacy in reducing anxiety and stress,[104] substance use,[105] depression, and smoking,[106] and in increasing positive emotions.[107] It's a promising technique that is gaining in scientific credibility, and it cannot be ignored.

Let's suppose for a minute that it's not all pie in the sky. Now put in your mind the hypothesis that mindfulness and other meditation techniques can really reduce violence. How might mindfulness and other meditation techniques work? What might be the mechanism of action? Recall that you are taught to become more aware of your own thinking.[108] You become increasingly conscious of when you are beginning to feel angry over a disparaging comment someone makes to you. You become better able to regulate your thoughts before you boil over in a rage. You become more attuned to the very first moments when, say, your partner made that critical comment that cascaded into a steady stream of unpleasant thoughts and associations. You become aware of your heart racing and your face flushing, and how negative emotions then rear their ugly heads. You are taught to become more accepting of these feelings, to control the urge to act, and to step back from your first instinctive emotional reactions. Because you have become adept at experiencing the negative thoughts and emotions that you felt when the argument started, you have learned to habituate or acclimate to them. That means you can better control your urge to lash out. By being more mindful of your anger at an early stage, you are better able to control and regulate it—at a point in time when your anger is more manageable and has not yet reached its crescendo.

If you think back at the neuroscience we looked at—the studies that document both the short-term and the long-term brain changes that occur with mindfulness—the effect of meditation begins to make some sense. Meditation enhances left frontal brain activity. That meshes with the fact that enhanced left frontal brain activation occurs when people experience positive emotions[109] and is associated with reduced anxiety.[110] It also increases frontal cortical thickness, and we know that this area is not just important in emotion regulation, but is also structurally and functionally impaired in offenders. Note also that meditation enhances brain areas important for moral decision-making as well as areas involved in attention, learning, and memory. We have seen that offenders have impairments in these cognitive functions. Meditation is improving brain areas involved in functions that are deficient in offenders, and that's why it may help.

Mind over brain matters. Brain over behavior matters. What matters to me is that hopefully in journeying through the anatomy of violence you have appreciated three important points. First, there is a basis to violence in the brain. Second, the biosocial jigsaw mix is critical. Third, we really can change the brain to change behavior.

In that third point we have options that run the gamut from concrete surgical castration to almost spiritual mind-over-matter training. In between these extremes we have prenatal nursing interventions, early environmental enrichment, medication, and nutritional supplements that can all make a difference.

Based on the biosocial model I've outlined here, we have promising techniques to block the foundational processes that result in the brain dysfunctions that in turn predispose an individual to violence. That has not been fully recognized within the traditional study of crime—and it really needs to be if we are to be sincere about stopping the suffering and pain associated with violence. We can wait until the milk is already spilled and we have to deal with the adult recidivistic offender who is so very hard to change. That's where we are today. Or we can invest in broad-based prevention programs that start in infancy and can benefit everyone—a public-health approach to violence prevention.

Ultimately, it is up to the public to make that decision. If you want my personal view—based on everything I have learned in my thirty-five-year career in research and practice—it would be this: the best investment that society can possibly make in stopping violence is to invest in the early years of the growing child—and that investment must be biosocial in nature. You cannot successfully intervene without addressing the brain.

Don't get me wrong. Biology is not the sole answer to stopping violence and never will be. Larry Sherman, a world-renowned experimental criminologist at Cambridge University, and others have marshaled systematic evidence from randomized controlled trials documenting that some traditional psychosocial and behavioral treatment programs *can* make a modest difference in offending.[III] What I am arguing here does not negate the positive work done to date by experimental criminologists. What I am saying, however, is that we can go one better with biological interventions that take into account the anatomy of violence—and break the mold that is today giving birth to violent offenders in droves. We have much research ahead of us to develop new and innovative biosocial interventions, but we now have a base on which to build—if we are willing to.

Imagine how society would change if for once we could cure crime. Can you picture a future where suddenly we crack the biological code to violence? How would that change how we think about violence? How would it affect our sense of culpability, punishment, and free will? Would it lead to changes in the law? We'll see in the next chapter that this future isn't so far away.

10

THE BRAIN ON TRIAL

Legal Implications

Michael—or Mr. Oft, as we will refer to him here—was pretty much your everyday, run-of-the-mill middle-aged American guy. In his early career he worked as a correctional officer, later earning a master's degree and becoming a schoolteacher in Charlottesville, Virginia. He liked teaching, and he liked kids. By all accounts he genuinely loved and cared for both his second wife, Anne, and his twelve-year-old step-daughter, Christina, whom he had known since she was seven years old. He got on fabulously well with her. Oft had no prior psychiatric history, nor any history of deviant behavior. He was not much different from you or me—until the clock moved toward the turn of the century in late 1999.

At the age of forty, his behavior slowly but surely changed. He'd never previously been interested in massages, but now he began to frequent massage parlors. He also began to avidly collect child pornography. Then the once-innocent act of putting his stepdaughter to bed changed in an unspeakable way.

As Christina recollects, Mr. Oft used to sing her lullabies before he tucked her in. But after his wife took a part-time job that kept her out of the house until ten p.m. two evenings a week, the usual bedtime practices became a little more sultry and sordid. Oft began to get into bed with Christina, and began to touch and fondle her.

Like many children suffering from abuse at the hands of a trusted relative, Christina was very confused. She knew that she loved her stepfather—but she also knew that what he was doing was wrong. She would argue with him over it, and it increasingly bothered her. But Michael's changes were growing. Normally engaging and likeable, he was becoming more short-tempered. On Thanksgiving Day in 1999 he pulled out some of his wife's hair in a fight. Oft was clearly a man in decline.

Finally, the tearful Christina spoke to a counselor about her stepfather's pedophilia. The counselor in turn spoke to her incredulous mother. Anne was shocked, horrified, and furious. She found "barely legal" child pornography in his possession—pictures of women supposedly of legal age but who looked thirteen or fourteen. She reported his behavior to the police.

Mr. Oft was legally removed from the home and duly charged with sexual assault. Diagnosed as a pedophile, he was found guilty of child molestation and given the option of either completing a treatment program for pedophiles or going to prison.

Naturally, Mr. Oft opted for the treatment program. But even during treatment Oft could not resist soliciting sexual favors from the female staff and other clients at the rehabilitation center. He was thrown out of the program. That was it. He now had to go to prison.

The very night before he was due to start the prison sentence, Mr. Oft went to the University of Virginia's hospital complaining of a headache. Unconvinced, the hospital turned him away, but as he was about to be discharged, he claimed he would kill himself if released, slipping in a threat to rape his landlady. The medical staff could not let someone go in those circumstances, so Mr. Oft was admitted to the hospital's psychiatry ward under the medical diagnosis of pedophilia. Naturally, one of the first things he did was to come on to the female hospital staff and request carnal medical attention.

That probably would have been his undoing if he had not also urinated on himself. The odd thing was that he did not seem concerned. He also began to walk a little unsteadily. A very astute neurologist—Dr. Russell Swerdlow—put two and two together and ordered a brain scan. The scan showed that Mr. Oft had a massive tumor growing at the base of his orbitofrontal cortex, compressing the right prefrontal region of his brain.[1] Brain surgeons resected the tumor, and a most remarkable change came about. Mr. Oft's emotion, cognition, and sexual activity returned to normal. The pangs of guilt and remorse at what he had done to his stepdaughter at last set in. He no longer sexually propositioned female staff members. He no longer felt the urge to rape his landlady or to commit suicide.

Mr. Oft was a changed man.[2] He was released from the hospital and went back to therapy. This time he successfully completed the twelve-step Sexaholics Anonymous program that he had previously failed so spectacularly. His behavior was now entirely appropriate. Seven months later he returned home to be reunited with his wife and stepdaughter to once more lead a normal life. It was a near-miraculous recovery—and it should have been a case of living happily ever after. But what appeared to be a medical miracle was a mirage. The headaches came back.

After several months of normal behavior, Mr. Oft again began to collect child pornography. Suspecting a relapse, one night his wife checked his computer and found the offending material—and once again Mr. Oft was in hot water. Yet thanks to a foresightful re-scan of his brain, his neurologist, Dr. Swerdlow, found that the tumor had grown back. In 2002 the tumor was resected for the second time.[3] Once again Mr. Oft made a complete recovery, and this time for six years after the second resection of his tumor his sexual urges and general behavior have been totally appropriate.

The case of Michael Oft is remarkable because it comes as close as one can get to demonstrating a *causal* link between brain dysfunction and deviant behavior. A double reversal of fortune over time. Going from normality to a growing tumor to the development of pedophilic urges then back to normality after the tumor is resected—with the pendulum swinging back yet again to repeat this tumor to pedophilia to resection to normality progression. The temporal ordering of events is very telling. But the powerful evidence suggesting that this man's antisocial behavior was due to an uncontrollable growth in his skull also raises a profound legal question: Was Mr. Oft legally responsible for his deviant behavior?

Some of the debates in life seem ancient and eternally fixed in time, like the frozen figures encircling a Grecian urn. Here we have such a debate. On one side of the urn we have Themis, the Greek goddess of law and justice. Themis wants no sob stories—she wants no excuses. Justice and retribution reign, and criminals have to be held responsible for their actions.

On the other side of the urn, we have the pleading figure of Mr. Oft and others like him, victims, in ways we are still trying to understand, of complex biosocial forces—forces frequently beyond our control.

In this penultimate chapter we will take a critical look at how research on the biology of violence is not just affecting the judicial system—but is also raising questions about core human values including free will. The new subdiscipline of "neurolaw" is playing a critical role in shaping our perspectives on this issue. Within this context we'll take a focused look at criminal responsibility, and, as one would expect in a legal context, we'll evaluate the cases for and against the relevance of neuroscience research on violence. Finally, we'll return to the question of Mr. Oft's responsibility—and examine the credibility of the current legal response.

HOW FREE IS FREE WILL?

We have been witnessing so far a myriad of biological, genetic, and brain factors that conspire together to create violence and crime. A number occur even before a child is born. A child does not ask to be born with birth complications or a shrunken amygdala, or to have the gene for low levels of MAOA. So if these factors predispose some innocent babies to a life of crime, can we really hold them responsible for what they eventually do—no matter how heinous the crime? Do they have free will in the strict sense of the word? That's the key question we must address.

At one extreme, many theologians, philosophers, social scientists—and likely yourself—would argue that barring exceptional circumstances such as severe mental illness, each and every one of us has full control over our actions. Theologians argue that we have a choice as to whether to let God into our soul, that we choose whether to commit sin or not, and consequently our criminal actions—our sins—are a product of a will that is under our full control.

At the other extreme, some scientists eschew the idea of a disembodied soul that has its own free will and take a more reductionist approach. Francis Crick, who won the Nobel Prize for the discovery of the structure of DNA, for example, believed that free will is nothing more than a large assembly of neurons located in the anterior cingulate cortex, and that under a certain set of assumptions it would be possible to build a machine that would believe it has free will.[4] Such a view harks back to our discussion of evolutionary perspectives. Perhaps we are indeed merely gene machines that con ourselves into believing we have choices in life.

I might argue for a middle ground between these two extremes. Free will likely lies on a continuum, with some people having almost complete choice in their actions, while others have relatively less. Rather than viewing intent in black-and-white, all-or-nothing terms, as the law does, with a few exceptions, I see shades of gray. Most of us lie between these extremes. Think of the free-will concept like IQ, extraversion, or temperature, which are dimensional in nature. There are degrees of free will, and we all differ on that dimension of agency.

What determines the extent of free will? Early biological and genetic mechanisms alongside social and environmental factors play substantial roles. For some, free will is significantly constrained early in life by forces far beyond their control. Let's look into the life history of one murderer and rapist I worked with to illustrate my argument. I'll first present his life circumstances that his defense team argued constrained his free will—but I'll return to his case later to offer a more retributive perspective from the prosecution.

Donta Page was born on March 28, 1976. His mother, Patricia Page, was only sixteen years old at the time of his birth. She had gonorrhea during her pregnancy. Her own mother was fourteen when she gave birth to Patricia, so Donta's mother was raised by her aunt and uncle, both of whom physically abused her, forcing her into an eight-year-long incestuous sexual relationship with the uncle that started when Patricia was four. Donta himself did not have a father at home, but he did inherit a family history from his father's side of crime, drug abuse, and mental illness.

Throughout his early childhood, Donta was a frequent visitor to the local emergency room. He had *five* recorded admissions before the age of two. When he was just nine months old he was taken there after he supposedly "fell" out of a car window—but in all likelihood the unwanted child was thrown out. A scar on his head as an adult is the only external sign of what likely resulted in a very significant closed-head injury. Due to lack of close parental supervision, he was also knocked unconscious by a swing, and when six months old he fell out of the top of a bunk bed. So before he was even two years old he had a substantial history of head injury—and very likely brain impairment.

When Donta was three he was moved into one of the worst neighborhoods in Washington, D.C. His defense attorney reported that as he walked in the area where Donta grew up, he could see that every fourth or fifth house was burned or abandoned. During this time Donta was bounced around from his mother to his great-aunt and back again, experiencing sustained instability in bonding and a normal family life. Frequently he was left at home alone to fend for himself for the whole day. Things were so bad with his abusive mother that by the time he was ten, Donta often preferred to sleep in an abandoned building rather than face the abuse at home.

Given how Donta's mother had herself been raised, it was not surprising that she physically abused him. His grandmother testified that as an infant he was vigorously shaken on repeated occasions for crying. At the age of three, he was punched in the head by his mother so hard that it caused him headaches. At six he was being beaten with an electrical cord that caused bleeding. He was beaten for wetting himself. He was beaten for getting bad grades. He was beaten for any minor misbehavior. When his teacher told his mother that she suspected Donta had ADHD, his mother went home to beat him because he had a childhood disorder. It was documented that by age ten he was being hit by his mother with a closed fist. Donta was also burned with cigarettes, leaving him with dark black spots on his arm that remain in adulthood— alongside scars on his thighs, back, flanks, arm, and chest that bear testimony to the bombardment of abuse.

That abuse was dealt out not just by his mother but also by neighboring predators. When he was ten he was violently raped by a next-door neighbor. Back in the local emergency room, it was documented that he had rectal bleeding. It was further suspected that he was also bleeding internally. Despite all the physical evidence of rape, the hospital never referred the matter to child protection agencies. Donta was sent back to live in the same house, across the way from the same rapist, likely to be raped again. He was given neither counseling nor one ounce of understanding. Neither the family nor the hospital cared about the safety of a small, unsupervised boy from predatory neighborhood rapists.

The abuse escalated. By age thirteen, he was yet again back in the ER because his mother had hit him very hard on the side of his head with an iron. The attending physician documented welts on his arm where his mother had struck him with an electrical cord and the swelling to his temple where he had been hit with the iron. This was clear documentation of child abuse,[5] but no action was taken, and Donta was ultimately returned back home to his mother.

As one might have predicted, Donta was committing property crimes by the age of sixteen and was duly sent to a juvenile detention center. When he was later on trial as an adult for homicide, his attorney carefully pointed out that by the age of eighteen, Donta had been referred by teachers and probation officers for psychological treatment an amazing nineteen times. Astonishingly, he never received even one treatment session. Eight of these attempted referrals were before he had committed a single criminal act.

Given the complete absence of any form of intervention, it is unsurprising that Donta quickly fell into a criminal lifestyle, committing robberies and burglaries that when he was eighteen resulted in a sentence of twenty years in prison and ten suspended sentences. He'd only served four years, however, before being let out on parole and sent to a halfway house on Stout Street in Denver, Colorado, in October 1998. The respite from prison did not last long. He assaulted one of the other residents, and on February 23, 1999, he was told he would be sent back to Maryland to serve the rest of his sentence. It was on the following day, when he was due to be returned to prison, that he robbed and killed Peyton Tuthill in Denver.

Before the trial began I had been contacted by James Castle, Donta's defense attorney, who had heard of my brain-imaging work on murderers. He believed that Page's abominable social history would likely have consequences for brain functioning, and that this in turn would have consequences for behavioral control. I frequently get requests of this sort and usually turn them down, but after considering the details Jim Castle presented me with, I believed Page's case deserved a closer look.

We arranged to have Donta Page brought across state lines from Colorado to California, so he could be brain-scanned in the same PET scanner that I had used in our study of murderers—using the very same methodology. I presented Donta's brain scan at his trial as an expert witness and compared it to fifty-six normal controls. I gave my opinion to the judge and jury: Donta Page showed clear evidence of reduced functioning in the medial and orbital regions of the prefrontal cortex, as well as the right temporal pole.

You can see Donta's brain scan alongside the normal controls in Figure 10.1, in the color-plate section. In the top half of the figure you are looking head-on and slightly up at the individual. Look at the normal controls on the right, and you can see a lot of warm red and yellow colors in the top half of the brain—the prefrontal cortex—indicating relatively normal prefrontal activity. If you look at the top left, at Donta Page's brain, you can see those cooler green patches that indicate reduced glucose metabolism in the frontal poles.

Now look at the lower half of the figure. You are looking down on the brain—a bird's-eye view of a slice through the ventral part of the brain. At the top of the illustration you are looking at the frontal cortex. You can see that the controls are showing good activation in the medial region of the frontal cortex and at the two sides that make up the orbitofrontal cortex. In contrast, Donta Page demonstrates a distinct lack of medial and orbitofrontal functioning. It's not far off being a black-and-white difference. Page clearly shows brain functioning that is quite different from that of normal people.

By now you will have picked up on the significance of these brain regions. You'll recall that the brain regions found to be impaired in Phineas Gage are critically important for cognitive, emotional, and behavioral control. The medial prefrontal cortex—especially the frontal pole—is involved in behavioral control, as well as moral decision-making, empathy, social judgment, and insight into oneself.[6] The ventral prefrontal cortex, including the orbitofrontal cortex, is critically involved in emotion regulation and impulse control—as well as fear conditioning, the ability to switch behavioral response strategies, compassion and caring for others, and sensitivity to others' emotional states.[7] Neurological patients with damage to these regions show impulsivity, loss of self-control, immaturity, lack of tact, inability to modify and inhibit inappropriate behavior, poor social judgment, loss of intellectual flexibility, and poor reasoning and problem-solving skills, as well as psychopathic-like personalities and behavior.[8] We've seen how these processes—when turned off—are important predispositions to violent and antisocial behavior. You will also recall that prefrontal dysfunction is especially characteristic of impulsive killers.[9]

When you place this scientific knowledge in the context of Donta Page's behavior, his actions become more explicable. He had not planned to rape and kill Peyton Tuthill—he just wanted to raid her house for any money he could get. It was not too dissimilar from the case of Antonio Bustamante, who impulsively burglarized a home for money and then battered an old man to death. As we saw in chapter 3,

Bustamante's PET scan similarly revealed orbitofrontal dysfunction. When Peyton surprised Page by unexpectedly coming back home, Donta acted impulsively. Once in full control of this beautiful young blond woman in the bedroom, his emotions and sexual instincts overcame him and he perpetrated on her the ugly act that had been perpetrated on him when he was young and vulnerable—rape.[10] He lacked self-regulation and emotional control. He also lacked the ability to empathize with his victim or to be sensitive to her fear, and when she fought back, he stabbed her. He was angry at how his life, which had been at the brink of being turned around, had fallen back into old patterns. He was angry that he was being sent back to prison that very day, and he took that blazing anger and frustration out on his victim. Given his lifelong history of serious childhood abuse, it's quite likely that either at a conscious or unconscious level, this was redirected aggression—dealing out to Peyton the abuse he had been on the receiving end of as a child.

Nobody can deny that Page's acts were abhorrent—and some would even say evil. But can you deny the predisposing factors that led him down the road to violence?

A significant aspect of Page's brain scan is that the most salient areas of damage included the orbitofrontal cortex and the temporal pole—the frontal tips of both brain regions. These are the areas that are most susceptible to head injury due to the way they sit in the brain. And this damage can result from events far less insidious than the shocking head injuries that resulted in Page being taken to the hospital in infancy and toddlerhood.

We know from family members' testimony that Page's mother would vigorously and repeatedly shake baby Donta—simply because he cried too much. When that occurs, the brain of the baby rocks backward and forward inside the skull, with the orbitofrontal and frontal-temporal pole areas rubbing up against bony protuberances on the inner surfaces of the skull—and getting damaged. So the brain impairments that we saw in his PET scan are quite consistent with the social history of very significant and severe child abuse.

There were more elements of Page's history that struck me. He was enuretic and encopretic until he was ten—he could not control his bladder and bowels in bed. For that he was beaten by his mother. You might see this in children at age three or four years, but the fact that it went on till the age of ten illustrates the anxiety, fear, and tension that the young Donta Page must have experienced in his unbearably traumatic upbringing. He clearly had a very disturbed and harrowing childhood.

At a neuropsychological level, Page performed poorly on the Wisconsin card-sorting task, a classic measure of executive functions—what we'd expect given the results from the PET scan that showed a lack of regulatory prefrontal functioning. Donta also flunked three grades as a child, a clear indication of learning disability.

At a psychophysiological level, his resting heart rate was 60 beats per minute. I compared that to a demographically matched sample of males his age, and this would place him in the bottom 3 percent of the distribution. We've seen earlier that low resting heart rate is one of the best-replicated biological correlates of antisocial behavior—a marker of fearlessness and an indicator of low arousal that can give rise to stimulation-seeking behavior.

At a cognitive level there was a striking difference between his verbal and spatial IQ scores—a gap of 17 IQ points, with the spatial "right hemisphere" score being much lower than the left, suggesting relatively more impairment to the "emotional" right hemisphere. Neuropsychological testing also revealed memory impairments in both auditory and visual modalities, consistent with his history of head injury to the temporal region of the brain.

Three experts documented that Donta suffered from some form of mental illness, likely organic in nature. Given also a familial history of mental illness on his father's side, to say nothing of the aberrant social family history on his mother's side, it is likely that genetic factors also played some role in predisposing him to a dysregulated and impulsive lifestyle, including violence.

Let us not forget that the social environment can have profound "biosocial" effects on the brain, and the fact that Donta as a young baby had a rejecting, uncaring, and callous mother who severely neglected her son. In chapter 8 we talked about interaction effects between biological and social factors. We have few details of Donta's birth, but we do know that his mother had gonorrhea at the time of pregnancy. This can result in obstetric complications, including premature rupture of the membrane that surrounds the baby in the uterus, infection of the amniotic sac and fluid, and early onset of labor. Donta himself could have even contracted a sexually transmitted disease when he passed through the birth canal at the time of delivery.[11] We have seen earlier that maternal rejection of the child, when combined with obstetric complications, triples the likelihood of adult violence.[12] Given the abject poverty in which he lived, the growing Donta was very likely undernourished as a baby and young infant, another important factor that can negatively influence the developing brain.

We've seen how the term "biosocial" can also be viewed in a different light—social factors giving rise to biological risk factors for violence. At the level of environmental toxins, Donta's great-aunt reported that he ate paint debris in the house as a toddler. The old housing in which Donta lived had lead-based paint, and we've seen how lead is neurotoxic, resulting in brain damage. Donta had little in the way of food, and children at any age when hungry will try to eat whatever comes their way—including paint chips—as they crawl around and put their fingers in their mouth. We've seen that poor nutrition is associated with later antisocial behavior—a social risk factor that impairs brain functioning. So at both levels, the social adversity that Donta experienced likely produced brain deficits that in turn contributed to his evolution into a violent offender.

All of these processes—social and biological—can shape further risk factors for violence. His first-grade teacher documented "emotional disturbances" when Donta was six and a half years old. She clearly saw that the young Donta was completely out of it, and that something was profoundly wrong with him. Similarly, his grandmother viewed him as seriously troubled and depressed at ages five and six, and also as being distractible, impulsive, and hyperactive. These clinical behavior problems are again well-documented risk factors for later antisocial and violent behavior.[13]

Let's summarize the case of Donta Page here. Teenage pregnancy. Potential birth complications combined with uncaring, callous mother. Total absence of father. Impoverished neighborhood. Vigorous shaking as an infant that likely resulted in a disconnection between the frontal cortex and the limbic system. Sustained and severe physical and sexual abuse, including rape resulting in scarring and rectal bleeding. Total neglect. Early head injuries and multiple visits to the emergency hospital in the first two years of life. Neurotoxic lead exposure. Poor nutrition. A complete lack of supervision. Learning disability. A family history of mental illness and signs of depression, ADHD, and conduct disorder as early as elementary school. Impaired executive functioning and memory. Low physiological arousal. Poor functioning of the orbitofrontal and medial prefrontal cortex in addition to reduced temporal-pole functioning.

This shopping list of risk factors looks as if it was freshly plucked from a neurocriminological recipe book for creating a recidivistic violent criminal. Donta Page was a walking time bomb waiting to go off. He was totally unloved and uncared for right from the moment he popped out of his gonorrhea-infected mother's womb. It was Peyton Tuthill's dreadful bad luck to be in the wrong place at the wrong time when Page blew up in her face.

Page himself wrote lucidly on his life and the perspective of the jury in a letter read out to the court before sentencing:

> All they see is a black man that killed a white woman. Nobody took the time to ask why but rather who. I've been asking for help for years. Nobody cares until I hurt someone, then they wanted to give me medicine, but when I went home nothing until I got in trouble again. . . . I don't see what I really have to live for. I'm 24 years old. I never had a chance to live. Now it's over.[14]

"I never had a chance to live."[15] He was a 300-pound African-American who had raped and killed a pretty young blond woman. This interracial rape and homicide is rare. Most violence—about 90 percent—is intra-racial.[16] The racial dynamic must surely have ratcheted up the retribution factor in the minds of the jurors. When they returned after three days, they found him guilty of first-degree deliberate murder and rape—and a prime candidate for the death penalty.

The jury took time to answer the question of "who?" but spent much less time to answer Page's own more pertinent question of "why?," a question so childishly simple that it is almost impertinent. Yet we sometimes need to ask an impertinent question to find our way to the pertinent answer. We need to understand the "why"—the causal factors that explain the crime—if we are ever going to prevent horrific crimes like the one poor Peyton Tuthill had to suffer.

Page is also essentially correct on the remainder of his letter. Very early behavioral signs of disturbance flagged him immediately. He was crying out for intervention. Eight documented referrals for treatment *before* he had even committed a single crime—and heaven knows how many undocumented referrals. He desperately needed an expert to defuse the toxic mix of risk factors that was thrust upon him so early in life. These were life circumstances he had no control over whatsoever.

Looking at the free-will continuum as a totem pole, Donta is down at the bottom, where destiny lies. He was always in the red zone. Anyone could have seen that; indeed, they did see it. If you want to lay the blame on someone, blame his psychopathic-like mother for the wretched life she knowingly and uncaringly thrust on her son. Blame the indolent bystanders who witnessed what was going on and did nothing to intervene. Blame the social services for a complete and abject failure to act in a case that was crying out for intervention. Blame society for not doing more to protect once-innocent lives.

But don't blame Cain. Donta's case shows that free will is not as free as law and society would like to believe.

MERCY OR JUSTICE—SHOULD PAGE BE EXECUTED?

Should we execute Page? He was eventually found guilty and was facing the death penalty. We strongly suspect that brain damage made him significantly more likely to commit violent acts. We have also ascertained that the likely cause of damage occurred early in life for reasons beyond his control. Of course we have to protect society, and unless we can treat this brain dysfunction we may need to keep him in secure conditions for the rest of his life. But does Page deserve more punishment? Should he lose his life, given the early constraints on his free will?

One argument rests on the belief that we all have free will and agency even in the face of risk factors. It's almost a religious belief. Surely we all have a choice? If I were to ask you to explain why you are reading this book right now, you'd say something like, "Well, I wanted something to read today and decided to pick up your book. I've always been fascinated by violence, and these days we're hearing a lot more about the brain and biology. So here I am now."

Sounds reasonable, doesn't it? You can choose. You have free will. I was not standing beside you with a gun to your head coercing you to buy it, was I? Surely this has to be full-bodied proof of free will? No, it's not.

You did not choose to read this book. Your brain made you do it. You likely had "risk factors" for buying this book, whether you are conscious of them or not. You may have been a victim of crime. You may have yourself bordered on committing a crime—and always wondered where the line between offenders and good citizens lay. Alternatively you were born good, giving you the fascination for the bad seed that you are not. You may have been exposed to domestic violence and abuse. If you are a woman, we know you are more attracted to books on crime than men—likely because you have a greater fear of being a victim. These factors produce a causal chain of events that predisposed you to read this book. You saw the bold title and colorful cover. In milliseconds it triggered a chain of past emotional memories and associations that made you pick up the book and start reading its contents.

You want so desperately to believe that you determine things in your life, yet that belief has no true substance. It floats like a ghost in a mind machine forged by ancient evolutionary forces. You were as helpless in deciding to buy this book as I was in writing it.

Even if you decide to put this book down right now to prove me wrong, it wasn't you that chose to close it. It was your Bolshie brain that was programmed to be oppositional and defiant when challenged. Free will is sadly an illusion—a mirage. I wish it were not, because I too find this perspective unsettling. But there we have it.

Here's another example. We know that alcoholism is a disease state that has a substantial genetic component. If we sit an alcoholic and a nonalcoholic in front of a glass of beer and tell them not to drink it—then yes, in some sense they do indeed "choose" to drink it or not. But in a probabilistic sense we also know that the alcoholic is going to be less able to resist drinking from the glass. In this situation, the alcoholic's freedom of will has been constrained in large part by genetic, biological, and, to be sure, environmental forces beyond his control. Offenders like Donta Page are no different.

Okay, you say, so Page has a whole bunch of risk factors for violence. Sure, he got a rough deal in life. But he's still as responsible as anyone. If an individual possesses characteristics that make him disproportionately more likely to commit violence, then he has to take responsibility for those predisposing factors. Just as an alcoholic knows he has a drinking problem and must seek out treatment, so the person at risk for violence needs to recognize those risk factors and take preventive steps to ensure that he doesn't harm others. He has a choice, and he needs to act. He is responsible.

This makes good practical sense, but there is a problem with this argument. Responsibility and self-reflection are not disembodied, ethereal processes but are instead rooted firmly in the brain. Functional imaging research has shown that the medial prefrontal cortex is centrally involved in the ability to engage in self-reflection.[17] And it is this very area of the brain that has been repeatedly found to be structurally and functionally impaired in antisocial, violent, and psychopathic offenders. Similarly, patients who have damage to the medial prefrontal cortex are known to become irresponsible, to lack self-discipline, and to reflect less on the consequences of their actions. The very mechanisms that subserve the ability to take responsibility for one's actions were impaired in Donta Page. If you take a look at Figure 10.1 in the color-plate section, you can see very clearly the reduced medial prefrontal cortical functioning. He is less capable than the rest of us to reflect on his behavior, to recognize factors that place him at risk for violence, and to take responsibility for those risk factors and seek treatment.

Let's step back and consider the counter to my own court testimony. Aren't we treading into legal quicksand if we accept the biosocial argument for clemency to Donta Page? Let's concede that genes place the bullets in the gun. I'll admit that the environment cocks the trigger. But surely it is your choice whether to pull the trigger?

Scientifically, I take a more deterministic—and some would say pessimistic—perspective. If there are people stumbling around with a loaded, cocked gun all the time, somebody for sure is going to get shot. We cannot prove that brain impairments cause violence, but as with Page, we can come close.

But your retort is that these offenders must have *some* degree of insight into their loaded-gun condition, and must know there's something just not right with them. Based in part on the four years I worked with prison inmates, I'm not so sure. Most prisoners whom I suspected to have brain dysfunction simply had no idea that anything was wrong with them. This is not entirely surprising when you consider the neurodevelopmental basis to violence, with brain mechanisms not developing normally throughout childhood and adolescence. In many cases these offenders grew up with brain dysfunction, so it has always been an intrinsic part of them. Even when their biological dysfunction is pointed out to them, like many of the general public they believe that the causes of violence nevertheless lie squarely in social factors like poverty, unemployment, bad influences, poor parenting, and child abuse. That's what they have grown up to believe. I think that these offenders and some of you think that way because poverty and bad parenting can be objectively seen and recognized, and are consequently very salient—

whereas biological risk factors are invisible to the naked eye. Yet the neurobiological reality is that many offenders, like Phineas Gage, and individuals with Alzheimer's disease, have brain impairments and cannot objectively evaluate their own minds.

But even if offenders knew they were at risk for violence, the way society is constructed precludes them from doing anything about it. Even if Donta Page had been able to recognize and comprehend the implications of the many factors that placed him at high risk for impulsive violence, what was he going to do about it? Go to the police and tell them he felt like raping someone?[18] We know what the societal response to that would be, and you cannot blame an individual for not wanting to be locked up for a long time in prison. There are no self-help groups for foresightful criminals.

In reading over the case of Donta Page, you may have been reminded of a friend, an acquaintance, or even a family member who might have had some biological and social risk factors for crime, and yet they did not succumb. So you say, surely there must be something profoundly wrong with this actuarial approach of weighing degrees of risk for violence.

The counterargument? The concept of protective factors. That person sticking out in your memory with all those risk factors for crime likely had positive influences on their lives—factors *protecting* them from future crime in the face of the biosocial bogeymen. For example, positive family functioning can protect a child from antisocial behavior in the face of living in a community with a high level of violence.[19] Or, conversely, I have shown that good fear conditioning[20] and high levels of arousal[21] serve as biological factors that protect a child from adult crime even when that child was antisocial during the teenage years. These protective factors helped them along a different course, but not necessarily because they had exerted "free will."

There is a side of me that would argue that Page should not have been punished as fully as he could have been in the eyes of the law. There are limitations to his free will that we should take into account when sentencing criminals like him. We are not all the same.

RETRIBUTION REIGNS

Let's now argue the other side of the case we have before us. There is a compelling reason that we should be unwilling to let Page off the hook, despite all the risk factors he had against him. Retribution—the mainstay philosophy within the legal system for justifying the punishment of an offender. Peyton Tuthill had her throat cut and died in a pool of her own blood after enduring a horrific rape. Should not the victim's cries for justice be heard and a pound of flesh rendered?

You almost certainly have been a victim of crime at some point—a burglary, a robbery, a theft, or an assault. Do you remember the outrage and injustice you felt? The unwillingness to forgive? The instinct to demand an eye for an eye? Justice exists to address a victim's powerful psychological need for retribution. If we were to take tough retributive justice away and replace it with softer sentencing, would that not leave a bitter aftertaste of injustice in the mouths of the victims?

I've presented to you the case for clemency for Donta Page, but now let us go through the hard facts of the rape and murder. This will not be as vivid as it would be were you sitting in the jury box at the trial, facing the photographs and forensic testimony, but perhaps it will give you pause before rendering your verdict—and help you better understand the retributivist's position.

First and foremost, Peyton Tuthill was a truly wonderful young woman. As an undergraduate at the College of Charleston, in South Carolina, she had been a cheerleader, athlete, lifeguard, and sorority president. She worked as a drug-abuse peer counselor. She volunteered in a convalescent home for the elderly. She had an intense sense of social responsibility toward the less fortunate. She worked selflessly during her studies to help underprivileged minorities—mentoring children from very poor homes and organizing the "adoption" of five of them by her sorority. When she left college she moved out to Denver to eventually attend the Colorado Institute of Art. While she waited, she registered at a temporary employment agency for work—I know all too well what that is like. Ironically, she had even visited the Stout Street Foundation, where Donta Page lived, and spoken to officials there about drug and alcohol rehabilitation. She was considering volunteering for them and perhaps might have even helped in Page's rehabilitation. More ironically, they had reassured her that where she lived was quite safe, and that if she ever needed help she should get in touch.

On February 24, 1999, she went to an interview with the Cystic Fibrosis Foundation. Meanwhile, Donta Page was back at the Stout Street treatment center waiting for a lift to the bus station for his one-way trip back to prison in Maryland. He had two hours to kill before his ride, and impulsively decided to burglarize a nearby home.

Returning from her interview, Peyton parked her car outside her duplex. When she entered the house she encountered Page. Terrified, she ran upstairs. Page chased after her, catching her at the top of the stairs, where he proceeded to punch her several times in the face. He hit her hard on the head with the butt end of the knife he had taken from the kitchen drawer. Blood splatters on the railing, floor, and wall showed that she was cut here. As her dog barked loudly in one closed upstairs room, Page dragged Peyton into another bedroom. He tied her hands with cord, and asked where her money was. She told him it was in her purse in her car outside.

Page went out for the money. Peyton, meanwhile, got her hands loose and ran downstairs, seemingly free of her ordeal. But she encountered Page for a second time, as he was coming back up the stairs. With no way out, she ran into the bedroom again. Page followed. He stripped her of her blouse and panties, and raped her on the bed. He raped her vaginally, then he raped her anally. Blood marks down the wall indicate that her head, bleeding from the wound she had received on the stairs, was banging up against the wall in what must have been a truly horrendous ordeal for her.

In his confession tape, Page revealed that Peyton's terrible screams ultimately drove him to kill her.[22] He pulled her to the edge of the bed into a sitting position, took the kitchen knife, and cut her throat. Blood gushed from the wound—but she still screamed, desperately fighting for her life. Bravely she struggled against a man more than twice her size. She grabbed the knife, but it severed the webbing between her thumb and forefinger. Page tried to silence her again—this time by plunging the knife twice into her chest.

She still would not give up. Standing up valiantly against her assailant, she suffered two more knife wounds. One ran deep, with the blade plunging eight inches into her chest, cutting major blood vessels around her heart. Peyton staggered forward two or three steps, and then collapsed. The coroner testified that it likely took another minute before her wretched ordeal was over and she died in a pool of her own blood. Page returned to Stout Street just in time to catch his 1:30 p.m. bus.

The mother of Peyton Tuthill would later say that her daughter was not killed, but that she was "butchered"—like an animal. Should we really excuse Page after he slaughtered this wonderful, charitable woman who was only just beginning her life? She had given unceasingly to underprivileged minority children—and, paradoxically, it was an underprivileged minority child who as an adult paid her back with this bestial treatment. Her life was snuffed out in hideous fashion by a vicious thug. Imagine Peyton as your best friend, your girlfriend, your sister, or your daughter. Can you imagine the pain, fear, and humiliation she must have suffered? If a defendant ever deserved what is a justifiable legal punishment under the law, then surely Page deserves it. Even that punishment would be far more humane than what Peyton was forced to undergo.

Let's take another example. I'll call him Fred Haltoil. Fred was brought up in an abusive household and, according to his sister, was thrashed by a bad-tempered father who had little if any understanding of his son. His home life was traumatic, with four of his siblings who didn't survive beyond childhood. The antagonism between father and son was long-standing and bitter. His family moved repeatedly. Like many offenders he failed in school—having been expelled from one—and left education at the age of sixteen without a diploma. He joined the military, where he proved to be a fearless soldier who fought courageously for his country during wartime. Fred took up one of the most dangerous positions—as a message runner—and was gassed in the process. Hospitalized, he was blind for a month and suffered post-traumatic stress disorder for his near-death experience.[23] Perhaps not surprisingly, like many war veterans his emotional compassion for others was blunted as a result of his traumatic war experiences.

Demobilized, Fred was unemployed and slept part of the time in shelters for the homeless, moving around from place to place.[24] Lacking education and useful life skills, he had no true sense of direction or ambition. His social dysfunction was such that he was never able to develop an intimate physical relationship with another person. His repeated attempts to normalize his life by unrealistic applications for art school and architecture were inevitably unsuccessful, given his lack of training and true talent. He was on a downward spiral. After serving a five-year prison term,[25] he, like Page, went on to become a killer after his release.

Given the same option that the judges of Donta Page had between the death penalty and life in prison without the possibility of parole for this murderer, would you as a juror spare Fred the death penalty? I think many of you would. He had a lot of the risk factors for violence—child abuse, negative home background, traumatic life events with the early illnesses and deaths of his siblings, school failure and expulsion, unemployment and occupational failure, homelessness, and major trauma exposure. Like Page, does he not deserve some degree of clemency?

Perhaps not for Fred Haltoil—alias Adolf Hitler—who was responsible for the deaths of 6 million Jews and many millions more. There's no question that Hitler was not a good man. His best defense lawyer would have had to admit that he pushed the envelope a bit when it came to social policy. Like Page, he was at best a flawed character, and at worst an inhuman monster. For any other killer, we might show mercy. But could you ever excuse Adolf Hitler?

In case you are willing to show mercy to Hitler and those like him who perpetrated genocide—Idi Amin, Pol Pot, Joseph Stalin—bear in mind that American society is wired differently than you are. James Castle, the defense attorney of Donta Page, offered to enter the plea of guilty on all charges and receive a life sentence without the possibility of parole *before* the trial began. Page would never again be free to terrorize anyone outside of prison. Despite this, the prosecution pressed for the death penalty and went to trial—at great expense to you. Clearly this mind-set goes well beyond the protection of society and into the realm of costly retribution.

Are we wired for retribution? I believe that we evolved to have inside us deep-rooted feelings of retribution and rage at those selfish psychopaths who cheat on our civilized rules of social engagement, and who ruthlessly exploit our charity and trust. Without that powerful emotional mechanism to motivate rage and righteous indignation against these offenders, our current-day civilized society would not exist. If we forgave psychopaths we would be overrun by them. We need to hold a grudge. There is surely something to be said for simmering retribution as a mainstay of our society.

You may alternatively have bought into the risk-factors argument for clemency I have given you. You may stand unswayed by the retributive argument. Others will feel differently. I can understand—I used to feel just like you. Why do people differ in their views? If you, unlike others, feel in favor of clemency, perhaps unlike Peyton Tuthill you have not had your throat cut recently.

You'll recall from the Introduction my own feelings of being a victim of violence and the Jekyll-and-Hyde debate I have with myself today. That scientifically trained alter ego has spent his life trying to stop crime by working out what causes it and then developing treatments. He's spent four years of his life holed up in top-security prisons helping the dregs of society, running the gauntlet of the prison hierarchy from murderers and bank robbers at the top to pedophiles at the bottom. He's even argued that recidivistic crime is a clinical disorder and that we should go easier on those that we hit the hardest. And he is resolute in his belief—based on the body of scientific evidence that has been amassed—that early risk factors beyond the individual's control help launch some into criminal careers. He urges all of us to take a hard look at the scientific evidence, and not to let our instincts and emotions hijack our rational thinking.

And yet—can I really forgive? Can I forget? Can I let slip for just once my evolutionary instincts that yearn for revenge and retribution? The Amish apparently could when Charles Roberts shot ten of their little girls in a schoolhouse in Lancaster County, down the road from me in Pennsylvania. That community's response to this despicable act was:

> I don't think there's anybody here that wants to do anything but forgive and not only reach out to those who have suffered a loss in that way but to reach out to the family of the man who committed these acts.[26]

The Amish visited the killer's family to express their forgiveness and even set up a fund for them. I was brought up a Catholic and always admired Jesus Christ, so why can't I have his sense of forgiveness and resolve to turn the other cheek? And if you find it hard to believe the response of the Amish, can you more easily believe that others criticized their response as misguided and tantamount to denying the existence of evil?[27, 28]

So I argue back and forth with myself on this perspective, first arguing one side, and then the other. It sounds a little crazy, but it's really all right to talk to yourself—as long as you don't interrupt! And perhaps there is a bit of Jekyll and Hyde inside many of us. The ultimate challenge arises in how to reconcile these conflicting perspectives within ourselves to develop a compromise position. We'll return to this issue further when we turn to the future of neurocriminology in the next chapter. But right now let's return to our starting point—the two case studies that may help shape our perspective and judgment on the Jekyll-and-Hyde debate.

TURNING BACK A PAGE TO OFT

Some of us have felt the double-edged sword that neurocriminology offers up to us. Peyton Tuthill forcefully felt the sharp edge of the blade. I felt the same edge, but far more lightly. Tuthill's mother, Pat, vents against the violence done to her daughter. My Mr. Hyde rages for revenge.

Yet is there a blunter edge to the blade that can soften these retributive feelings, and give us pause for thought on punishment? Perhaps the medical model, with its Hippocratic oath of doing no harm, can help render a more benign judgment on this tortuous issue. Let's look back both at Donta Page and also at our point of departure, Mr. Oft.

The medical information on Donta Page's early life—as well as his brain scan in adulthood—did not deter the jury from finding him responsible and rendering him guilty of first-degree deliberate murder, first-degree felony murder, first-degree sexual assault, first-degree burglary, and aggravated robbery against Peyton Tuthill. But would it make a difference in deciding whether he should live or die? In Colorado, on February 20, 2001, this question was decided at Page's sentencing hearing by a panel of three judges who had to weigh the evidence and make a fateful decision. Would Page be held fully responsible for his acts and be executed by lethal injection? Or would they accept the biosocial argument that factors early in his life, beyond his control, led him down the path to violence? Should these facts mitigate the punishment, resulting in prison without the possibility of parole?

The panel decided not to execute him. They accepted the argument that a toxic mix of biological and social factors mitigated, to a degree, Page's responsibility. It is what I and the defense team had argued for. But is that the right decision? Or is it nothing more than a slippery slope down to a future lawless society that knows no bounds and where all evil acts have some type of "excuse"? Where no one is responsible for anything?

Retributivists can be reassured that Page was found to be legally responsible for what he did. But what about Mr. Oft? *Should he be held responsible for his actions?* Would *you* hold him responsible? Bear in mind that in Donta Page's case, we are talking about a correlation—not causation—between brain dysfunction and later violence. Yet in Mr. Oft's case we come much closer to causality—the dramatic temporal sway of orbitofrontal disturbance with the sexual swing of his pedophilic passion. What is your verdict? Take a moment to render your judgment.

I put this very question to an assembly of fourteen federal and state judges in the Federal Courthouse in Philadelphia on a cold November morning in 2011. It was a seminar organized by the AAAS—American Association for the Advancement of Science—aimed at bringing neuroscientists together with the judiciary.[29] I suggested to them that Mr. Oft *was* legally responsible for his pedophilia. Every one of these judges agreed. It's not that I have any expertise in law—unlike my good colleague Stephen Morse, at the University of Pennsylvania Law School, who is an international expert on criminal responsibility and who educated me on the case.

How can that decision possibly be reached when we have a clearcut case of a medical condition—far beyond the individual's control, let alone his wishes—that hijacks his brain control center and turns him into a sexual predator? Pedophilia in itself is so "unnatural" that it smacks of a clinical disorder even if there were no corroborative medical evidence.[30] How can you turn a blind eye to the on-off tumor growth and the on-off pedophilia?

The legal answer is relatively simple. In American law, legal responsibility is defined in terms of mental capacity—specifically, the capacity for rational thought.[31] Let's assume that you clearly committed a criminal act. In order not to be held responsible, you need an "affirmative defense." Here you "affirm" that the crime took place—you did it—but your defense is that you are not culpable or worthy of blame because you lacked "rational capacity." You could lack rational capacity because you were suffering from a serious mental illness such as schizophrenia, or because you were mentally retarded, or because you were just a young, irresponsible child.[32] If you could be shown to lack normal capacity for rational thinking, you would not be held responsible for the crime you committed, even if you freely admit to committing the crime.[33] In these cases, you lack substantial capacity to appreciate the wrongfulness of your act.

To translate this legalese into common parlance, rational capacity requires two basic conditions. First, you knew what you were doing. Second, you knew that what you were doing was wrong. How does Mr. Oft's mental state line up with these two conditions?

On the first condition, Mr. Oft knew what he was doing. He freely admitted to the fact that he knew he was going to bed with his twelve-year-old stepdaughter and molesting her. On the second condition, he knew that what he was doing was wrong. It was almost as if he, like me, had a Dr. Jekyll and a Mr. Hyde inside him, with Mr. Hyde having a more telling influence. In reflecting on his pedophilic action with his stepdaughter he comments: "Somewhere, deep, deep in the back of my head, there was a little voice saying 'You should not do this.' But there was a much louder voice saying 'What the heck? Why not?'"[34]

No matter what you would like to believe—or what you think others should believe—there is no hiding from the legal fact that Mr. Oft is responsible for his pedophilic acts. He was fully aware of his action at a cognitive level.

Yet how in the course of justice would you compare Mr. Oft to a pedophile who did just the same act—but did not have that whopping orbitofrontal tumor clouding his moral sense and propelling him to those under-the-sheets illicit activities? Are they one and the same? If you agree, Mr. Oft would beg to disagree: "Now, whether I should be held as accountable for it as someone without a tumor? No, I don't think so."[35] Nevertheless, under current law in the United States, they are both viewed as legally responsible for their acts.

Mr. Oft knew what he was doing. Yet, at another level—at the affective, emotional level—there was something amiss in Oft. As his wife, Anne, comments when she discusses how she confronted Mr. Oft on what he did: "It seemed as though he got that what he was doing was wrong, but he just didn't seem to get it. He just sort of had this look of 'What?'"[36]

Yes, Mr. Oft knew at a cognitive level that what he was doing was wrong, but did he have the *feeling* that it was wrong? When he wet his pants after admission to the hospital, he did not experience the secondary emotions of embarrassment and shame. That lack of feelings arises after damage to the ventral orbitofrontal cortex.[37] Similarly, he did not experience a sense of shame and remorse when committing acts of pedophilia.

We can place this affective deficit in the context of offenders more broadly. We saw earlier, in chapter 3, how psychopathic offenders fail to show activation of the brain's emotional circuitry when contemplating moral actions, and we've seen how the ventral orbitofrontal cortex is also structurally impaired in offenders. Mr. Oft's case is just the tip of the iceberg of a much larger group of offenders in whom the brain contributes to crime.

This in turn leads to a broader and perhaps more troubling question. If you agree that Mr. Oft was not responsible for his actions because of his orbitofrontal tumor, what judgment would you render on someone who committed the same act as Mr. Oft but, rather than having a clearly visible tumor, had a subtle prefrontal pathology with a neurodevelopmental origin that was hard to see visually from a PET scan? Because such a pathology consists of a slowly evolving maldevelopment of this self-control region, there is no rapid switching from brain abnormality to behavioral abnormality. An individual with this kind of pathology lacks self-control from an early age and is always viewed by those around him as a "bad egg." He will grow up to be your archetypal evil monster. How should we view him with respect to responsibility? If you cut Mr. Oft some slack, why not individuals like that? And if, on further reflection, you would not cut them some slack, would their case make you feel differently about how you view Mr. Oft?

Regardless of this latter issue, you might view Mr. Oft as not responsible not just because his tumor "caused" his pedophilia, but also because the tumor could be resected and return him to normality. He could be quickly and convincingly treated, unlike most offenders with more subtle brain impairment. His treatability is making you think differently about his culpability—it's altering your moral evaluation of his act. And yet you would view today's untreatable offenders with volume reductions in their prefrontal cortex and amygdala as more responsible and worthy of punishment? How could we ethically condone such a difference in our evaluation? Today's brain-impaired offenders cannot help the fact that we cannot currently reverse that brain impairment in the way we could with Mr. Oft. Would we call that difference in our opinion "justice"?

Perhaps the majority of you may agree that Mr. Oft was not responsible for his pedophilia. Some will disagree. All I will say for now is that currently the law holds him responsible, standing almost agnosticly to neurocriminology. But what does the future hold for the application of neurocriminology to the law? Stephen Morse has argued that severe psychopaths just do not get the point of morality—just as Mr. Oft could not when questioned by his wife. They are blind to moral concerns and have no capacity for conscience. As such he believes they should be excused from crimes that violate the moral rights of others in society.[38]

If we were to agree with this leading expert in criminal responsibility, might there be some basis for applying a similar line of thinking to

Mr. Oft? Should the law be changed in the light of what we are learning not just in a case like Mr. Oft, not just in severe psychopaths, but also in recidivistic violent offenders who also lack this moral sense and feeling of what is right and wrong? And yet we have seen in chapter 5 there is initial evidence for a neurobiological basis to even white-collar crime. Will there come a day when the Bernie Madoffs of the world plead that it's not their fault—that they were just as biologically predisposed to white-collar crime as Mr. Oft was predisposed to pedophilia?

This issue on the future applications of neurocriminology brings us to the final chapter, where I will give you my own guarded perspective not just on this issue, but also on other societal values that may have to be reevaluated in the new light of neurocriminology. What does the future hold for us?

II.

THE FUTURE

Where Will Neurocriminology Take Us?

Can you remember Kip Kinkel? Probably not. He's easily forgotten amid all the other mass killers in America and elsewhere.

You certainly won't recall Howard Unruh, who shot thirteen people in New Jersey in 1949. I doubt you'll recall the tragic murder of sixteen Scottish primary-school children in 1996. You've probably never heard of One Goh, the Korean-American who killed seven people at his Christian college in Oakland, California, in April 2012. You might remember the twelve students and one teacher killed in Columbine High School by Eric Harris and Dylan Klebold—or Seung-Hui Cho, the Korean-American who killed thirty-two at Virginia Tech in 2007. You'll very likely recall James Holmes killing twelve people during the midnight showing of *The Dark Knight Rises*. It may be some time before you forget Adam Lanza's gunning down of twenty schoolchildren at Sandy Hook Elementary School in Newtown, Connecticut, on December 14, 2012. But the rest become a blur—it's really hard to keep track of them all, and they go back a long way. They are outrageous and completely unacceptable in any society. But heinous killings are not going to go away—unless we take fairly radical steps.

It's in this context that I want to explore with you the possible directions neurocriminological knowledge may take us in the future—for better or for worse—in preventing these and other tragedies. I want to explore how a public-health approach to violence can help create a healthier future for us all. But before we begin our exchange of perspectives, I must first refresh your memory on Kip.

Kip Kinkel was a fifteen-year-old schoolboy in Springfield, Oregon, and he loved guns. That's not uncommon, especially in the rural American Northwest. So he was delighted when his father, Bill, bought him a 9-mm Glock semiautomatic handgun. Bill bought it because he had difficulty connecting with his son, and he thought maybe a gun would help. He'd already given Kip a .22 rifle, and enrolled him in gun-safety courses so that his son could safely channel his enthusiasm for firearms. Kids like Glocks because they are easy to fire, lightweight, and stylish. But Bill never thought Kip would bring a gun to school. Kip was caught with a loaded, stolen handgun in his locker. In England it's cell phones that bother teachers in the classroom. In the United States it's guns. Kip was suspended from school and faced expulsion.

His parents were absolutely distraught. Both Bill and his wife were highly respected teachers in their middle-class community, and now their son had been arrested on a felony charge. Bill collected Kip from the Springfield Police Station, where he had been booked, and they drove together to their secluded rural home. It was the middle of the afternoon. Bill sat at the kitchen counter drinking coffee, no doubt contemplating what could be done for Kip, wondering what on earth would happen next with his son.

What happened next was that Kip delivered a single bullet to the back of Bill's head, behind his right ear, using a rifle that he had retrieved from his bedroom. Kip then waited anxiously for about two hours for his mother, Faith, to come home from work.[1] As she walked into the house Kip first told her that he loved her. Just as Adam Lanza shot his mother four times in the face before killing twenty schoolchildren, Kip fired two bullets into the back of his own mother's head. But she was still alive. So Kip fired three bullets into her face, one into her forehead above her left eye, one through her left cheek, and one close up in the center of her forehead. Yet she still moved. Kip put the sixth and final bullet into her heart.

Kip then put the theme song from the 1996 movie *Romeo and Juliet*, which had starred Leonardo DiCaprio, onto continuous play. He had watched this classic romantic tragedy in his English class. The next morning, on May 21, 1998, he drove to his high school dressed in a trench coat and armed with an arsenal of weapons. Kip walked into the cafeteria of Thurston High, where 150 students were having breakfast. Shooting from the hip with his semiautomatic rifle he got off forty-eight rounds in one minute and very quickly killed one teenager and wounded twenty-six others. One of those wounded later died at the hospital. He would have killed more, but as he was reloading, a wounded member of the high school wrestling team, enraged that his girlfriend had been shot, tackled him. Kip quickly got out his Glock and managed to fire just one more round before six other students fought him down to the ground. He was arrested and charged with four counts of aggravated murder and twenty-six counts of attempted murder.[2]

Kip's attorneys had a dilemma on their hands. They could have entered an NGRI plea—not guilty by reason of insanity—because there was evidence that Kip was mentally ill. Yet a jury might not easily accept going soft on a wayward teenager who had killed so many in cold blood.

Instead, the defense decided to cut a deal with the prosecution: Kip would plead guilty to murder and attempted murder. But while he normally would receive twenty-five years for each of the four murders, the prosecution agreed to recommend that the sentences run concurrently instead of consecutively. That way he would get a maximum of twenty-five years. Thus, with the support of the prosecution, Kip could be out at the age of forty. The defense had found the presiding judge, Jack Mattison, to be fair, reasonable, and rational. They were confident about their case. Because Kip pleaded guilty there was a six-day hearing at Lane County Circuit Court instead of a trial by jury.

Speaking for the defense of Kip Kinkel was Richard Konkol, the chair of pediatric neurology at Kaiser Permanente and also adjunct professor in neurology and professor of pediatrics at Oregon Health & Science University. Konkol had conducted a functional brain scan on Kinkel and documented poor functioning in several areas of the brain.[3] Konkol convincingly pointed out that the most striking dysfunction were "holes" that appeared in the ventral or underside of the prefrontal cortex. These were not physical holes but areas of poorer functioning.[4] Both sides of the orbitofrontal cortex showed much-reduced functioning, but the right orbitofrontal cortex was particularly impaired.

Dr. Konkol buttressed the brain-scan findings with his own neurological examination of Kinkel, which revealed multiple signs of neurological disorder. His examination included tests of cranial nerve functioning, neuromotor functioning, tone and muscle functioning, reflexes, sensory functions, and neurocognitive functioning. He testified that the neurological findings concurred with the imaging findings of frontal and temporal lobe abnormalities, and argued that the impairment was neurodevelopmental in nature. The prosecution elected not to cross-examine Dr. Konkol.

Psychiatric experts also testified for the defense. Kip had suffered from depression the year before the killings and had had nine sessions with a therapist. His mother, Faith, had been concerned with his temper and obsessive interest in guns, knives, and explosives. He also had police reports for shoplifting and throwing a rock at a car from an overpass. The therapy focused on depression and anger management. After the sixth session he was put on Prozac. Prozac worked so well in lifting Kip's depression and emotional problems that after three months his therapist, his mother, and Kip jointly decided he could be taken off it. That may have been a well-meaning mistake.

It was after the seventh therapy session that Bill bought his son the Glock semiautomatic. In hindsight, it sounds like a really irresponsible thing to do, but Bill was a sensible and rational man who was desperately trying to improve the strained relationship he had with his son. He was careful to create very strict operational guidelines for its use and storage. This cherished parental present was to become one of the guns that Kip took to school to execute his murderous plan.

Several psychiatrists testified that Kip was suffering from paranoid schizophrenia at the time of the homicides and that he heard voices resulting in command hallucinations. One voice told him to "Shoot him!" when he had arrived back home with his father. Another voice said "Go to school and kill everybody. Look what you've already done" after he had killed his father.[5]

It was also revealed by psychiatrists that Kip suffered from delusions. He believed China was going to invade the United States, and in preparation he kept explosives under the house. Disney was going to take over the world, with Mickey Mouse's effigy stamped on the new world's currency. Experts testified about his learning disability, particularly with respect to reading and spelling. Kip was dyslexic. He first started hearing a voice when he was eleven, a voice that had told him, "You are a stupid piece of shit. You aren't worth anything." Another psychiatrist documented that there were multiple cases of mental illness in Kip's family history, including schizophrenia.

The prosecution took only four hours to present its case. It did not contest any of the psychiatric and neurological evidence. It was going to be up to the judge to agree or disagree with the prosecution's and the defense's joint recommendation that the sentences should run concurrently for a total of twenty-five years.

When Judge Mattison rendered his judgment, his reference point was a change to the constitution of Oregon two years earlier, which had placed the rationale of punishment away from reforming the individual and toward both the protection of society and also personal responsibility. In this context he argued:

> To me, this was a clear statement that the protection of society in general was to be of more importance than the possible reformation or rehabilitation of any individual defendant. . . . [M]y focus must be much broader than the possible reformation or rehabilitation of Mr. Kinkel.[6]

On November 10, 1998, he sentenced Kip to 111 years in prison without the possibility of parole. Kinkel became the first juvenile to serve a life sentence in the state of Oregon. He could never be free again.

We now move into the future. We pluck the same Kip from 1993 and skip him forty years ahead in time to 2039. He is now a ten-year-old schoolboy, five years before the fateful killings. A new school screening program has identified him as a potential killer. He obtains residential state-of-the-art treatment that successfully tackles the neurodevelopmental factors placing him at risk for future violence. He is later released and lives out a normal life as a crime-free citizen and functional father. Bill and Faith become doting grandparents, two other children live out their lives instead of dying a harrowing death, and twenty-five more people are no longer life-scarred victims of deadly assault.

That's a future I will suggest to you in this final chapter—Lombroso's legacy. Stopping crime before it starts with advanced prediction and treatment efforts. Addressing, with modern technologies and scientific techniques, this grave public-health problem that kills so many globally. Can we improve our approach to crime prevention to create a society that is both more civilized and safer, where a belief in actual rehabilitation trumps the retributive instinct that dominates our justice system today? I believe we can. But before that happens, we need to take a fresh look at the causes of the violence that infect our society and cultivate a more compassionate perspective, not just for the victims, but also for the perpetrators who live on.

FROM SHADOWS TO SUNSHINE—
VIOLENT CRIME AS A CLINICAL DISORDER

I'd first like to share something personal with you—my sister. Roma was like a mother to me. The years have passed since I last saw her, but my memories of her are clearly etched in my mind. I remember her perching me on the countertop in the kitchen and putting on my socks and shoes. Or the day she sat on the sofa in the living room with me on her lap in my new trousers. She fussed over me like a mother hen. I remember the soft touch of her hand as she walked me down the street early one evening as the sun was fading, stretching out our elongated shadows into the remains of the day. I remember her holding me in her arms and telling me how lovely I was. I could feel her caring, her warmth, and her tenderness. Roma to me always looked so special, so serene, so beautiful. Right now I can see her beautiful face, her gorgeous dark curly hair, and her understanding eyes.

Roma left school at sixteen and worked for a while in Binns, the main department store in Darlington, our hometown in the northeast of England. She was a natural-born caregiver who always wanted to help people—just as she cared for me. So she became a nurse at Darlington Memorial Hospital. What happened to my sister next, when she was just eighteen, is narrated by her nursing colleague and friend Clare Fitzgibbon, who won the Macmillan Gold Medal for nursing.

Clare recounts in her book *Sunshine and Shadows* her time being a nurse working on the ward with Roma. Her close friend had become pale and tired, and had continuous sore throats. Roma eventually collapsed on duty and was taken to an infectious-disease unit on the edge of town. Clare was wondering one day what was wrong with Roma when the ward sister told her to quickly prepare a side room on the Florence Nightingale Ward. A new leukemia patient was being admitted.

The pale patient was being wheeled along the corridor and through the swing doors on a trolley with a blood-transfusion bottle rocking precariously on a short pole. Clare was in shock to see that the new cancer patient was Roma.

She nursed Roma through her final days, surprised at how quickly her dear friend had faded. As she recounts in her book, a very moving narrative on caring for cancer victims:

> her dark Italian eyes seemed to have taken over her face, her clear, pale skin was now ashen, framed by her dark, beautiful hair. . . . Roma looked straight at me. "I'm dying" she said simply, clutching my hand. Blood was trickling down both her nostrils. "Please tell them I love them" she gasped, "my mother, my father," again she gasped for air, "all my family," she managed a half smile, her face now colourless, "and you." . . . Tears were streaming down my face. "And we all love you too" I got out, as she died in my arms.[7]

The bright sunlight of my sister's radiant life was overshadowed by a particularly acute form of leukemia. On September 18, Roma's life was snuffed out—perhaps mercifully—in just two weeks,[8] although that's painfully long compared with the victims of most acts of violence. We all miss her, just as Clare does to this day.

I have reflected a great deal on Roma, and her death has profoundly affected my thinking. The other cancer that bloodies the lives of so many more people—violence—is to me as much medical as the sickness that killed my sister. For me, Roma's death is a metaphor for how I think we need to treat violence. It requires more compassion, less retribution, and a new clinical perspective that I want to move you toward considering.

As a psychology undergraduate in the 1970s I had been fascinated by the psychosomatic approach to illness—mind-over-body causation. Susan Sontag wrote provocatively twenty years after Roma's death about how cancer—the paradigmatic disease for much of the twentieth century—was wrongfully viewed as something to be ashamed of, something to be covered up.[9] The psychosomatic perspective on illness considered the *person* to have caused their own cancer. Their internal aberrant personality, hallmarked by inhibition and anger suppression, caused a somatic disease, and psychotherapy was offered as an alternative treatment. The person was responsible, not any outside agent.

I believe we currently view the cause of violence in a similar way. Don't you think some offenders are just plain evil? It's the serial killer's own internal demon that caused him to kill. Two world-leading academic clinical psychologists have in the past provoked me to consider that possibility—and it is indeed provocative. Perhaps there are no external biological or social causes—instead it's evil. Could that really be?

Perhaps. But my concern is that if we begin to think in that almost spiritual way, we have regressed to how crimes were explained in medieval days—by an evil spirit. Surely we have progressed further, scientifically and rationally? Cancer is not a punishment for our sins but a disease produced by external biological and social forces that can be treated. I would ask you to not only consider violence as a public-health problem, as a disease that affects our society—but also to think about it rationally and clinically, not inflected by ideas of sin and evil. I sense that that was the essence of Sontag's point on the illness my sister died from—the same cancer that Sontag herself would die from—and it's the same point I want to make to you about the nature of violence.

Just as our perspectives on cancer have now radically changed, so too, I believe, are our perspectives on violence about to change. Like Clare, I've been on a ward with my own patients, being up close and personal with them for four years—working with them as a psychologist in top-security prisons and caring for them in therapy. For thirty-five years I've been trying to understand what causes their illness. We've given up on lifers in much the same way that doctors had to give up on my terminally ill sister, moving Clare Fitzgibbon to hatred for them when all they could say was, *"It's time you called the priest. We've done all we can."*[10] It was seemingly time for Roma to confess her sins and take responsibility for causing her cancer. How in future years can we turn the dark shadows of prisons into sunshine? How can we cure this violent cancer?

Before moving into the future to provide an answer I need to explain my own perspective on violence further. Let's drift back twenty years, to a book I wrote in 1993, six years before Kinkel's conviction, called *The Psychopathology of Crime: Criminal Behavior as a Clinical Disorder*. I argued that repeated violent offending is a clinical disorder[11] in just the same way that cancer, depression, and anxiety are viewed today. In viewing violence this way, I'm not referring to someone who loses his temper one day and slaps someone, but to the class of violent criminal offenders who repeatedly perpetrate significant criminal violence upon others. I would also include nonviolent criminal offenders—those who are recidivistically antisocial. I believe there are good grounds for this view.[12]

Fundamental to this idea is the definition of clinical disorder as a "dysfunction."[13] Essentially, something is not working right in the individual. The DSM—the *Diagnostic and Statistical Manual of Mental Disorders*—is used by psychiatrists and clinical psychologists to diagnose all clinical disorders.[14] It is a veritable bible for psychiatry. Let's see how a revision proposed for the next edition, which represents 36,000 mental-health physician leaders, defines what a disorder is and how it fits recidivistic violence. The proposed definition in *DSM-5* is as follows:

> A Mental Disorder is a health condition characterized by significant dysfunction in an individual's cognitions, emotions, or behaviors that reflects a disturbance in the psychological, biological, or developmental processes underlying mental functioning. Some disorders may not be diagnosable until they have caused clinically significant distress or impairment of performance.[15]

Do violent offenders have abnormal functioning in terms of how they think, feel, and behave? Yes, they certainly do. Does this "dysfunction" have a biological basis? Is something not going right in their development? I have argued that crime germinates early in life from a neurodevelopmental and genetic base. I've suggested that there is a heck of a lot that is just not working right in violent offenders. They are also impaired in how they perform in life—whether at school, at home, or at work. Violence certainly causes distress to others, and the offender himself is frequently in a distressed state. Repeated violent offending *is* a clinical disorder.[16]

In the field more broadly there are at least nine different criteria for judging whether a certain condition is a clinical disorder—such as statistical infrequency, deviation from the social norm, and deviation from ideal mental health.[17] Recidivistic crime is relatively infrequent. It deviates from the social norm. And we know that offenders are not the picture of ideal mental health. Combine this with distress and suffering to others and self; impairments in social, occupational, behavioral, educational, and cognitive functioning; and the host of biological and brain impairments we have documented already, and the case is fairly complete. Of course, most individual criteria of what constitutes psycho-pathology have significant weaknesses, but when combined together they help describe a gestalt picture of psychopathology against which violent crime may be viewed. Recidivistic offending meets these criteria just as well as most disorders listed in the *DSM*, and, indeed, it fits better than some already listed.[18]

What will be the critical turning point that will lead to this radical way of thinking? It will be the development of new treatments that conclusively stop violence in its tracks. Once that happens—once we can "treat" offenders successfully—retributive justice will seem archaic. We'll witness a significant change in society's perspective, driven particularly by how judges decide to deal with defendants in sentencing.

For that to happen, of course, we'll need some really big break-throughs. But even today there are signs of progress, many coming from advances in other medical disciplines. Let's take a closer look at leukemia, as a current-day example of what could happen tomorrow with violence. Leukemia very likely results from a genetic mutation in DNA that produces protein abnormalities that make too many white blood cells. Normally these white cells are produced in the bone mar-row and protect us from viruses. But the new white cells produced by the illness are immature, and they crowd out the healthy cells, damp-ening the immune system and reducing the number of red blood cells that provide oxygen. That results in anemia, pallor, and shortness of breath, as it did with Roma. The reduction in blood platelets, which normally aid clotting, resulted in Roma bleeding from almost every ori-fice in her body. The immune-system suppression results in unremit-ting infections, such as Roma's repeated sore throat and infected tonsils, and eventually death.

For one form of leukemia, called chronic myelogenous leukemia—CML—we have an understanding of its genetic basis. Genes on two chromosomes normally regulate white-blood-cell growth. In leukemia, the ends of these two chromosomes get switched around, with one get-ting shorter. This shortened chromosome is named after the city I cur-rently work in—the Philadelphia chromosome—and was discovered in 1960, just three years after Roma's death. It now contains a new hybrid gene that uses a molecule called ATP that activates other proteins and causes the cancerous growth that produces the excessive white blood cells. How can ATP be blocked? By using a drug called imatinib, sold under the name Gleevec.[19]

Okay, you say, this is all fine and dandy for a cure for cancer, but crime and violence is only half genetic and it's just not so clear-cut. Yet the reality is that while some cancers show heritability at a similar level to crime and violence, many cancers are *not* heritable, even though they have a biochemical genetic basis.[20] So what's happening here?

Duing the time you read the paragraph on Roma's death, hundreds of changes had taken place in your genome. Hundreds of thousands occur every day, but we have natural repair mechanisms that reverse this genetic damage.[21] When these repair mechanisms go awry, mutations can result in gene abnormalities, producing defective proteins that in turn result in faulty physiological functioning and impaired health. What can cause some interruption to the natural correction that normally takes place? Think back to the concept of epigenetics that we discussed in chapter 8. Environmental experiences alter gene expression. That's why many cancers have little or no heritability and yet they operate through genetic processes.

For that reason I fundamentally believe that what we see today in cancer can happen tomorrow for violence. Mutations can be repaired with medication. The speed with which science made progress on the human genome project is just one example of the rapidity of change that is possible. I ground this prediction on what I have seen in the past thirty-five years of my research career on crime. I've seen how breakthroughs come first in physical clinical conditions and the development of new medicines—not infrequently from research on cancer. Those conceptual breakthroughs tend to filter down to other medical illnesses. Advances in medication get applied to psychiatric illness. Then from psychiatry there is invariably a trickle-down effect to violence and crime. Take cognitive-behavior therapy, pioneered by Tim Beck at the University of Pennsylvania. It was first developed for depression, and now it is one of the best and most used interventions for adolescent and adult antisocials alike.[22] Take the application of medications for epilepsy, psychosis, and ADHD, which are being used today for aggressive children and adolescents. Very slowly—but very surely—I see it happening.

Why am I sure this change will occur? Because the theoretical framework and science are in place right now, and because treating the physical causes will work more quickly and effectively than repairing the complicated social factors that also contribute to criminal behavior. Bad neighborhoods basically don't change much over decades,[23] and the cycle of poverty is equally resilient. You now know that the environment critically interacts with biological and genetic risk factors in shaping violence.[24] You now know that there is a significant genetic basis to crime, aggression, and violence. You now know about epigenetics—that changing the environment changes gene expression. You know that current medications can attenuate aggression and violence. You know that a new generation of cancer medications has the capacity to reverse gene mutations. We could have the capacity to change violent behavior more quickly through biological interventions.

From a practical standpoint, can we stop the social causes of crime? John Laub and Rob Sampson are prominent criminologists who argue for the importance of the neighborhood in crime causation.[25] Improving neighborhoods will help reduce crime, and we should certainly do more for that goal. They also persuasively argue that daily situational contexts and experiences can be turning points that either start or stop crime, whether it's getting married, getting a job, or even joining the army. I believe they are right. Yet the problem remains that it's going to be darn near impossible to control people's daily social interactions and experiences. After all, our lives can turn on a dime, with a chance meeting. We won't be able to predict and control these chance fluctuations. Not now—nor in thirty-five years' time.

Yet we also know that environmental and even chance events can promote genetic and biological alterations through the process of epigenetics. Can we control the physiological effects that give rise to basic cognitive, emotional, and behavioral risk factors that spawn violence? In theory we could, by developing drugs in the same way they are currently being developed to treat some forms of cancer. The future promise is that a new generation of medications can be developed to block the functioning of the faulty proteins that will be identified in the future as the genetic and biological bedrock for violence. We first need to identify which structural genetic mutations give rise to which specific faulty proteins that in turn give rise to the biological risk factors for violence. It will take time—a long time—but the *theoretical* potential is there if we have the courage and conviction to pursue that path. So far we have not.

It may be even less of a choice—and more of a future sociopolitical tipping point that we have seen so many times before—that will bring about a change. Let's now delve into that future.

THE LOMBROSO PROGRAM

It's 2034. The past decades have seen enormous efforts spent on reducing crime through social programs to increase equality. But it's not working. The Internet, which so effectively democratized knowledge, has inadvertently resulted in a much smarter breed of crooks who, though failures at school, have succeeded in home-schooling themselves on high-tech ways to evade the surveillance of global CCTV. Clearance rates for homicide have moved from a national high of 65 percent in 2010[26] to 38 percent in 2034—arrests of suspects were dropping precariously. Serial killings are on the rise. Prisons are not just full to capacity, they are bursting at the seams. Back in 2012, the United States made up 5 percent of the world's population but was incarcerating 24 percent of the world's prisoners. That number has grown to 31 percent. Police are working around the clock on overloaded portfolios of unresolved cases.

The public is growing enraged at decades of failure and the increasingly intolerable condition of living under stifling and ineffective public surveillance. People are fed up with the long legacy of attempted rehabilitation efforts, and alarmed at well-publicized accounts of furloughed criminals committing fresh crimes. But it's more than that. The economic cost of crime is now astronomical. Back in 2010, the cost of homicide in the United States was estimated at over $300 billion— more than the combined budgets of the Departments of Education, Justice, Housing and Urban Development, Health and Human Services, Labor, and Homeland Security.[27] Way back in 1999, it was estimated to consume 11.9 percent of GDP,[28] but in 2034 it is gobbling up 21.8 percent. The more crime got out of control, the less the government could spend on education, health, and housing—and that just fed into more and more crime.

The tipping point came in 2033, when one "low-risk" mentally ill offender was released early on supposedly supervised medication to help relieve the massive prison overcrowding. Through an administrative oversight his dangerousness assessment report had been mixed up with that of another offender. He was high-risk—not low-risk. Just two weeks after his release he held up a store in Washington, D.C., during which a young woman was killed in cross fire between the ex-con and the police. By sheer bad luck the victim was the U.S. attorney general's daughter.

This incident, combined with the mounting economic and public concern, now leads the government to launch the LOMBROSO program—Legal Offensive on Murder: Brain Research Operation for the Screening of Offenders. The logic behind LOMBROSO is surprisingly simple. Back at the turn of the century, in 2006, it was known that 22 percent of all those arrested for murder were probationers and parolees—those who had been released from prison.[29] Criminologists in 2009 had then used early machine-learning statistical techniques to predict which parolees would go on to commit homicide. They had only basic demographic and prior-crime data to work with then, and yet they were still able to correctly classify 43 percent as likely to be charged with homicide only two years after their release.[30] Of course there was still the false-positive problem—those who were predicted to commit homicide but who did not.[31] But a replication study with a longer follow-up period provided better results. By the 2020s, interdisciplinary neurocriminologists, statisticians, and social scientists improved the predictive power of this model by adding brain, genetic, and psychological risk factors into the equation. By the early 2030s they took it a step further by developing algorithms for violence in the community at large. Then, in 2034, the LOMBROSO program was put into place.[32] It was a chance for a failing government to reverse its declining popularity in the polls.

Under LOMBROSO, all males in society aged eighteen and over have to register at their local hospital for a quick brain scan and DNA testing. One simple finger prick for one drop of blood that takes ten seconds. Then a five-minute brain scan for the "Fundamental Five Functions": First, a structural scan provides the brain's anatomy. Second, a functional scan shows resting brain activity. Third, enhanced diffusion-tensor imaging is taken to assess the integrity of the white-fiber system in the brain, assessing intricate brain connectivity. Fourth is a reading of the brain's neurochemistry that has been developed from magnetic resonance spectroscopy. Fifth and finally, the cellular functional scan assesses expression of 23,000 different genes at the cellular level. The computerization of all medical, school, psychological, census, and neighborhood data makes it easy to combine these traditional risk variables alongside the vast amount of DNA and brain data to form an all-encompassing biosocial data set.

All those convicted of homicide in the United States have been assessed on the Fundamental Five Functions. This was going on for research purposes well before the homicidal tipping point arrived. An equal number of noncriminals was drawn from the community as a comparison group. Fourth-generation machine-learning techniques looked for complex patterns of linear and nonlinear relationships between these predictor variables and the homicide-control grouping. One conceptual advance that was learned in the previous decade and that enhanced the accuracy of violence prediction was the critical importance of factoring in the *interaction* between social and biological variables. The samples of murderers and controls were randomly divided into three separate pools of data. The first pool of murderers and controls was used as a training set—allowing machine-learning techniques to "learn" how to predict homicide. The second pool of data was used to test out the prediction formula to see if it held water. After further refinement, the formula was tested and finalized on the third data set.

The result is not perfect prediction, but it is pretty darn good— good enough for an outraged society. Those tagged as LP-V (Lombroso Positive—Violence) as a group have a 79 percent chance of committing a serious violent offense within the next five years. Those classified as LP-S (Lombroso Positive—Sex) have an 82 percent chance of committing either rape or pedophilic offenses. Finally, those classified as LP-H (Lombroso Positive—Homicide) have a 51 percent chance of killing someone in the next five years. Some have dual designations.

The program works like this: those who test positive—the LPs— are held in indefinite detention. In light of the administrative lapse that originally sparked LOMBROSO when test results were mixed up, LPs are given the legal right to challenge the findings and be retested by an independent authority. The detention centers are highly secure, but are not the harsh holding bays of decades gone by. They are equipped as a home away from home. Conjugal visits are allowed on weekends, albeit under surveillance that is a bit too close for comfort for the partners concerned. There are full recreational and educational services. They are allowed to vote. The LPs have full communication access to their family and even friends—after appropriate security checks on those concerned. It sounds quite cushy, but remember that the LPs have not actually committed a crime. Perhaps the main drawback is who they live with, housed as they are in facilities full of other LPs—time bombs waiting to explode.

Every LP is reassessed every year, as the changes brought about by the detention environment and treatment can bring about significant epigenetic change and hence a change in their LP status. They can be downgraded to tagged probation where they will be back in the community and kept under continuous auditory and visual scrutiny. With time they could entirely lose their LP status, while others could also eventually age out of their LP designation.

Release is also possible, and long-term detention can be avoided. The LP-S group, for example, can elect to have surgical castration and will be set free immediately, although they have to continue to undergo mandatory weekly testosterone checks to ensure that they are not taking hormone-replacement therapy. Others, depending on their bio-profile, can also be placed on mandatory medication and tested at halfway houses. Most releases, however, are the result of the intensive treatment programs implemented in the LOMBROSO centers.

These are scientific interventions, deriving from the experimental criminology movement beginning in 1998 espousing practice based on randomized controlled trials.[33] Society accepted that serious recidivistic crime was a clinical disorder when new biological treatments were shown to work. State-of-the-art biopsychosocial treatments are intensively explored for all LPs, but are tailored to their unique biosocial profile. Alongside more traditional cognitive-behavioral therapy sessions, treatments range all the way from sophisticated derivatives of the earlier deep-brain stimulation[34] and noninvasive transcranial-magnetic-stimulation techniques[35] to next-generation medications that enhance prefrontal functioning. Sophisticated nutritional programs that include omega-3 as well as mindfulness training that incorporates fMRI biofeedback are also options.

What has created the most consternation to the public is LP-P status—Lombroso Part-Positive. Risk assessment is essentially dimensional—there are degrees of risk. LP-Ps are not exactly high-risk, but they are not low risk either, and need careful monitoring. In the event of a serious offense occurring that cannot be cleared reasonably quickly, law-enforcement agencies have access to the identities of those in the pool of LP-Ps to help narrow their search. They effectively become prime suspects. Politicians skillfully negotiated a solution to the protests that broke out over the invasion of civil liberties and the potential threat to employment and insurance. There is quadruple encryption of the data to protect identities, with only senior police officials having the ability and authority to decrypt the LP-P database on a case-by-case basis.

At first there were remonstrations over excessive government control and breach of civil liberties. But the government has been able to come up with scientific backing for the validity of its policy. Back in 2009, the importance of science and evidence-based practice had transformed the attorney general's office through the pioneering efforts of Laurie Robinson, then the assistant attorney general.[36] The government argued that, just as we screen for cancer to prevent deaths, we should also screen for violence to prevent loss of life. Critics railed against the enormous expense of the new program, but the government ingeniously issued bonds that were bought by private investors to help finance it. If it works—and the evidence suggests it will—private speculators will get a handsome return on their investment. With increasing political debate, it was argued that the only people who really had anything to worry about are those at high risk for committing homicide. That shut the protestors up.

THE NATIONAL CHILD SCREENING PROGRAM

It's now 2039, and five years after the introduction of the LOMBROSO program. An independent analysis was conducted on the efficacy of the government's program. After years of gradual increases, the homicide rate has been cut nearly 25 percent. Similar reductions have been seen for rape, pedophilia, and serious crime. Government spending on health, education, and housing have increased, given the savings on the cost of crime that they shared with private investors. Civil libertarians are flabbergasted by the fact that a scheme they thought would be racially prejudicial actually resulted in a *lower* proportion of minorities being detained as LPs. The jury system of the 2010s was undoubtedly racially biased, with a black offender more likely to be convicted of the same crime as a white offender.[37] LOMBROSO, in contrast, is scrupulously objective and data-driven, and the results have pleased civil libertarians and minority leaders alike. After all, it was known all along that minorities are disproportionately the victims of violence,[38] and now they are disproportionately benefiting from violence reduction.

Everyone feels discernibly safer. Oddly enough, many LPs are not too dissatisfied with their lot. Conditions are fairly reasonable. The food is quite good and nutritious. Those with partners have sex every weekend but without the social obligations and hassles that go with it. Their kids are not around to have screaming arguments with. There is no work to produce work pressure. They have TV, movies, books, gym, swimming, basketball, and other recreational activities. There is less stress all around. Even the treatment is not a problem, and in fact the therapy sessions are stimulating and provocative and something they look forward to. Ironically, what they least like is being around people like themselves, the other LPs. Overall, though, it isn't all that bad—a bit like being in a summer camp but without having to pay. Or like resting up in the hospital but without feeling ill.

The astonishing success of the program was one of the reasons for the reelection of the party that had initially introduced LOMBROSO. And yet there is still a significant level of serious teenage violence, with two separate mass killings in shopping malls in the same year involving young teenagers. Homicide rates are also not as low as they were in the good old days of 2013, even though they have come down. The government and its scientific advisors sat back from the glow of the independent review that lauded the program as a breakthrough. They hunched around a conference table and thought it through. "It's never too late to prevent violence" had been the mantra of the scientific advisors in 2034. Now, in 2039, they have a new prevention mantra—"It's never too early to stop the rot." If LOMBROSO is working well with a screening at eighteen years of age, then why not screen earlier?

In 2040, the National Child Screening Program (NCSP) is announced. All children ten years of age are given a comprehensive medical, psychological, social, and behavioral evaluation that incorporates all prior school, social, and medical-record data. Anxiety and stress in youngsters are on the rise, just as autism was at the turn of the century, together with obesity, depression, and a host of other medical and psychiatric conditions. The screening program is ostensibly an evaluation of dyslexia and learning disabilities, allergies, vision, and obesity—indeed, all physical and mental health problems that go along with children entering puberty earlier than they used to. What is also included in the health screening under the rubric of "behavior problems" are "emotion-regulation problems" and "violence potential." After all, violence is now widely viewed as an international public-health problem.

Prospective longitudinal studies are increasingly documenting the biosocial package of early factors giving rise to adult crime. Together with advanced machine-learning statistical techniques, they are doing a decent job of predicting future crime from childhood data. Not as well as LOMBROSO did at eighteen, because it's harder to predict crime from an earlier age—but with persuasive predictive power nonetheless.

Under the new NCSP, parents of some ten-year-olds are informed that their child is a rotten apple. The NCSP determines that little Johnny has a 48 percent chance of developing into a serious violent offender in adulthood, and a 14 percent chance of committing homicide. That's the bad news.

The good news, however, is that the NCSP has developed residential treatment programs that should be successful in cutting these odds by more than half, to 18 percent for serious violence and 6 percent for homicide. It does, of course, mean that Johnny will have to be taken away for two years for intensive biosocial therapy, but after that he will be back home.

Yes, it is true that it is not a perfect solution. There will still be a chance that he will become an offender anyway, even if his parents do opt for the residential treatment. And yes, the overall odds that he will become a serious violent offender without intervention are a fraction less than half. But there you have it—it's your choice. What will you decide for your little Johnny?

What would *you* decide if you were Johnny's mother or father? Put yourself in their situation. Do you want *your* child whisked off to an institution for treatment and branded as a potential future offender? What are you going to tell your relatives and friends and neighbors? Think of the stigma. What about Johnny losing his friends? And what bad new friends will he make in this residential program for criminals-in-the-making that might make real a self-fulfilling prophecy?

On the other hand . . . are you just going to stand by and do nothing? You know full well that Johnny has a very significant chance of ruining not just his own life, but your life, and the lives of innocent victims. These are lives *you* could save if you only act.

On balance, the majority of parents give up their children for residential treatment. Bill and Faith Kinkel decide to put their son Kip into treatment—it is, if nothing else, a welcome break from their endless struggle to get him back on the rails. Yes, in the NCSP even good parents like the Kinkels have children who are identified as violence-prone—it's the well-off as well as the underserved who are affected.

In 2042 there is a controversial change to the NCSP initiative after two eleven-year-old schoolchildren coldheartedly tortured and killed a three-year-old child, having abducted him from a shopping mall while his mother was distracted. The act was caught on the global CCTV network. It turned out that both of the killers had been identified by the NCSP the previous year as being in dire need of residential treatment, but their respective parents had elected to decline intervention. Analysts argued that children in the red zone likely have parents who do not have the best interests of their children at heart. They are not responsible parents and not good decision-makers—reasons their child is in the red zone in the first place. NCSP officials now need to act "in loco parentis"—to step into the parents' shoes and make the decision. The treatment now becomes compulsory.

Just two years later, in 2044, research analysts on the LOMBROSO program make another recommendation to the government that results in a further addendum to the National Child Screening Program. If a child is in the red zone, isn't his biological father likely a bad apple too? What's he up to these days? After all, like father, like son. Perhaps he missed his LP screen when he was eighteen. His new status as the biological parent of the offspring identified in the NCSP needs to be factored into the equation. He is now brought into detention pending reevaluation of his LP status; 2044 is slowly but surely sounding all too like 1984.

THE MINORITY REPORT

It's now 2049 and the fifteenth anniversary of the LOMBROSO program. The nation is nine years into the NCSP. Together these programs are undeniably making a dent in the rates of juvenile and adult violence. They have also significantly reduced nonviolent crime. It has unquestionably been a dicey game to play, but cost-benefit analyses clearly document the winnings, which are invested back into welfare programs and have gained bipartisan political support for the program. The government is popular, but the opposition is ever present. Fortunately, the government's research analysts have another card up their sleeve.

LOMBROSO and NCSP are certainly costly prevention programs even with investment from the private sector. There could be even greater savings. An avant-garde cadre of research analysts and neuro-

criminologists propose a controversial program that is outvoted by
other advisors. But a minority report is written and submitted alongside
the majority vote for senior government officials to consider. Following
in the traditions of LOMBROSO and NCSP initiatives, the minority
report proposes to stop crime before it starts. But this time it proposes
that citizens get a license before they even have a child. After a very long
and heated debate, there is a small majority vote in favor, and the policy
becomes law.

The train of thinking in the minority report goes something like
this: Poor parenting has undeniably been linked to later violence.
Genetic studies documented not just that antisocial parents transmit
their bad genes to their children, but that the negative social experience
of having a bad parent is also a causal factor for antisocial behavior.
The issue is not to use eugenics as a final solution to crime, advocates
argue, but to create a social policy to promote positive behavior. Better parents, better children. The minority report's perspective focuses
on children's rights—minors need to be protected and better treated,
and would-be parents need to be responsible. They must report in for
licensing.

Cars can be killers, and so you need a license before you can drive.
Kids can be killers too. So the logic goes that you should also have a
license before you can have a child. Just as you need to document practical skills in driving a car and also knowledge of the right way to drive,
you also need to show theoretical and practical proficiency in rearing a
child. It's only right for the child and society.

Civil-rights activists remonstrate loudly against the minority report,
claiming it is taking away a fundamental human right. In response, the
government adds the caveat of compulsory classes in parenting skills in
all schools. Now everyone has the potential to pass the licensing exam,
they say. No child left behind. No more excuses.

Classes are structured to be age-appropriate and to start at a relatively early age. They teach children everything from the basics of
reproduction to prenatal nutrition, stress reduction, the early needs
of a developing baby, providing structure and support for the growing child, negotiation skills with teenagers, what psychological problems teenagers have, and how to help them. The broader context is
on becoming a responsible citizen, with the curriculum covering
knowledge-acquisition, social skills, decision-making, and emotion-
regulation. The examination covers practice as well as theory, just like

a driving test. What to do—and what not to do. The large majority of children pass and get their license.

Some parents are opposed, but what wins the day is that kids actually enjoy the one-hour Friday afternoon class far more than Monday morning's matrix algebra. The teenagers love to talk about sex, intimate relationships, dealing with drugs, and peer-group pressures—all the stuff they are going through and will have to deal with in their own child. They enjoy the "good parent—bad kid" role-play pairings in which one of them acts as the good parent while the other one acts—well—at being basically themselves.

Some teenagers never knew that vigorously shaking a baby when it cries cuts the white fibers connecting the prefrontal cortex with the limbic system. They did not know that babies have to be fed in the middle of the night. They never knew the long-term financial cost of having to bring up a kid. They not only learn about how to be a better parent, but they also learn social skills that help them manage their current relationships with their parents, boyfriends, and girlfriends, as well as academic skills on human development, brain development, and behavioral control. Schoolkids like it, teachers like it, and parents actually learn a useful thing or two from their kids about parenting that they did not know. The kids themselves are actually becoming more manageable and understanding of their parents' position. It is an all-around winner.

Yet the licensing program still has significant opposition from human-rights advocates. Civil liberty advocates remonstrate that the government is taking away the right to have children and essentially criminalizing pregnancy. The government's retort is that any woman can become pregnant—she just has to pass the licensing exam before she gives birth.[39] To make it enforceable, there have to be sanctions for illegal parenting—just as there are sanctions for dangerous driving. If she is unlicensed, a mother caught with a baby has her child taken away into a foster home but is also offered a crash course on parenting and the opportunity to take the examination. If she passes, her baby will be returned—although there are inevitably yearly follow-ups on her parental skills, given her documented lack of responsibility and law-breaking behavior. DNA banks also allow the biological fathers to be tracked and sanctioned if they are not licensed.

Opponents argue vociferously that the program is inherently eugenic, as those with learning disabilities are less able to pass the examination. The government has countered by arguing that only a small minority will fail, and, as in a driving test, they will be given a second chance. They can learn the skills if they really want to. There is also a surprising number of more privileged kids who in pilot testing showed themselves to be pretty clueless at parenting—it isn't just the poor kids who have problems. In fact, quite a number of underprivileged kids have done very well on the exam—because they have already taken on the role of parenting their younger siblings. They know all the ropes of parenting already.

Despite strident debate, the majority of the public feels on balance that there is something inherently sensible in the government's plan. Most people recognize that parents are not perfect and laud efforts to reduce child abuse, improve parenting skills, and prevent future violence. The school authorities are surprisingly oppositional. It turns out that they want as much class time as possible for traditional academic subjects because school evaluations are based on that. The government puts paid to that objection by mandating school evaluation based partly on grades in parenting—and school authorities are then suddenly in strong support. In 2050 the Parental License Act is passed.

In the first few years, parenting skills go up and unwanted pregnancies go down. Juvenile delinquency declines too, as adolescents achieve a greater sense of responsibility, empathy, and agency alongside slightly improved relationships with their parents. There are long-term reductions in child abuse and later adult violence as teenagers grow up to be more responsible parents. The result is a new generation of children more cared for and loved by their parents. It is a winner with the public, and the government continues to win its war on violence—and its battle with the opinion polls.

Let's now step back from Big Brother and the impending glare—or glitter—of these hypothetical programs. Consider two quite different questions on the three future programs I have outlined. Could they happen? Should they happen? The practical, and the philosophical.

THE PRACTICAL—COULD THIS HAPPEN?

LOMBROSO could certainly come about in practice in twenty years, or something quite like it. Let's face it, elements are already in place right now. The prison at Guantánamo Bay is just one example of how indefinite detention is being used by countries throughout the world in the name of national security. Indefinite imprisonment for dangerous criminal offenders—or "preventive detention," as it is neatly packaged—is common in many countries.

You also know that all it takes is one tinderbox crime to set off a new law to protect society. That happened with Megan's Law, which required the public registration of sex offenders after the rape and murder of seven-year-old Megan Kanka in 1994 by a man with prior convictions for sexual assaults against young girls.[40] It also happened with Sarah's Law, in England, after the murder of eight-year-old Sarah Payne in 2000 by a sex offender named Roy Whiting. As we learned earlier, physical castration is offered right now in Germany and some other countries as a treatment option for sex offenders—we don't have to wait two decades for that to happen.

Society over the years is also becoming more controlling, with enhanced safety and security at all levels. I can check the Megan's Law Web site for where I live with my wife and two boys, and I can see pictures of all the convicted sex offenders living near me, together with their addresses and what their offenses were. There are sixty-nine in my zip code right now.

On the other side of the fence there are ever-stricter safety and security measures in place. My boy Andrew asked me to bring him a potato gun back from England, as I had told him I had one when I was a kid. But now I find out that they are not sold anymore for health and safety reasons. My sister Sally, in Darlington, tells me she needed an Enhanced Criminal Records Bureau check so she could monitor children's examinations at the school near her—a check on whether she has any registrations for offenses under the Protection of Children Act. You just never know what my sister could get up to with kids—although I've checked her convictions certificate and she seems clean. Kids cannot play conkers anymore at school for safety reasons.[41] Are we too concerned about children's safety? Are we wrapping them in plastic bubbles and not allowing them normal life experiences where they can grow? Or are we not being safe enough? In any event, society is certainly becoming more controlling over time—and that control can be subtly extended.

We also know all too well the political "something must be done" brigade. They never hesitate to introduce new laws supposedly to solve society's problems and win power. Just look at what happened in relatively liberal societies like the United Kingdom in recent years. Tony

Blair in 1997 won a landslide victory for the center-left Labour Party with his mantra to be "tough on crime and tough on the causes of crime." In 2003, Blair's party launched the Criminal Justice Act, which set in motion Imprisonment for Public Protection—the IPP program. Under the act, judges can sentence offenders to life in prison even though the crime they committed would not normally receive a life sentence. If they have previously committed one of a list of 153 offenses, and if they have currently committed a "serious" offense on that list, and if the judge feels they might commit another serious offense in the future, they will get life.[42] In fact, the judge is *legally compelled* to give a life sentence if an offender meets these criteria.[43] Judges are also required to say what sentence they would have given if they had not viewed the offender as potentially dangerous. In about a third of cases the sentence they would have received, the "tariff" sentence, is only *two years* in prison—and yet the offender will now get life unless a parole board decides to release him.

Crimes on the list are quite interesting. They range from "serious" offenses such as taking an indecent photograph of a child to attempting to procure a girl under the age of twenty-one.[44] It covers quite a lot of ground. Prisons swelled, and 5,828 had been given the IPP life sentence by 2010. Even though about 2,500 of them had served their tariff sentence, only ninety-four—or 4 percent—were released. Even then, of this tiny number of released offenders, a quarter were dragged back into prison after initial release.[45] They had served their time and yet are locked up for life.

Will we lock up offenders in the future longer than their "just deserts" if we feel there is a chance they might commit another violent or sexual offense? Of course we will—we do it now! Did the public kick up a fuss with IPP? No, they didn't! If you think the legislation that launches LOMBROSO in 2034 is a tad hasty and not all that well thought out, bear in mind that IPP has been lauded as "one of the least carefully planned and implemented pieces of legislation in the history of British sentencing."[46] More bungled legislation can follow even sooner than 2034.

My socialist country went one better than IPP. In 2000, magicians in the government conjured up from nowhere the label of "dangerous and severe personality disorder"—in the face of overwhelming opposition from psychiatrists.[47] Under this new legislation, the police have the power to whisk potentially dangerous people off the streets and into holding institutions for further assessment and treatment—even if they have committed no crime. More commonly prisoners who have served out their sentences can be detained further "for the public good." The practice is still ongoing, with the British government contemplating increasing and diversifying its operations.[48]

Forensic psychiatrists in both the United Kingdom and the United States, meanwhile, are remonstrating strongly against the increasing pressure to use forensic psychiatry to protect the public.[49] Yet the public doesn't seem to mind, and my family in England did not even know about the existence of these programs when I asked them. The essence of the LOMBROSO program has been essentially alive and well for years in countries like England, which has far less of a retributivist stance than the United States, China, or Singapore, all of which impose the death penalty. Yet, paradoxically, it was not tough enough for judicial officials, with the Lord Chief Justice complaining in 2004 that Tony Blair had not been tough enough on the causes of crime.[50] Blair slipped up—he really should have launched the LOMBROSO program if he had wanted to stay in power.

Using neuroscience to aid risk assessment has its advocates in the most foremost intellectual circles. The Royal Society in the United Kingdom commissioned leading academics to examine whether neuroscience technologies now or in the future could help law courts decide the fates of offenders. The ensuing report was appropriately cautious, yet at the same time suggested that neurobiological markers might indeed be shown to be useful, in conjunction with other risk factors, to identify risk for violence when making decisions about probation or parole.[51] It further suggested that neuroscience may be used more widely in the future to decide which potentially dangerous offenders should be detained to protect society. Let's reflect on this. If the scientific potential is being envisioned in 2011, it's not entirely unreasonable to imagine the field moving futher, albeit precariously, in that future direction.

What about the National Child Screening Program? Could that nefarious venture come about? Let's look back to Kip Kinkel. Just after his killings, in June 1998, President Clinton toured the school corridors and cafeteria at Thurston High, where Kip had gunned down his classmates. It was not too dissimilar to President Obama's visit to Newtown after the Sandy Hook Elementary School tragedy. He met with the surviving victims and gave them more than presidential comfort. Clinton instructed the attorney general to generate a new school guide entitled "Early Warning, Timely Response" that would help keep kids out of harm's way. Scientists and practitioners got in on the act too, with the American Psychiatric Association announcing "22 warning signs" of dangerous kids.[52] Do you see some of these same signs in your own child or younger sibling? Things like:

- angry outbursts
- depression
- social withdrawal and isolation
- peer rejection
- fascination with guns
- poor school performance
- lack of interest in school

Kip had them all—and a lot more besides, including cruelty to animals, attention deficit, and recorded juvenile delinquency.[53] There is almost always a reasoned sociopolitical response to national tragedies, and such tragedies will continue to cultivate new policies out of current-day events. The Minnesota Department of Health, in conjunction with the Minnesota Department of Education, has a brief and simple screening program to identify not just health problems in children, but also social and emotional problems like emotion regulation difficulties.[54] Instead of starting at ten years it starts very early, screening children aged zero to six years. It's an excellent program, there are many like it, and neither I nor anyone else is complaining. Yet can we not see this and other screening programs like it creeping further along as violence is already viewed as a public-health problem by the World Health Organization[55] and the Centers for Disease Control and Prevention?[56]

Is it too much of a stretch of the imagination to conceive that private investors would actually foot the bill for a LOMBROSO program, as I suggested? Not if they are already doing it. Tracy Palandjian is the charismatic chief executive officer of Social Finance, a nonprofit organization that is drawing in investment capital to finance social benefits like stopping crime. In 2010, Social Finance[57] launched the first Social Impact Bond, aimed at preventing male prisoners from re-offending upon their release in Peterborough, England. If it reduces re-offending by more than 7.5 percent, the financial savings get returned to investors. So far, savings range from 2.5 percent to 13 percent.[58] President

Obama in 2012 slated $100 million for Social Impact Bonds, and Boston is currently the first to show interest in helping juvenile offenders successfully transition into productive lives.[59] If the capital-cost side of crime-prevention programs is being handled by the private sector right now, why not in twenty years' time for the LOMBROSO prevention program?

As for parental licensing, this has been debated in both the popular press[60] and the academic press[61] for some years. Articles point out that poor parenting is a well-replicated risk factor for adult violence. Indeed, some governments have already acted to do something about it. In May 2012, Prime Minister David Cameron, leader of the Conservative-Liberal coalition party in the United Kingdom, committed over $5 million for a state Web site to advise parents on how to raise their children. Cameron argued:

> It's ludicrous that we should expect people to train for hours to drive a car or use a computer but, when it comes to looking after a baby, we tell people to just get on with it. . . . We've all been there in the middle of the night, your child won't stop crying, and you don't know what to do.[62]

How long will it be before the state tells us what to do by initiating compulsory parenting classes in school, arguing that some parents don't know that shaking their crying baby at night causes brain damage now and violence later, and that it is "ludicrous" to allow an unlicensed adult to be a responsible parent? We may not be there yet, but today's daydream can easily become tomorrow's nightmare.

Other forces that can lead to these future programs include our sense of retribution and the power of politics. As I argued when discussing the evolutionary basis to violence, the retributivist stance is ingrained in every one of us, part of our evolutionary heritage to deter the cheats. It's not going to fade away too easily, and it is particularly alive and well in the United States and other countries with the death penalty, as well as inside me.

We saw with Kip Kinkel how retribution trumped rehabilitation. Consider the four legal philosophies that justify punishment: deterrence, incapacitation, rehabilitation, and retribution. We can also add a fifth—reelection. In the future, when do-gooder efforts to stop the rot have failed, society certainly may wonder whether it's time to get to the heart of the matter, protect ourselves and our children, and halt the moral decay with a tough political party willing to get going when the going gets tough. It's a new landscape not far from the Queen of Hearts' "off with their heads" call for law and order in *Alice's Adventures in Wonderland*.

Politicians will continue to overreact to isolated tragic events in order to quell the public outcry and try to solve society's problems. With more water under the bridge, scientific advances in knowledge, and a much broader, multidisciplinary perspective to crime causation that incorporates neurocriminology, the ability to predict—and pre-emptively act—will, I believe, become more probable, not just possible. These things can happen. You can debate that particular conclusion later, but right now let's move to a more poignant point—do you *want* programs like LOMBROSO?

THE NEUROETHICS OF NEUROCRIMINOLOGY: SHOULD THIS HAPPEN?

That's a question for all of us to consider. It sends shivers down my spine to think I could be convicted without committing a crime. It would send shivers down your spine too if you had a brain scan like mine that looks like a serial killer's, together with low resting heart rate, birth complications, minor physical anomalies, early vitamin B deficiency, and a past that included bootlegging and gambling by the age of eleven. But let's hear all sides on the neuroethical issues surrounding neurocriminological research, and where we may or may not be taken to in the future. Neuroethics is a new subdiscipline of bioethics championed by my colleague Martha Farah at the University of Pennsylvania. It concerns ethical issues surrounding the brain and mind and the ways neuroscience affects society for better or for worse.[63] Let's take a look at the three futuristic programs highlighted above in the worldview of neuroethics and our broader attempt to understand humanity.

Of course there are civil-liberties issues in detaining people before they have committed a crime. But as I alluded to earlier, are there not civil-liberty issues involved in *not* doing anything when you know someone has a 79 percent chance of committing a serious violent act—and you can do something to stop that happening? Yes, some people will be detained who may not pose a risk—yet the harsh reality of daily life is that we have to balance risks with benefits.

Think back to the case of the young graduate Peyton Tuthill, whose life was snuffed out by the cancer of violence, in the person of Donta Page, just as a different kind of cancer snuffed out my sister Roma's life. Recall that Donta Page had been out of prison for just four months, and while he had been sentenced to twenty years for robbery, he was released after serving just four years. What if I had been asked to assess him before he was prematurely released? I would have said exactly what I said in court when defending him. All the biosocial boxes were checked. He was a walking time bomb waiting to explode. He was at heightened risk for committing violence for reasons beyond his control. It wasn't exactly destiny, but he was much more likely to be impulsively violent than not. Even if he had not previously been convicted of robbery, my conclusion would have been broadly the same.

That was back in the 1990s. In 2034 we will be even better placed not just to identify such individuals before they act, but to help them. In the same way that I wish Roma could have benefited from a cure for her deadly cancer that didn't exist at the time, I wish Donta Page could have been in a future LOMBROSO program and been treated at all levels, including with innovative drugs that would have blocked the bad chemistry that in part creates violence. If we could just have vaulted Donta ahead in time into the far more sophisticated risk-assessment mechanism of the LOMBROSO program of 2034, we would identify him as an LP-H or an LP-S, or both, by age eighteen.

Frankly, his fate under the LOMBROSO regime would be better than it is now. Should we not have some feeling for offenders as well as their victims? He would have had some chance of release and of living out his life in humane conditions. Right now he is resigned to living in a hellhole for the rest of his life. More important, young Peyton Tuthill would be alive and well today, enriching the lives of others. We can stand in the way of future progress because of our ethical fear of the frightful risks, yet let us not forget that in doing so there are benefits that will surely be lost, including lives that could have been saved. At what cost civil liberties?

On the early identification of potentially dangerous children there is no question that there are important neuroethical issues that have to be recognized. At the same time, both the public and scientists alike have an honest and growing interest in what to make of the anatomy of violence. The issues are perhaps best summed up by Philipp Sterzer, a neuroscientist and researcher of psychopathic behavior from the Department of Psychiatry in Berlin, who wrote an editorial on Yu Gao's identification of poor fear conditioning at age three as a putative bio-marker for crime at age twenty-three. His critical evaluation, entitled "Born to Be Criminal? What to Make of Early Biological Risk Factors for Criminal Behavior" ended with the following summary paragraph:

> If not handled with great caution, neurobiological markers can easily be misused to stigmatize individuals who are perceived as a potential threat to society. With the increasing availability of data that help us prevent, diagnose, and treat antisocial behav-ior early in life, we also need a public debate on how to use this information and, even more important, how to avoid its mis-use. Neurobiological research offers a great chance to further our understanding of antisocial and criminal behavior. This understanding should be used to benefit those children who are at greatest risk for a criminal career and to design interven-tions that are tailored to their needs.[64]

We certainly have more data available to us than in past genera-tions, and that will only increase. What to do with it does require due caution, protection against misuse, and minimization of risks. Yet the potential benefits exist, and there should be discussion about them. Jon-athan Kellerman had the courage in 1999 in the wake of the Jonesboro middle school shooting to voice his views on the issue. He argued that we already know the warning signs of troubled children, we should take them very seriously, and preventive custody with appropriate treat-ment is a solution worth implementing for a small minority.[65] Just as Philipp Sterzer argued, we should use new knowledge to benefit chil-dren who need help, and create new individualized interventions for them. Leading scientists have been arguing for some years that the U.S. government should save children from a life of crime by establishing a national program of risk-focused prevention—identifying those at risk and intervening early.[66] But should we? If we go out too far, do we run the risk of falling through thin ice? How do we know that the bad old days of eugenics are really over?

On parental licensing, is it really a moral right to have a child or not? Should it instead be considered a privilege that needs to be earned? Even today we take away parental rights. Parents who lack the capacity for care and nurturing, and instead hurt their child, lose their paternal rights—just as we saw in chapter 5 in the case of the Russian-roulette boy. Their child is taken away from them into care. It's not too far a leap to go one step further by conducting preventive intervention to preclude harm to the child occurring in the first place.

In a future where every individual is assessed not just on his or her parental capacity but on his or her risk quotient for child abuse, would such preventive intervention not be in the best interests of both parent and child? We adults have our human rights, but what about the rights of the child—whether they are born or not? Do today's children at least have a right to minimal standards of care and upbringing? Do you really wish to deny to Donta Page, Henry Lucas, Carlton Gary, and many other killers who suffered horribly at the hands of their abhorrent parents the right to an upbringing that is not a total affront to human dignity? Even if you could have just ensured that these killers had been treated slightly worse than your average pet dog when they were growing up, you would very likely have prevented many homicides. Is that too much to ask for the care of an innocent young baby—as these killers once were?

Is being a parent any less responsible an activity than being a doctor? Would you go to a doctor if she was not licensed to practice? Parenting a child is not so very different from a therapist caring for a client. If anything, parenting requires much more responsibility. Parents need to care for their children far more than do licensed therapists. We are very ready to protect our own turf—so why don't we protect the next generation of people like us? No one has the capacity to harm a child more than its parent, and, indeed, 80 percent of all child abuse is perpetrated by parents.[67] We protect ourselves from inept therapists by requiring a license for them to practice—why not protect future children from inept parents?

You may reasonably remonstrate against licensing. I did when I first encountered the idea. It just did not *feel* right to me for reasons I could not entirely put my finger on. My reaction was typically the System 1 thinking elucidated by the Nobel Prize winner Daniel Kahneman—emotional, fast, and intuitive.[68] It was a gut reaction. Licensing just smacked of the sneering superiority of the privileged classes. I thought, Surely we all have a right to reproduce?

Perhaps you think and feel the same way I did. If you had that same negative feeling that a parental license is just not right, let's try to get a better handle on our reaction. Is it because we feel that everyone knows how to be a parent? Animals get on with it pretty well, don't they? Surely we are better than animals? Yet consider adoption. Not everyone is automatically assumed to be a good enough parent to look after a child. Potential parents are scrutinized very carefully by the state on background and financial circumstances to ensure that the child will enter a loving and stable home. Because of that competency screening, the rate of child abuse in adoptive homes has been argued to be less than half of that for children reared by their natural biological parents.[69] We ensure standards for unwanted children—so why not apply such screening to us all to help *every* child in society and cut child abuse?

Legal and spiritual perspectives may help to partly explain our negative reaction to parental licensing. English common law has historically treated children as their father's property. That proprietary right of parents to possess their child as chattel stills holds a subconscious sway today in the most educated sectors of society. The God-given right to procreate certainly lies at the heart of many peoples' objections to parental licensing—but is it a good ethical reason to reject the idea?[70]

What is ultimately the most potent force fueling that undefined feeling against curtailing parental rights? I think the answer lies in evolutionary forces, a powerful drive built into us to reproduce at all costs. We've discussed how we are essentially gene machines whose primary mission is to reproduce and be represented in the next gene pool. Without that powerful drive, we would not be here right now to debate this ethical issue. So I sense evolution instills in us the *feeling* that licensing is wrong, motivating us to invoke counterarguments, whether they hold water or not. Arguments like the one saying that parental licensing is a subtle form of eugenics to create a master race, or that it may discriminate against some sectors of society, or that being a parent is a natural, God-given right that should not be taken away. Are these arguments merely a specious by-product of the genetic, instinctive need to reproduce? Are we capable of rising above our genetic heritage and our instinctive "feeling" that parenting is a right—while licensing is wrong? Or are we destined to remain the instinctive animals that we are?

Perhaps we have become numb to child abuse. Wretched parenting is rampantly common. I was visiting my family in Darlington and discussing parental licensing with my sister and brother-in-law over a cup of tea—a split vote there—and they showed me an article in their daily newspaper. An eleven-year-old boy from Blackpool was forced by his parents to live in a filthy, windowless outhouse, a coal bunker with a concrete floor. With no heat and very little light, he was locked up in this barren room every night with a potty-chair for a toilet. His parents bullied him and half-starved him.

Why did they force him to live for a year in the bunker? They told the police that it was punishment for taking some food from the refrigerator. The gross ill-treatment only came to light after his school became concerned that he was constantly hungry, leading social workers to visit the home and discover the conditions. Doctors found him to be stunted due to malnutrition. It was reported at court that he was traumatized and psychologically damaged by the experience, with the judge calling the bunker "akin to a prison cell from a third-world country."[71] His parents were jailed for two years. And yes, you've guessed it from our evolutionary perspective in chapter 1, the "father" was the stepfather.

Of course this story is barely considered newsworthy—it was buried on page 19 of the paper. After all, child abuse happens so often, it's just not news. So what's more important? The front page was devoted to David Cameron stuffing a meat pie into his mouth and society's sigh of relief that a new tax on the Cornish pastie was withdrawn. Like the parents of the Blackpool boy, we appear to care more about how much food we can shove into our own mouths than what a child can minimally eat to stay alive. We adults count, our children are chattel, and we do with them what we will behind closed doors and in underground bunkers. We care more about the cost of a pie, which is why if LOMBROSO saves society substantial money it really will take root.

One of the most difficult neuroethical challenges that neurocriminology will give rise to in the future is undoubtedly the sensitive balance between protecting society and protecting our civil rights. The futuristic National Child Screening Program that I described is based on a docudrama I took part in for the BBC in December 2004 in the United Kingdom, a documentary about what we know and don't know about the biology of violence, interlaced with a fictional drama about how an NCSP could go horribly wrong in the future. Immediately after the screening I took part in a studio debate with Jeremy Paxman, a forceful, assertive, witty, and very astute TV interviewer known for his incisive and unyielding questioning of politicians. It also featured Shami Chakrabarti, a very intelligent and likeable civil-liberties leader in England. We chatted together before we went on the air, and I was very impressed with her sincerity and thoughtfulness.[72] During our debate Paxman put a provocative question to Chakrabarti that highlights the

tension between violence prediction to protect society and the violation of human rights:

> PAXMAN: If science could predict with 100 percent certainty
> who was going to commit a violent crime, would it be
> legitimate to act before they commit that crime?
> CHAKRABARTI: I would have to say that in a liberal society of
> human beings, and not animals, my answer to your question
> would be "No."
> PAXMAN: So someone would have been potentially killed
> by this person despite the fact that that life could have been
> saved. Even if science can do it 100 percent, you still say it
> would be wrong?
> CHAKRABARTI: We also have to look at the kind of society
> that we live in, and even while the risk-free society, drama
> and illusion that it may be, is touted by popular politicians . . .
> there is a huge cost to our way of life and to the kind of liberal
> democracy that I say we want to live in.[73]

Shami had understandable difficulty with that particular question,
which challenged the civil-libertarian perspective. It seems that in the
name of a liberal democracy and human rights, we would wave good-
bye to a life we could have saved, even in the face of perfect prediction, as in the movie *Minority Report*. It is always a question of balance
in weighing protection and civil liberty, never a question of absolutes.
In striking out for liberal democracy we must also look down at the
blood we have on our hands—the blood of innocent lives that could
have been saved had we only chosen to act. Would you really agree
with Shami Chakrabarti?

Let me attempt to defend Chakrabarti's point of view. Once we
begin to slip on our democratic principles, we can end up on a scrap
heap of human-rights violations. Before we vote, politicians tempt us
with the illusion of the risk-free society that we say we want. But isn't
that just a charlatan's call echoing in an immoral wilderness? Is it not a
mirage, a future that we dearly want to see, but will never have unless
a huge price is paid, the price of gross injustice to the innocent who are
wrongly accused?

I think some of you may disagree with Chakrabarti's perspective.
You may conclude that in the face of perfect prediction, perilous though

the ethics may be, we must act. Yet if even one human right is violated, can we in good conscience live with that policy? That is Chakrabarti's provocative point. Is that moral sense the reason in a previous chapter you would not push the corpulent man off the footbridge to stop a runaway trolley from killing five railway workers? You object to the principle of utilitarian moral decision-making—the greater good of the greater number. Well, let's push the envelope on that issue one nudge further.

Consider Adolf Hitler.[74] Hitler, as we discussed, was by anyone's standards a flawed character—but he was also a human being and he had the right to live. Yet would you or Chakrabarti not have killed Hitler in 1933 to save the lives of 6 million Jews and 60 million German, British, Russian, American, and other international civilians and soldiers?

Imagine yourself standing beside Hitler on March 23, 1933, in the Kroll Opera House in Berlin. He is giving his speech just before the Enabling Act, the law that would make him a dictator with absolute power. He talks about the "decision to carry out the political and moral cleansing of our public life."[75] You have a gun in your pocket. You can predict the future and you know for certain you will save 66 million lives if you put the gun to the back of his head near his right ear—as Kip did with his father—and shoot him. No harm would come to you, and the world would be a better place. Would you kill Hitler?

Think it through. Sixty-six million lives and countless suffering to many millions more. Dreadful though the dilemma may be, I think that is the particular trigger I would be prepared to pull. Is doing that really living like an animal rather than a civilized human being? Is there not a huge cost to pay in *not* taking this particular life, even if it comes with the huge moral cost of murder?

And yet once we take that step, where will this journey lead us? Let me walk you through the valley of darkness and into the barren desert of just deserts. The question comes down to where exactly in the shifting sands of sensible reasoning you are willing to judiciously draw the line that delineates the protection of society on one side and the invasion of civil liberties on the other. The overall risks weighed against the overall benefits. The difference between right and wrong—between life and death. Between acceptance of the neurocriminological knowledge we are rapidly gaining—and the social concerns we all have over equity, ethics, and liberty.

Where exactly on that sliding scale of violence prediction will you be prepared to act? There will never be perfect prediction, not the 100 percent in Paxman's scenario. But what if it was 90 percent? Or 80 percent? Would you enact LOMBROSO, or something like it, at 79 percent? I know that we are all going to draw different lines. Can we agree on a consensus—the average of all the lines we have drawn?

You may be unwilling to draw any line. You may feel as ethically outraged as Shami Chakrabarti was about where neurobiological research on violence may be taking us. But if the idea of programs like LOMBROSO and NCSP give you pause, consider this: they at least give offenders a chance for deliverance. Criminals would not be stripped of the basic human rights that we deny them today. Under LOMBROSO they could vote, whereas those with criminal convictions cannot vote in the United States and many other countries. They would have conjugal visits. Most prisoners today do not.

Do you realize that we currently practice passive eugenics on our prisoners in forty-four out of the fifty United States? Male prisoners are not allowed to send their sperm out. Female prisoners are not allowed to send their eggs out or receive sperm. If you are serving life without the possibility of parole, your genes will not reproduce. You are a loser in the evolutionary game of reproduction. That line was drawn in the judicial sand long ago.

This glaring fact is extraordinarily hush-hush. Have you ever thought about it yourself? When I raised this issue with some of my criminology colleagues, their response was that it had not crossed their minds. When I spoke to over 200 correctional staff in Trenton, New Jersey, in 2009, they admitted they had never thought about it. When I have raised this issue on several occasions in academic talks and lectures, it is followed by universal silence.

There is irony here. Genetic researchers in the 1990s were accused of fostering a eugenic "final solution" to stopping crime. That accusation was demonstrably false. But let's be sure about one thing: our current policy of what I call "passive eugenics" on criminals did not emerge from genetic or biological research. It was a direct product of social policy. Although some well-intentioned people believe that genetic research on crime should be stopped because it could lead to eugenics, there has been no similar call to halt social-science or public-policy research on crime. And yet through such policy we are effectively reducing the genetic fitness of the most serious offenders and limiting their genetic material in future gene pools.

Social scientists may have decried Lombroso's nineteenth-century thinking in branding criminals as evolutionary throwbacks, but in many ways our current thinking and our passive-eugenics policy are still stuck in the nineteenth century. Prisoners are today viewed as little more than Lombrosian subhuman savages who are not fit to reproduce. We practice passive eugenics, don't we? They shoot horses, don't they?

Consider the counterpoint. Losing the right to have children is just part and parcel of committing crime. Prisoners lose their freedom. They lose their right to vote. So why not the right to give life, especially for those who have taken life already? Retribution and deterrence are the rules of the legal game we play with prisoners, and disenfranchisement and passive eugenics are most regrettably the costs that those dealt losing hands in life simply have to pay. And yet . . . I was always brought up to believe that eugenics was a bad thing.

CONCLUDING COMMENTS—ENDING UP AN OSTRICH

Kip Kinkel can't have kids. Not with 111 years in prison without the possibility of parole. It's ironic that the logic of neurocriminology asks us to cut offenders like Kip some slack, to assist in their defense, not to punish them so harshly because reasons beyond their control constrained their free will. It's ironic because biological researchers on crime have in the past been accused of having the worst intentions for criminal offenders. Have we gone wrong somewhere, and do we need to change our perspective? If we compare some salutary events from the recent past with where we stand today, I think a shift has already occurred in our thinking. We are on the cusp of crossing into new territory.

Wouter Buikhuisen was a criminologist at Leiden University in the Netherlands in the 1970s and '80s who believed that there was a psychophysiological basis to crime. That perspective resulted in his being hounded like a wild animal and torn to pieces in the Dutch popular press.[76] His position was debated in parliament, and he ultimately had to resign his position as chair of the criminology department at Leiden in 1988. It was intolerable at that time to think of crime and criminality being anything other than a social construction caused exclusively by social forces. As a young scholar, I visited Wouter at Leiden in 1987. We had met the previous year in Italy—he wanted to bring me on board in a faculty position at Leiden. Instead I went to Los Angeles, where I hoped the academic atmosphere would be more liberal. But was it?

In 1994 I presented my research findings from Denmark at the annual meeting in San Francisco of the American Association for the Advancement of Science. I showed that a combination of birth complications interacted with early maternal rejection in predisposing babies to be violent offenders eighteen years later.[77] An article in *Science* in March that year published a figure illustrating my main findings under the headline WAR OF WORDS CONTINUES IN VIOLENCE RESEARCH.[78] It reported my own hope that this new biosocial research could lead to "feasible, practical, and benign ways" of preventing violence. Nevertheless, as *Science* reported, it was subjected to "a unified and outspoken assault" by other scientists at the meeting, who characterized my findings as "racist and ideologically motivated."[79] My sample was all white, so targeting minorities was not the issue. Instead, the findings suggested that biology worked in concert with social influences—and that was intolerable. Twelve years earlier, in 1982, I had to take a chapter on biosocial influences out of my thesis at the insistence of the external examiner in order to obtain my PhD—even though I had published that work two years earlier in a scientific peer-reviewed journal.[80]

Twenty years has seen an enormous change in the political landscape of an anatomy of violence. Back in 1994, suggesting an interaction between biological and social factors in predisposing individuals to violence was anathema. Today it is totally passé. Of course such biosocial interactions occur, what's all the fuss about? In the Netherlands, Wouter Buikhuisen has now been exonerated and given an apology for his persecution,[81] and in my experience the Netherlands today has more interest in neurocriminology than any country outside of North America.

Yet the very beginning sentence of that article in *Science* on violence still rings like a gunshot in my ears:

> There are few certainties in life, but here's one: The uproar surrounding attempts to find biological causes for social problems will continue.[82]

Will the emerging science of neurocriminology and the double-edged sword that it wields continue to remain bogged down in a minefield of unproductive diatribes? One of the continuing problems is that this research field borders on the politically incorrect. The left doesn't like it, and the right doesn't like it either. Liberals and center-left parties fear that the research will be used to stigmatize individuals and take attention away from social problems, the true causes of crime. Conservatives and the center-right are concerned that it will be used to let offenders off the hook and take away responsibility and retribution. There is no question that neurocriminology is a difficult terrain to tread, and some would wish it did not exist at all. Are we certain that the uproar will continue—or is the tide turning?

Critics will further contend that neurocriminology raises the ominous specter of violence being reducible to a physical neural cause, the erosion of the concepts of individual accountability and free will, an abandonment of social injustice as an explanation of crime, and the consequent derailment of social intervention programs for underserved populations. Attacking the law's freedom-of-will assumption with a deterministic-sounding neurobiological excuse could lead to a "throw away the key" solution because we feel biology cannot be changed. Would it be the start of a slippery slope toward the dissolution of responsibility, an increase in unbridled violent offending, and the implosion of civilization, as Shami Chakrabarti feared?

That ever-feared slippery slope. It's a common refrain surrounding the moral implications of my work. *If we take these steps, what quagmire do they slide us into?* Far too often the slippery slope argument is presented at the end of a discussion. *Well, there's a slippery slope, so let's play it safe and tread no further.* That's a cop-out, and when it comes to the active suppression of new knowledge or the ignorance of silence, it generally stems from the desire of certain groups to maintain the status quo. It turns out that most slopes aren't so slippery after all if we care to confront our fears and cautiously weigh the risks and benefits of action. There is firm ground underfoot and ample opportunities up and down that slope to choose where we stand—*if* we have the courage to do so.

Neurocriminology is now providing the foundations upon which to not just dissect the future Hannibal Lecters and Donta Pages, but to potentially prevent their very occurrence in the first place—if we act early. In the wake of the Newtown shootings, many officials and citizens were quick to point not only to guns as the culprit but to our general lack of mental health services. Can we do more for those all too often underserved children like Donta Page and prevent future disasters? After all, what's so heinous about investing resources in better pre- and postnatal nutrition and care for the underserved, better elementary school nutrition, reducing lead exposure, implementing education on parenting skills, and identifying children with serious behavior problems for benign interventions? Investing resources—costly though it may be—in the next generation of adolescents at risk for violence is not just a place on that slope where I am prepared to stand, but it's where I hope you'll stand with me.

Most positive societal advances somehow involve a so-called slippery slope. Can we not find ways to collectively and humanely move forward to reduce violence? Neurocriminology and a more profound understanding of the early biological causes of violence can help us take a more empathic, understanding, and merciful approach not just to the victims of violence but also to the prisoners themselves. In that process, would not the standing of all of us in an allegedly civilized society be raised?

As I sit writing here in a room in Churchill College, Cambridge, reflecting on our outlook on prisoners, my mind inevitably turns to Winston Churchill, who himself had been a prisoner during the Boer War. More than a hundred years ago, Churchill, as home secretary, stood up in the House of Commons and gave his perspective on how we should treat criminals:

> The mood and temper of the public in regard to the treatment of crime and criminals is one of the most unfailing tests of the civilization of any country. A calm and dispassionate recognition of the rights of the accused against the state, and even those of convicted criminals against the state, a constant heartsearching by all charged with the duty of punishment, a desire and eagerness to rehabilitate in the world of industry all those who have paid their dues in the hard coinage of punishment, tireless efforts towards the discovery of curative and regenerating processes, and an unfaltering faith that there is a treasure, if you can only find it, in the heart of every man—these are the symbols which in the treatment of crime and criminals mark and measure the stored-up strength of a nation and are the sign and proof of the living virtue in it.[83]

That was more than a century ago, and yet today how calm and dispassionate have the most civilized countries in the world become on this issue? Are we tirelessly pursuing curative and regenerative processes for the cancer of crime? Do we genuinely desire rehabilitation? Or does our mood and temper move us in anger to the costly coinage of retribution that we saw served out to Kip Kinkel, and societal protection above all cost? How would Churchill view us today if he could see where we currently stand in our treatment of prisoners?

We look back 200 years and are aghast at an age when mentally ill patients were kept locked in fetters and chains, and treated little better than animals because of their unacceptable behavior. In a society that was in its time at the pinnacle of world knowledge, such treatment of patients seemed totally appropriate. It was a radical and revolutionary approach for the physician Philippe Pinel to free mentally ill patients from their shackles in Paris in 1793 and place them under more humane conditions. Today the inhumane treatment of the mentally ill seems unconscionable to us. The critical question for us to consider is whether less than a hundred years from now, a much more advanced society than the one we live in will look back aghast at our current conceptualization of violence and our concomitant incarceration and execution of prisoners with the same incredulity with which today we look back at the earlier treatment of mental patients. They may well wonder how society could have countenanced such practices and overlooked the glittering gems—small though they may be—in each and every offender who had the potential to contribute positively to society.

In a wider context, others concur with Churchill's early vision of the potential for living virtue in our society. As Steven Pinker eloquently outlined in his book *The Better Angels of Our Nature,* our society is moving us to be more empathic, better able to control our impulses, and to reason rather than react. The result, he argues, is that over the course of history, despite periodic swings, violence has slowly declined.[84] The history of the world has also shown that as society becomes more ennobled and sophisticated, physical and mental disabilities such as epilepsy, psychosis, mental deficiency, and alcoholism cease to be viewed in a moral or theological context and become perceived more in the humanitarian context of treatment.[85] Just as mental disorders were once viewed as a product of evil forces, will the evil behavior of violent offenders eventually be reformulated as treatable clinical disorders? Society may deny this perspective in the short term, but I believe that a future generation with a calmer and more dispassionate perspective will indeed take this conceptual leap.

Extreme views certainly require due caution, but we must not forget that extreme views can be appropriate, and that moderate views can be erroneous. During the witchcraft hunts of the Reformation era in Europe, a moderate view would have been to wake up one morning and decide not to burn too many witches that day. An extreme view would have been to wake up and decide not to burn *any* witches. The notion of recidivistic violence as a clinical disorder may currently seem ludicrous to you. We must, however, face the possibility that if we close the door to even considering this perspective, we open the gates to tragedy—that breakthrough advances in remediation and treatment of crime will be foreclosed or hopelessly stalled, and future lives will be lost. Some think the issues are too hot to handle, or, as one leading criminologist in good faith once confided to me, "No good can ever come of genetic research on violence." We must indeed be ever mindful of how neurobiological research findings are interpreted, as such research can be misused. Yet if we don't allow ourselves the opportunity to consider new approaches for a better society, are we not all diminished by our blindness?

We live today in the most scientific and intellectually advanced society in the history of the world. We aspire for heavenly knowledge and have formulated firm convictions that we hold to be true. History has shown, though, that societies at different stages in history with a similar thirst for science have made grievous misjudgments under the banner of absolute knowledge. We have to cure ourselves of that irritating itch for absolute knowledge and certainty. I must consider the possibility that I err in creating a bridge between crime and cancer. Violence may not be a clinical disorder. I do not have the answers on some issues; I am not even sure where I stand on others. Some of my scientific views are tinged with personal perspectives, and like all scientists I stand on the edge of error in my empirical research. In the same spirit of humility, I hope that in your own mind and heart you can at least consider this new zeitgeist.

What is the main message I want to leave you with? I want to suggest that society's willingness to firmly grasp the neuroethical nettles that entangle neurocriminology, and to sensibly and cautiously integrate innovative clinical neuroscience findings with public policy, will be a critical ingredient for our future success in violence prevention. Building further on a public health approach to violence truly has the capacity to create a healthier future. We can seize the day, change tomorrow, and create a safer world for the next generation. An open and honest dialogue on the issues raised here will prepare the public for future developments—whatever they may be—and help facilitate future success in violence prevention.

When we finally get to 2034, will it be utopia or dystopia? You may think that the future landscape I have painted has an Orwellian echo—but it need not have a bleak Orwellian ending. You may recall the chant from Orwell's *Animal Farm* of "Four legs good, two legs better," as the privileged pigs tottered around their underling animals on two trotters. Their propaganda had closed down the minds of their comrades and created a class-based, inequitable society. Winston Smith in the end of 1984 was reduced to doublethink—believing in two contradictory views. Perhaps the government's LOMBROSO program, which would tell us we can protect society and rehabilitate offenders at the same time is a similar contradictory double message. Yet if we retain an open dialogue on these issues we can prise apart doublethink and both keep our cake and eat it too.

I do believe that in tomorrow's world we can rise above our feelings of retribution, reach out for rehabilitation, and engage in a more humane discourse on the causes of violence. After all, while we may disagree on the finer points, I believe we can all agree on our priority of *preventing* future violence. We can have a braver new world where sunshine replaces shadows. You can either stay where you are in the dark with our retributivist perspective, as I myself have been, or you can move ahead into a new day. We do have a choice—and you can choose.

We cannot continue to maintain an uncompromising mind-set, where one perspective—social or biological—dominates the other in a stranglehold over who calls the shots in curbing violence. Amy Gutmann and Dennis Thompson in *The Spirit of Compromise* argue that in the polarized political arena, all sides need to give up ground in a mutual sacrifice for sound governance, adjusting long-cherished principles for the greater good.[86] Achieving this in academic criminology is an enormous challenge, requiring traditional social scientists to reverse-thrust on their long-held beliefs and embrace the anatomy of violence—a new body of knowledge that can be suffocating to some in its sophistication. Yet standing steadfast on social principles can equally stifle progress. It is up to you the reader today to help us scientists surface for air, and with your civic perspective move us forward in reevaluating where we should stand tomorrow on the seething hotbeds of violence prevention.

In the final analysis, you may decide to stand your ground and turn a blind eye to the science this book has summarized and the societal issues I have raised. You may want to believe that a biological basis to violence does not exist, or it's going to be explained away in some manner. Like an ostrich evading the hunter, you may decide to bury your head in the sand. But if we do not make a move and act on the anatomy of violence, I believe this cancer will continue. And you had better watch out—the ostrich may get shot.

My sincere hope is that you will not turn a blind eye to the science— I want those ostriches to be alive and well. Nevertheless, you may be completely convinced that the fundamental message of the anatomy of violence is profoundly misguided. But if you happen to be a Christian, consider the words of Oliver Cromwell when he spoke to the Church of Scotland against its intended alliance with King Charles II:

> I beseech you, in the bowels of Christ, think it possible that you might be mistaken.[87]

And if you are not a Christian, I beseech you in your own bowels—or any other part of your anatomy that you choose—to consider that we all have the capacity to be wrong. In dissecting the anatomy of violence, I have that capacity—don't you too? More important than persuasion and conviction is open discussion, laying forth scientific reality, and allowing society to judiciously choose how to act in the ensuing light. My sincere hope is that our discussion will continue in the forthcoming decades and move us all into a safer and more humane society.

Notes

PREFACE

1. Wolfgang, M. E. (1973). Cesare Lombroso. In H. Mannheim (ed.), *Pioneers in Criminology*, pp. 232–91. Montclair, N.J.: Patterson Smith.
2. Sellin, T. (1937). The Lombrosian myth in criminology. *American Journal of Sociology* 42, 898–99.
3. Kellerman, J. (1999). *Savage Spawn: Reflections on Violent Children*. New York: Random House.

INTRODUCTION

1. I was able to buy a replica of this knife at the Bodrum marketplace and realized it was a cheap knife, probably used more for threat and defense rather than as a serious weapon. I had it on my office desk at the University of Southern California as a memento until it was stolen by an office cleaner.
2. Wilson, J. Q. & Herrnstein, R. (1985). *Crime and Human Nature*. New York: Simon & Schuster.

I. BASIC INSTINCTS

1. Horn, D. G. (2003). *The Criminal Body: Lombroso and the Anatomy of Deviance*. New York: Routledge.
2. Gibson, M. (2002). *Born to Crime: Cesare Lombroso and the Origins of Biological Criminology*, p. 20. Westport, Conn.: Praeger.
3. Wolfgang, M. E. (1973). Cesare Lombroso. In H. Mannheim (ed.), *Pioneers in Criminology*, pp. 232–91. Montclair, N.J.: Patterson Smith.
4. Shakespeare, W. (1914). *The Tempest*, Act IV, Scene 1. London: Oxford University Press.
5. Gibson, *Born to Crime*.
6. Dawkins, R. (1976). *The Selfish Gene*. New York: Oxford University Press.
7. Trivers, R. L. (1971). The evolution of reciprocal altruism. *Quarterly Review of Biology* 46, 35–57.
8. Cleckley, H. C. (1976). *The Mask of Sanity*. St. Louis: Mosby.

9. Hare, R. D. (2003). *The Hare Psychopathy Checklist—Revised (PCL-R)*, 2nd ed. Toronto, Canada: Multi-Health Systems.

10. Harpending, H. & Draper, P. (1988). Antisocial behavior and the other side of cultural evolution. In T. E. Moffitt and S. A. Mednick (eds.), *Biological Contributions to Crime Causation*, pp. 293–307. Dordrecht: Martinus Nijhoff.

11. Lee, R. B. & DeVore, B. I. (1976). *Kalahari Hunter-Gatherers*. Cambridge, Mass.: Harvard University Press.

12. Murphy, Y. & Murphy, R. (1974). *Women of the Forest*. New York: Columbia University Press.

13. Harpending & Draper, Antisocial behavior and the other side of cultural evolution.

14. We should not take the parallel between the Mundurucú and psychopaths too far. The lifestyle of the male Mundurucú does not exactly parallel the Western male psychopath. Western psychopaths do not in general form long-term relationships with either sex, and do not coexist and engage in joint enterprises. In contrast, the male Mundurucú do engage in long-term interpersonal relationships with members of their own sex and engage in all-male cooperative efforts for the benefit of the whole settlement.

15. Hare, R. D. (1980). A research scale for the assessment of psychopathy in criminal populations. *Personality and Individual Differences* 1, 111–19.

16. Chagnon, N. A. (1988). Life histories, blood revenge, and warfare in a tribal population. *Science* 239, 985–92.

17. Hare, R. D. (1993). *Without Conscience: The Disturbing World of Psychopaths Amongst Us*. New York: Guilford Press.

18. Woodworth, M. & Porter, S. (2002). In cold blood: Characteristics of criminal homicides as a function of psychopathy. *Journal of Abnormal Psychology* 111, 436–45.

19. Centers for Disease Control and Prevention National Center for Injury Prevention and Control (2002). *WISQARS Leading Causes of Death Reports, 1999–2007*, http://webapp.cdc.gov/sasweb/ncipc/leadcaus10.html.

20. Overpeck, M. D., Brenner, R. A., Trumble, A. C., Trifiletti, L. B. & Berendes, H. W. (1998). Risk factors for infant homicide in the United States. *New England Journal of Medicine* 339, 1211–16. While the first year of life is the time when you are most likely to be killed, for some ethnic groups this is rivaled by the risk of being a victim of homicide during adolescence and early adulthood.

21. Ibid.

22. Ibid.

23. Daly, M. & Wilson, M. (1988). Evolutionary social psychology and family homicide. *Science* 242, 519–24.

24. Wadsworth, J., Burnell, I., Taylor, B. & Butler, N. (1983). Family type and accidents in preschool-children. *Journal of Epidemiology and Community Health* 37, 100–104.

25. Daly, M. & Wilson, M. (1988). *Homicide.* Hawthorne, N.Y.: Aldine de Gruyter.

26. Lightcap, J. L., Kurland, J. A. & Burgess, R. L. (1982). Child-abuse—A test of some predictions from evolutionary-theory. *Ethology and Sociobiology* 3, 61–67.

27. Daly & Wilson, Evolutionary social psychology and family homicide.

28. Ibid.

29. Ibid.

30. Gottschall, J. A. & Gottschall, T. A. (2003). Are per-incident rape-pregnancy rates higher than per-incident consensual pregnancy rates? *Human Nature* 14, 1–20.

31. Thornhill, R. & Palmer, C. (2000). *A Natural History of Rape.* Cambridge, Mass.: MIT Press.

32. Singh, D., Dixson, B. J., Jessop, T. S., Morgan, B. & Dixson, A. F. (2010). Cross-cultural consensus for waist-hip ratio and women's attractiveness. *Evolution and Human Behavior* 31, 176–81.

33. Ward, T., Gannon, T. A. & Keown, K. (2006). Beliefs, values, and action: The judgment model of cognitive distortions in sexual offenders. *Aggression and Violent Behavior* 11, 323–40.

34. Levin, R. J. & van Berlo, W. (2004). Sexual arousal and orgasm in subjects who experience forced or non-consensual sexual stimulation: A review. *Journal of Clinical Forensic Medicine* 11, 82–88.

35. For counterarguments to the notion that orgasm can facilitate fertility and has an evolutionary basis, see Lloyd, A. E. (2005). *The Case of the Female Orgasm: Bias in the Science of Evolution.* Cambridge: Harvard University Press.

36. Polaschek, D.L.L., Ward, T. & Hudson, S. M. (1997). Rape and rapists: Theory and treatment. *Clinical Psychology Review* 17, 117–44.

37. McKibbin, W. F., Shackelford, T. K., Goetz, A. T. & Starratt, V. G. (2008). Why do men rape? An evolutionary psychological perspective. *Review of General Psychology* 12, 86–97.

38. Thornhill, N. W. & Thornhill, R. (1990). An evolutionary analysis of psychological pain following rape, vol. 1, The effects of victim's age and marital status. *Ethology and Sociobiology* 11, 155–76.

39. Russell, D.E.H. (1990). *Rape in Marriage.* Indianapolis: Indiana University Press.

40. Buss, D. M. (2000). *The Dangerous Passion: Why Jealousy Is as Necessary as Love and Sex.* New York: Free Press.

41. Daly & Wilson. Evolutionary social psychology and family homicide.

42. Buss, D. M., Shackelford, T. K., Kirkpatrick, L. A., Choe, J. C., Lim, H. K., et al. (1999). Jealousy and the nature of beliefs about infidelity: Tests of competing hypotheses about sex differences in the United States, Korea, and Japan. *Personal Relationships* 6, 125–50.

43. Andrews, P. W., Gangestad, S. W., Miller, G. F., Haselton, M. G., Thornhill, R., et al. (2008). Sex differences in detecting sexual infidelity: Results of a maximum likelihood method for analyzing the sensitivity of sex differences to underreporting. *Human Nature: An Interdisciplinary Biosocial Perspective* 19, 347–73.

44. Goetz, A. T. & Causey, K. (2009). Sex differences in perceptions of infidelity: Men often assume the worst. *Evolutionary Psychology* 7, 253–63.

45. Gage, A. J. & Hutchinson, P. L. (2006). Power, control, and intimate partner sexual violence in Haiti. *Archives of Sexual Behavior* 35, 11–24.

46. Lalumiere, M. L., Harris, G. T., Quinsey, V. L. & Rice, M. E. (2005). *The Causes of Rape: Understanding Individual Differences in Male Propensity for Sexual Aggression.* Washington, D.C.: APA Press.

47. Baker, R. (1996). *Sperm Wars.* New York: Basic Books.

48. Buss, D. M. (2009). The multiple adaptive problems solved by human aggression. *Behavioral and Brain Sciences* 32, 271–72.

49. Daly, M. & Wilson, M. (1990). Killing the competition: Female/female and male/male homicide. *Human Nature* 1, 81–107.

50. Wilson, M. & Daly, M. (1985). Competitiveness, risk-taking, and violence: The young male syndrome. *Ethology and Sociobiology* 6, 59–73.

51. Buss, D. M. & Shackelford, T. K. (1997). Human aggression in evolutionary psychological perspective. *Clinical Psychology Review* 17, 605–19.

52. Tremblay, R. E., Japel, C., Perusse, D., McDuff, P., Bolvin, M., et al. (1999). The search for the age of onset of physical aggression: Rousseau and Bandura revisited. *Criminal Behavior and Mental Health* 9, 8–23.

53. Archer, J. (2009). Does sexual selection explain human sex differences in aggression? *Behavioral and Brain Sciences* 32, 249–311.

54. Ibid.

55. Bettencourt, B. A. & Miller, N. (1996). Gender differences in aggression as a function of provocation: A meta-analysis. *Psychological Bulletin* 119, 422–47.

56. Campbell, A. (1995). A few good men: Evolutionary psychology and female adolescent aggression. *Ethology and Sociobiology* 16, 99–123.

57. Zuckerman, M. (1994). *Behavioural Expressions and Biosocial Bases of Sensation Seeking.* New York: Cambridge University Press.

58. Campbell, A few good men.

59. Ibid.

60. Archer, Does sexual selection explain human sex differences in aggression?

61. Buss, D. N. & Dedden, L. A. (1990). Derogation of competitors. *Journal of Personality and Social Relationships* 7, 395–422.

62. Ibid.

2. SEEDS OF SIN

1. *60 Minutes: Murder Gene: Man on Death Row Bases Appeal on the Belief That His Criminal Tendencies Are Inherited* (2001). CBS television, February 27.

2. It is thought that this "malfunction" or spontaneous event of identical twinning occurs when a blastocyst collapses and splits the progenitor cells in two, with the same genetic material in both sides of the embryo, resulting in the development of two identical embryos.

3. Baker, L. A., Barton, M. & Raine, A. (2002). The Southern California Twin Register at the University of Southern California. *Twin Research* 5, 456–59.

4. Baker, L. A., Jacobsen, K., Raine, A., Lozano, D. I. & Bezdjian, S. (2007). Genetic and environmental bases of childhood antisocial behavior: A multi-informant twin study. *Journal of Abnormal Psychology* 116, 219–35.

5. Ibid.

6. The heritability of 98 percent that we obtain from our twin study is very high, and might be applying to children who are seen to be antisocial by all informants of their behavior. In contrast, other children may be antisocial, but their parents and teachers are not aware of their antisocial behavior.

7. Baker, L., Raine, A., Liu, J. & Jacobsen, K. C. (2008). Genetic and environmental influences on reactive and proactive aggression in children. *Journal of Abnormal Child Psychology* 36, 1265–78.

8. Burt, S. A. (2009). Are there meaningful etiological differences within antisocial behavior? Results of a meta-analysis. *Clinical Psychology Review* 29, 163–78.

9. Arseneault, L., Moffitt, T. E., Caspi, A., Taylor, A., Rijsdijk, F. V., et al. (2003). Strong genetic effects on cross-situational antisocial behaviour among 5-year-old children according to mothers, teachers, examiner-observers, and twins' self-reports. *Journal of Child Psychology and Psychiatry and Allied Disciplines* 44, 832–48.

10. Viding E., Jones, A. P., Frick, P. J., Moffitt, T. E. & Plomin, R. (2008). Heritability of antisocial behaviour at 9: Do callous-unemotional traits matter? *Developmental Science* 11, 17–22.

11. Grove, W. M., Eckert, E. D., Heston, L., Bouchard, T. J., Segal, N., et al. (1990). Heritability of substance abuse and antisocial behavior: A study of monozygotic twins reared apart. *Biological Psychiatry* 27, 1293–1304.

12. Christiansen, K. O. (1977). A review of criminality among twins. In S. A. Mednick and K. O. Christiansen (eds.), *Biosocial Bases of Criminal Behavior*, pp. 45–88. New York: Gardner Press.

13. Schwesinger, G. (1952). The effect of differential parent-child relations on identical twin resemblance in personality. *Acta Geneticae Medicae et Germellologiae.* Cited in ibid.

14. Grove, et al. Heritability of substance abuse and antisocial behavior.

15. Baker, et al. Genetic and environmental bases of childhood antisocial behavior.

16. Moffitt T. E. (2005). The new look of behavioral genetics in developmental psychopathology: Gene-environment interplay in antisocial behaviors. *Psychological Bulletin* 131, 533–54.

17. Bouchard, T. J. & McGue, M. (2003). Genetic and environmental influences on human psychological differences. *Journal of Neurobiology* 54, 4–45.

18. Mednick, S. A., Gabrielli, W. H. & Hutchings, B. (1984). Genetic influences in criminal convictions: Evidence from an adoption cohort. *Science* 224, 891–94.

19. Raine, A. (1993). *The Psychopathology of Crime: Criminal Behavior as a Clinical Disorder.* San Diego: Academic Press.

20. Moffitt, T. E., Ross, S. & Raine, A. (2011). Crime and biology. In J. Q. Wilson and J. Petersilia (eds.), *Crime and Public Policy,* 2nd ed. Oxford: Oxford University Press.

21. Ibid.

22. Ibid.

23. Ibid.

24. In contrast to twin studies, several adoption studies have not shown heritability for violence. One explanation is that adoption studies rely on convictions for violence as their measure, yet conviction data is a notoriously poor measure, as most people who are violent are never even arrested, let alone convicted. In contrast, twin studies have relied more on laboratory, parent, teacher, child, and adult ratings of aggressive and violent behavior, which assess *degree* of aggression and hence have a much broader, more reliable, and more systematic radar screen compared with conviction data, which offers a much simpler yes/no dichotomy.

25. Jacobs, P. A., Brunton, M., Melville, M. M., Brittain, R. P. & McClemont, W. F. (1965). Aggressive behavior, mental sub-normality, and the XYY male. *Nature* 208, 1351–52.

26. Voorhees, J. J., Wilkins, J., Hayes, E. & Harrell, E. R. (1970). Nodulocystic acne as a phenotypic feature of the XYY genotype. *Archives of Dermatology* 105, 913–19.

27. Lyons, R. D. (1968). Ultimate Speck appeal may cite a genetic defect. *New York Times,* April 22, p. 43. http://select.nytimes.com/gst/abstract.html?res=F20C10FA355D147493C0AB178FD85F4C8685F9.

28. Telfer, M. A, Baker, D., Clark, G. R. & Richardson, C. E. (1968). Incidence

of gross chromosomal errors among tall criminal American males. *Science* 159, 1249–50.

29. Davis, R. J., McGee, B. J., Empson, J. & Engel, E. (1970). XYY and crime. *Lancet* 296, 1086.

30. Witkin, H. A., Mednick, S. A., Schulsinger, F. et al. (1976). Criminality in XYY and XXY men. *Science* 193, 547–55.

31. Ibid.

32. Ross, J. L., Roeltgen, D. P., Kushner, H., Zinn, A. R., Reiss, A., et al. (2012). Behavioral and social phenotypes in boys with 47, XYY syndrome or 47, XXY Klinefelter syndrome. *Pediatrics* 129, 769–78.

33. Brunner, H. (2011). Do the genes tell it all? Invited address, Congress on *Crime and Punishment: A Case of Biology*, Organization for Biology, Bio-Medical Sciences and Psychobiology, University of Amsterdam, Netherlands, January 19.

34. Brunner, H. G. (2011). Personal communication, Amsterdam, January 19.

35. Brunner, H. G., Nelen, M., Breakfield, X. O., Ropers, H. H. & van Oost, B. A. (1993). Abnormal behavior associated with a point mutation in the structural gene for monoamine oxidase A. *Science* 262, 578–80.

36. Ibid.

37. Farrington, D. P. (2000). Psychosocial predictors of adult antisocial personality and adult convictions. *Behavioral Sciences & the Law* 18, 605–22.

38. Brunner, H. (1996). MAOA deficiency and abnormal behaviour: Perspectives on an association. *Ciba Foundation Symposium* 194, 155–64.

39. Cases, O., Seif, I., Grimsby, J., Gaspar, P., Chen, K., et al. (1995). Aggressive behavior and altered amounts of brain serotonin and norepinephrine in mice lacking MAOA. *Science* 268, 1763–66.

40. Caspi, A., McClay, J., Moffitt, T., Mill, J., Martin, J., et al. (2002). Role of genotype in the cycle of violence in maltreated children. *Science* 297, 851–54.

41. Kim-Cohen, J., Caspi, A., Taylor, A., Williams, B., Newcombe, R., et al. (2006). MAOA, maltreatment, and gene-environment interaction predicting children's mental health: New evidence and a meta-analysis. *Molecular Psychiatry* 11, 903–13.

42. Beach, S.R.H., Brody, G. H., Gunter, T. D., Packer, H., Wernett, P., et al. (2010). Child maltreatment moderates the association of MAOA with symptoms of depression and antisocial personality disorder. *Journal of Family Psychology* 24, 12–20.

43. Williams, L. M., Gatt, J. M., Kuan, S. A., Dobson-Stone, C., Palmer, D. M., et al. (2009). A polymorphism of the MAOA gene is associated with emotional brain markers and personality traits on an antisocial index. *Neuropsychopharmacology* 34, 1797–1809.

44. Eisenberger, N. I., Way, B. M., Taylor, S. E., Welch, W. T. & Lieberman, M. D. (2007). Understanding genetic risk for aggression: Clues from the brain's response to social exclusion. *Biological Psychiatry* 61, 100–108.

45. Guo, G., Ou, X. M., Roettger, M. & Shih, J. C. (2008). The VNTR 2 repeat in MAOA and delinquent behavior in adolescence and young adulthood: Associations and MAOA promoter activity. *European Journal of Human Genetics* 16, 626–34.

46. McDermott, R., Tingley, D., Cowden, J., Frazzetto, G. & Johnson, D.D.P. (2009). Monoamine oxidase A gene (MAOA) predicts behavioral aggression following provocation. *Proceedings of the National Academy of Sciences USA* 106, 2118–23.

47. The important caveat that has to be borne in mind in interpreting the link between the low MAOA gene and antisocial behavior is that it accounts for only a small proportion of the variance. This is also true of most genes that have been linked to personality or mental illnesses.

48. Maori violence blamed on gene (2006). *The Dominion Post* (Wellington, New Zealand), August 9, Section A3.

49. Lea, R. & Chambers, G. (2007). Monoamine oxidase, addiction, and the "warrior" gene hypothesis. *New Zealand Medical Journal* 120, U2441.

50. Gibbons, A. (2004). American Association of Physical Anthropologists meeting. Tracking the evolutionary history of a "warrior" gene. *Science* 304, 818.

51. Newman, T. K., Syagailo, Y. V., Barr, C. S., et al. (2005). Monoamine oxidase A gene promoter variation and rearing experience influences aggressive behavior in rhesus monkeys. *Biological Psychiatry* 57, 167–72.

52. Lea & Chambers, Monoamine oxidase, addiction, and the "warrior" gene hypothesis.

53. Merriman, T. & Cameron, V. (2007). Risk-taking: Behind the warrior gene story. *New Zealand Medical Journal* 120, U2440.

54. Crampton, P. & Parkin, C. (2007). Warrior genes and risk-taking science. *New Zealand Medical Journal* 120, U2439.

55. Lea & Chambers, Monoamine oxidase, addiction, and the "warrior" gene hypothesis.

56. United Nations (2006). Intentional homicide, rate per 100,000 population. Office on Drugs and Crime, http://www.unodc.org/documents/data-and -analysis/IHS-rates-05012009.pdf.

57. Brunner, et al. Abnormal behavior associated with a point mutation in the structural gene for monoamine oxidase A.

58. Eisenberger et al., Understanding genetic risk for aggression.

59. It should be noted that the MAOA–antisocial relationship has not been found in all cultures. Shih and colleagues did not observe such a relation-

ship with either antisocial personality disorder or antisocial alcoholism in participants from Taiwan: see Lu, R. B., Lin, W. W., Lee, J. F., Ko, H. C. & Shih, J. C. (2003). Neither antisocial personality disorder nor antisocial alcoholism is associated with the MAO-A gene in Han Chinese males. *Alcoholism-Clinical and Experimental Research* 27(6), 889–93. Furthermore, the interaction between abuse and low MAOA has not been found in African-Americans in one report: see Widom, C. S. & Brzustowicz, L. M. (2006). MAOA and the "Cycle of violence": Childhood abuse and neglect, MAOA genotype, and risk for violent and antisocial behavior. *Biological Psychiatry* 60, 684–89.

60. Williams, et al., A polymorphism of the MAOA gene is associated with emotional brain markers.

61. Cadoret, R. J., Langbehn, D., Caspers, K., Troughton, E. P., Yucuis, R., et al. (2003). Associations of the serotonin transporter promoter polymorphism with aggressivity, attention deficit, and conduct disorder in an adoptee population. *Comprehensive Psychiatry* 44, 88–101.

62. DeLisi, M., Beaver, K. M., Vaughn, M. G. & Wright, J. P. (2009). All in the family: Gene x environment interaction between DRD2 and criminal father is associated with five antisocial phenotypes. *Criminal Justice and Behavior* 36, 1187–97.

63. Lee, S. S., Lahey, B. B., Waldman, I., Van Hulle, C. A., Rathouz, P., et al. (2007). Association of dopamine transporter genotype with disruptive behavior disorders in an eight-year longitudinal study of children and adolescents. *American Journal of Medical Genetics Part B-Neuropsychiatric Genetics* 144B, 310–17.

64. Gadow, K. D., DeVincent, C. J., Olvet, D. M., Pisarevskaya, V. & Hatchwell, E. (2010). Association of DRD4 polymorphism with severity of oppositional defiant disorder, separation anxiety disorder and repetitive behaviors in children with autism spectrum disorder. *European Journal of Neuroscience* 32, 1058–65.

65. Couppis, M. H. & Kennedy, C. H. (2008). The rewarding effect of aggression is reduced by nucleus accumbens dopamine receptor antagonism in mice. *Psychopharmacology* 197, 449–56.

66. Sokolov, B. P. & Cadet, J. L. (2006). Methamphetamine causes alterations in the MAP kinase-related pathways in the brains of mice that display increased aggressiveness. *Neuropsychopharmacology* 31, 956–66.

67. Caspi, A., Hariri, A. R., Holmes, A., Uher, R. & Moffitt, T. E. (2010). Genetic sensitivity to the environment: The case of the serotonin transporter gene and its implications for studying complex diseases and traits. *American Journal of Psychiatry* 167, 509–27.

68. Gelernter, J., Kranzler, H. R. & Cubells, J. F. (1997). Serotonin transporter

protein (SLC6A4) allele and haplotype frequencies and linkage disequilibria in African- and European-American and Japanese populations and in alcohol-dependent subjects. *Human Genetics* 101, 243–46.

69. Hariri, A. R., Mattay, V., Tessitore, A., Kolachana, B., Fera, F., et al. (2002). Serotonin transporter genetic variation and the response of the human amygdala. *Science* 297, 400–403.

70. Hanna, G. L., Himle, J. A., Curtis, G. C., Koran, D. Q., Weele, J. V., et al. (1998). Serotonin transporter and seasonal variation in blood serotonin in families with obsessive-compulsive disorder. *Neuropsychopharmacology* 18, 102–11.

71. Brown, G. L., Goodwin, F. K., Ballenger, J. C., Goyer, P. F. & Major, L. F. (1979). Aggression in humans correlates with cerebrospinal fluid amine metabolites. *Psychiatry Research* 1, 131–39.

72. Moore, T. M., Scarpa, A. & Raine, A. (2002). A meta-analysis of serotonin metabolite 5-HIAA and antisocial behavior. *Aggressive Behavior* 28, 299–316.

73. Coccaro, E. F., Lee, R. & Kavoussi, R. J. (2010). Aggression, suicidality and intermittent explosive disorder: Serotonergic correlates in personality disorder and healthy control subjects. *Neuropsychopharmacology* 35, 435–44.

74. Crockett, M. J., Clark, L., Tabibnia, G., Lieberman, M. D. & Robbins, T. W. (2008). Serotonin modulates behavioral reactions to unfairness. *Science* 320, 1739.

75. Glenn, A. L. (2011). The other allele: Exploring the long allele of the serotonin transporter gene as a potential risk factor for psychopathy: A review of the parallels in findings. *Neuroscience and Biobehavioral Reviews* 35, 612–20.

76. Beaver, K. M., Wright, J. P. & Walsh, A. (2008). A gene-based evolutionary explanation for the association between criminal involvement and number of sex partners. *Biodemography and Social Biology* 54, 47–55.

77. Orgel, L. E. & Crick, F. H. (1980). Selfish DNA: The ultimate parasite. *Nature* 284, 604–7.

78. Biémont, C. & Vieira, C. (2006). Genetics: Junk DNA as an evolutionary force. *Nature* 443, 521–24.

79. *60 Minutes: Murder Gene* (2001). CBS television, February 27.

3. MURDEROUS MINDS

1. Kraft, R. "My Life," chapter 5, Dad and the Fire. Death Row, California. http://www.ccadp.org/randykraft.htm.

2. McDougal, D. (1991). *Angel of Darkness*. New York: Warner Books.

3. Raine, A., Buchsbaum, M. S. & LaCasse, L. (1997). Brain abnormalities in murderers indicated by positron emission tomography. *Biological Psychiatry* 42, 495–508.

4. Barrash, J., Tranel, D. & Anderson, S. W. (2000). Acquired personality dis-

turbances associated with bilateral damage to the ventromedial prefrontal region. *Developmental Neuropsychology* 18, 355–81.

5. Bechara, A., Damasio, H., Tranel, D. & Damasio, A. R. (1997). Deciding advantageously before knowing the advantageous strategy. *Science* 275, 1293–94.

6. Blair, R.J.R. (2007). The amygdala and ventromedial prefrontal cortex in morality and psychopathy. *Trends in Cognitive Sciences* 11, 387–92.

7. Damasio, A. (1994). *Descartes' Error: Emotion, Reason, and the Human Brain.* New York: GP Putnam's Sons.

8. Bechara, A. & Damasio, A. R. (2005). The somatic marker hypothesis: A neural theory of economic decision. *Games and Economic Behavior* 52, 336–72.

9. Yang, Y. L. & Raine, A. (2009). Prefrontal structural and functional brain imaging findings in antisocial, violent, and psychopathic individuals: A meta-analysis. *Psychiatry Research: Neuroimaging* 174, 81–88.

10. The specific subregions of the occipital cortex found to be overactivated in murderers were visual areas 17 and 18.

11. *Understanding Murder: An Examination of the Etiology of Murder* (2001). The Learning Channel and Cronkite-Ward Productions, August.

12. *People vs. Antonio Bustamante* (1990–91). Case number: CR13160, Imperial County, Calif.

13. Bechara, A., Damasio, H. & Damasio, A. R. (2000). Emotion, decision making and the orbitofrontal cortex. *Cerebral Cortex* 10, 295–307.

14. *Understanding Murder.*

15. McDougal, *Angel of Darkness.*

16. Bechara & Damasio. The somatic marker hypothesis.

17. Kray, R. & Kray, R. (1989). *Reg and Ron Kray: Our Story*, p. 90. London: Pan Books.

18. Raine, A., Meloy, J. R., Bihrle, S., Stoddard, J., Lacasse, L., et al. (1998). Reduced prefrontal and increased subcortical brain functioning assessed using positron emission tomography in predatory and affective murderers. *Behavioral Sciences and the Law* 16, 319–32.

19. It is not just that homicidal acts can have a mixture of proactive and reactive aggression. An offender's criminal lifestyle can at times be at odds with their killing. Ron and Reggie Kray, for example, were organized gangsters who ruled the underworld in east London in the 1960s and 1970s, and participated in planned armed robberies and protection rackets. So while Reggie's killing of Jack "the Hat" McVitie was reactive aggression in nature, his criminal lifestyle was predominantly proactive.

20. Shaikh, M. B., Steinberg, A. & Siegel, A. (1993). Evidence that substance P is utilized in medial amygdaloid facilitation of defensive rage behavior in the cat. *Brain Research* 625, 283–94.

21. Adamec, R. E. (1990). Role of the amygdala and medial hypothalamus in spontaneous feline aggression and defense. *Aggressive Behavior* 16, 207–22.

22. Elliott, F. A. (1992). Violence: The neurologic contribution: An overview. *Archives of Neurology* 49, 595–603.

23. Adamec, R. E. (1991). The role of the temporal lobe in feline aggression and defense. Special Issue: Ethoexperimental psychology of defense: Behavioral and biological processes. *Psychological Record* 41, 233–53.

24. Mirsky, A. F. & Siegel, A. (1994). The neurobiology of violence and aggression. In A. J. Reiss, K. A. Miczek, and J. A. Roth (eds.), *Understanding and Preventing Violence*, vol. 2, *Biobehavioral Influences* (pp. 59–172). Washington, D.C.: National Academy Press.

25. Amen, D. G., Hanks, C., Prunella, J. R. & Green, A. (2007). An analysis of regional cerebral blood flow in impulsive murderers using single photon emission computed tomography. *Journal of Neuropsychiatry and Clinical Neurosciences* 19, 304–9.

26. Despite the lack of functional imaging research on murderers, there is a small literature on structural imaging. See, for example, Yang, Y. L., Raine, A., Han, C. B., Schug, R. A., Toga, A. W., et al. (2010). Reduced hippocampal and parahippocampal volumes in murderers with schizophrenia. *Psychiatry Research: Neuroimaging* 182, 9–13; Puri, B. K., Counsell, S. J., Saeed, N., Bustos, M. G., Treasaden, I. H., et al. (2008). Regional grey matter volumetric changes in forensic schizophrenia patients: An MRI study comparing the brain structure of patients who have seriously and violently offended with that of patients who have not. *Progress in Neuro-Psychopharmacology and Biological Psychiatry* 32, 751–54.

27. Soderstrom, H., Hultin, L., Tullberg, M., Wikkelso, C., Ekholm, S., et al. (2002). Reduced frontotemporal perfusion in psychopathic personality. *Psychiatry Research: Neuroimaging* 114, 81–94.

28. Hoptman, M. J. (2003). Neuroimaging studies of violence and antisocial behavior. *Journal of Psychiatric Practice* 9, 265–78; Miczek, K. A., de Almeida, R.M.M., Kravitz, E. A., Rissman, E. F., de Boer, S. F., et al. (2007). Neurobiology of escalated aggression and violence. *Journal of Neuroscience* 27, 11,803–6.

29. Gur, R. C., Ragland, J. D., Resnick, S. M., Skolnick, B. E., Jaggi, J., et al. (1994). Lateralized increases in cerebral blood flow during performance of verbal and spatial tasks: Relationship with performance level. *Brain and Cognition* 24, 244–58.

30. Sakurai, Y., Asami, M. & Mannen, T. (2010): Alexia and agraphia with lesions of the angular and supramarginal gyri: Evidence for the disruption of sequential processing. *Journal of the Neurological Sciences* 288, 25–33.

31. Rubia, K., Smith, A. B., Halari, R., Matsukura, F., Mohammad, M., et al. (2009): Disorder-specific dissociation of orbitofrontal dysfunction in boys

with pure conduct disorder during reward and ventrolateral prefrontal dysfunction in boys with pure ADHD during sustained attention. *American Journal Psychiatry* 166, 83–94.

32. Soderstrom, H., Tullberg, M., Wikkelso, C., Ekholm, S. & Forsman, A. (2000): Reduced regional cerebral blood flow in non-psychotic violent offenders. *Psychiatry Research: Neuroimaging* 98, 29–41.

33. Kiehl, K. A. (2006). A cognitive neuroscience perspective on psychopathy: Evidence for paralimbic system dysfunction. *Psychiatry Research* 142, 107–28.

34. Muller, J. L., Sommer, M., Wagner, V., Lange, K., Taschler, H., et al. (2003). Abnormalities in emotion processing within cortical and subcortical regions in criminal psychopaths: Evidence from a functional magnetic resonance imaging study using pictures with emotional content. *Biological Psychiatry* 54, 152–62.

35. Amen, D. G., Hanks, C., Prunella, J. R. & Green, A. (2007). An analysis of regional cerebral blood flow in impulsive murderers using single photon emission computed tomography. *Journal of Neuropsychiatry and Clinical Neurosciences* 19, 304–9.

36. Raine, A., Ishikawa, S. S., Arce, E., Lencz, T., Knuth, K. H., et al. (2004). Hippocampal structural asymmetry in unsuccessful psychopaths. *Biological Psychiatry* 55, 185–91.

37. Raine, A., Moffitt, T. E., Caspi, A., Loeber, R., Stouthamer-Loeber, M., et al. (2005). Neurocognitive impairments in boys on the life-course persistent antisocial path. *Journal of Abnormal Psychology* 114, 38–49.

38. Boccardi, M., Ganzola, R., Rossi, R., Sabattoli, F., Laakso, M. P., et al. (2010). Abnormal hippocampal shape in offenders with psychopathy. *Human Brain Mapping* 31, 438–47.

39. Swanson, L. W. (1999). Limbic system. In G. Adelman & B. H. Smith (eds.), *Encyclopedia of Neuroscience*, pp. 1053–55. Amsterdam: Elsevier.

40. Gregg, T. R. & Siegel, A. (2001). Brain structures and neurotransmitters regulating aggression in cats: Implications for human aggression. *Progress in Neuro-Psychopharmacology & Biological Psychiatry* 25, 91–140.

41. Kiehl, K. A., Smith, A. M., Hare, R. D., Mendrek, A., Forster, B. B., Brink, J. & Liddle, P. F. (2001). Limbic abnormalities in affective processing by criminal psychopaths as revealed by functional magnetic resonance imaging. *Biological Psychiatry* 50, 677–84.

42. Rubia, K., Halari, R., Smith, A. B., Mohammed, M., Scott, S., et al. (2008): Dissociated functional brain abnormalities of inhibition in boys with pure conduct disorder and in boys with pure attention deficit hyperactivity disorder. *American Journal of Psychiatry* 165, 889–97.

43. New, A. S., Hazlett, E. A., Buchsbaum, M. S., Goodman, M., Reynolds, D., et al. (2002): Blunted prefrontal cortical (18)fluorodeoxyglucose posi-

tron emission tomography response to meta-chlorophenylpiperazine in impulsive aggression. *Archives of General Psychiatry* 59, 621–29.

44. Maratos, E. J., Dolan, R. J., Morris, J. S., Henson, R.N.A. & Rugg, M. D. (2001). Neural activity associated with episodic memory for emotional context. *Neuropsychologia* 39, 910–20.

45. Mayberg, H. S., Liotti, M., Brannan, S. K., McGinnis, S., Mahurin, R. K., et al. (1999). Reciprocal limbic-cortical function and negative mood: Converging PET findings in depression and normal sadness. *American Journal of Psychiatry* 156, 675–82.

46. Ochsner, K. N. et al. (2005). The neural correlates of direct and reflected self-knowledge. *NeuroImage* 28, 797–814.

47. Fagan, J. (1989). Cessation of family violence: Deterrence and dissuasion. In L. Ohlin & M. Tonry (eds.), *Family Violence: Crime and Justice: A Review of Research*, pp. 377–425. Chicago: University of Chicago.

48. Wilt, S. & Olson, S. (1996). Prevalence of domestic violence in the United States. *Journal of American Medical Women's Association* 51, 77–88.

49. Guth, A. A. & Pachter, L. (2000). Domestic violence and the trauma surgeon. *American Journal of Surgery* 179, 134–40; Hamby, J. M. & Koss, M. P. (2003). Violence against women: Risk factors, consequences, and prevalence. In J. M. Leibschutz, S. M. Frayne & G. M. Saxe (eds.), *Violence Against Women: A Physician's Guide to Identification and Management*, pp. 3–38. Philadelphia: American College of Physicians.

50. Pihlajamaki, M., Tanila, H., Kononen, M., et al. (2005). Distinct and overlapping fMRI activation networks for processing of novel identities and locations of objects. *European Journal of Neuroscience* 22, 2095–105.

51. Sevostianov, A., Horwitz, B., Nechaev, V., et al. (2002). fMRI study comparing names versus pictures of objects. *Human Brain Mapping* 16, 168–75.

52. George, D. T., Phillips, M. J., Doty, L., Umhau, J. C. & Rawlings, R. R. (2006): A model linking biology, behavior, and psychiatric diagnoses in perpetrators of domestic violence. *Medical Hypotheses* 67, 345–53.

53. Ibid.

54. Babcock, J. C., Green, C. E., Webb, S. A. & Graham, K. H. (2004). A second failure to replicate the Gottman et al. (1995) typology of men who abuse intimate partners . . . and possible reasons why. *Journal of Family Psychology* 18, 396–400.

55. We are not the only group to be thinking along these lines. Others have hypothesized that spouse-abusers are hypersensitive to emotional stimuli that could be interpreted as threatening, such as slights and signs of disapproval, resulting in increased negative emotionality and reacting out of proportion to the social context. See George, D. T., Rawlings, R. R., Williams, W. A., Phillips, M. J., Fong, G., et al. (2004). A select group of perpetrators of domestic violence: Evidence of decreased metabolism

in the right hypothalamus and reduced relationships between cortical/ subcortical brain structures in position emission tomography. *Psychiatry Research: Neuroimaging* 130, 11–25; also Babcock et al., A second failure to replicate the Gottman et al. (1995) typology.

56. Babcock, J. C., Green, C. E. & Robieb, C. (2004). Does batterers' treatment work? A meta-analytic review of domestic violence treatment. *Clinical Psychology Review* 23, 1023–53.

57. Twain, M. (1882). *On the Decay of the Art of Lying.* Boston: James R. Osgood and Company.

58. Very sadly, Sean Spence died prematurely, at the age of forty-eight, on Christmas Day, 2010, after suffering a long illness. He was a highly creative and energetic scientist that many of us miss.

59. Lee, T.M.C., Liu, H. L., Tan, L. H., Chan, C.C.H., Mahankali, S., et al. (2002). Lie detection by functional magnetic resonance imaging. *Human Brain Mapping* 15, 157–64.

60. Spence, S. A., Farrow, T.F.D., Herford, A. E., Wilkinson, I. D., Zheng, Y., et al. (2001). Behavioural and functional anatomical correlates of deception in humans. *NeuroReport* 12, 2849–53.

61. Langleben, D. D., Schroeder, L., Maldjian, J. A., Gur, R. C., McDonald, S., et al. (2002). Brain activity during simulated deception: An event-related functional magnetic resonance study. *NeuroImage* 15, 727–32.

62. Mackintosh, N., Baddeley, A., Brownsworth, R., et al. (2011). *Brain Waves Module 4: Neuroscience and the Law.* London: The Royal Society.

63. Greene, J. D., Sommerville, R. B., Nystrom, L. E., Darley, J. M. & Cohen, J. D. (2001). An fMRI investigation of emotional engagement in moral judgment. *Science* 293, 2105–8.

64. Koenigs, M., Young, L., Adolphs, R., Tranel, D., Cushman, F., et al. (2007). Damage to the prefrontal cortex increases utilitarian moral judgments. *Nature* 446, 908–11.

65. Moll, J. et al. (2002). The neural correlates of moral sensitivity: A functional magnetic resonance imaging investigation of basic and moral emotions. *The Journal of Neuroscience: The Official Journal of the Society for Neuroscience* 22, 2730–36.

66. Heekeren, H. R., Wartenburger, I., Schmidt, H., Prehn, K., Schwintowski, H. P., et al. (2005). Influence of bodily harm on neural correlates of semantic and moral decision-making. *NeuroImage* 24, 887–97.

67. Kumari, V., Das, M., Hodgins, S., Zachariah, E., Barkataki, I., et al. (2005). Association between violent behaviour and impaired prepulse inhibition of the startle response in antisocial personality disorder and schizophrenia. *Behavioral and Brain Research* 158, 159–66.

68. Kiehl, K. A., Smith, A. M., Mendrek, A., Forster, B. B., Hare, R. D., et al. (2004). Temporal lobe abnormalities in semantic processing by criminal

psychopaths as revealed by functional magnetic resonance imaging. *Psychiatry Research: Neuroimaging* 130, 295–312.

69. Yang, Y. L., Glenn, A. L. & Raine, A. (2008). Brain abnormalities in antisocial individuals: Implications for the law. *Behavioral Sciences & the Law* 26, 65–83.

70. Raine, A. & Yang, Y. (2006). Neural foundations to moral reasoning and antisocial behavior. *Social, Cognitive, and Affective Neuroscience* 1, 203–13.

71. Veit, R., Lotze, M., Sewing, S., Missenhardt, H., Gaber, T., et al. (2010). Aberrant social and cerebral responding in a competitive reaction time paradigm in criminal psychopaths. *NeuroImage* 49, 3365–72; Kiehl, K. A. (2006). A cognitive neuroscience perspective on psychopathy: Evidence for paralimbic system dysfunction. *Psychiatry Research* 142, 107–28.

72. New et al., Blunted prefrontal cortical (18)fluorodeoxyglucose positron emission tomography response.

73. Lee, T.M.C., Chan, S. C. & Raine, A. (2009). Hyper-responsivity to threat stimuli in domestic violence offenders: A functional magnetic resonance imaging study. *Journal of Clinical Psychiatry* 70, 36–45.

74. Rule, A. (2009). *The Stranger Beside Me*. New York: Pocket Books.

75. Vronsky, P. (2007). *Female Serial Killers: How and Why Women Become Monsters*. New York: Berkley Books.

76. Ibid.

77. Ibid., p. 132.

78. Ibid.

79. Bowlby, J. (1969). *Attachment and Loss*, vol. 1, *Attachment*. New York: Hogarth Press; Rutter, M. (1982). *Maternal Deprivation Reassessed* (2nd ed.). Harmondsworth, U.K.: Penguin.

80. Vronsky, *Female Serial Killers*.

81. Hare, R. D. (2003). *The Hare Psychopathy Checklist—Revised (PCL-R)*, 2nd ed. Toronto, Canada: Multi-Health Systems.

82. Crime: Chronic Murder. August 29, 1938. *Time*. http://www.time.com/time/magazine/article/0,9171,789132,00.html.

83. Glenn, A. L., Raine, A. & Schug, R. A. (2009). The neural correlates of moral decision-making in psychopathy. *Molecular Psychiatry* 14, 5–6.

84. Vronsky, *Female Serial Killers*.

85. Blair, The amygdala and ventromedial prefrontal cortex.

86. Raine & Yang. Neural foundations to moral reasoning and antisocial behavior.

4. COLD-BLOODED KILLERS

1. Chynoweth, C. (2005). How do I become a bomb disposal expert? *The Times* (London), February 24, http://business.timesonline.co.uk/tol/business/career_and_jobs/graduate_management/article517604.ece.

2. Elder, R. K. (2008). A brother lost, a brotherhood found. *Chicago Tribune,* May 17, http://www.chicagotribune.com/news/nationworld/chi-una bomber-story,0,7970571.story.

3. Forty-three years after his first IQ test, at age eleven, Ted Kaczynski was retested, for a score of 138. The drop from 167 is likely due to mental illness, which developed in early adulthood.

4. Eisermann, K. (1992). Long-term heart rate responses to social stress in wild European rabbits: Predominant effect of rank position. *Physiology & Behavior* 52, 33–36.

5. Cherkovich, G. M. & Tatoyan, S. K. (1973). Heart rate (radiotelemetric registration) in macaques and baboons according to dominant-submissive rank in a group. *Folia Primatologica* 20, 265–73; Holst, D. V. (1986). Vegetative and somatic compounds of tree shrews' behavior. *Journal of the Autonomic Nervous System,* Suppl., 657–70.

6. One reason it is hard for people to believe that low heart rate can predispose an individual to antisocial behavior is the idea that exercise reduces resting heart rate and we view people who exercise in a favorable light. Although this is technically true, surprisingly the effect is much smaller than people imagine. Even twenty weeks of endurance training lowers resting heart rate only by two beats per minute. The type of moderate exercise some of us regularly engage in has even smaller effects. See Wilmore, J. H., Stanforth, P. R., Gagnon, J., et al. (1996). Endurance exercise training has a minimal effect on resting heart rate: The HERITAGE study. *Medicine and Science in Sports and Exercise* 28, 829–35.

7. Raine, A. & Venables, P. H. (1984). Tonic heart rate level, social class and antisocial behaviour in adolescents. *Biological Psychology* 18, 123–32.

8. Raine, A. & Jones, F. (1987). Attention, autonomic arousal, and personality in behaviorally disordered children. *Journal of Abnormal Child Psychology* 15, 583–99.

9. Ortiz, J. & Raine, A. (2004). Heart rate level and antisocial behavior in children and adolescents: A meta-analysis. *Journal of the American Academy of Child and Adolescent Psychiatry* 43, 154–62.

10. The overall "effect size" was -0.44. Effect sizes tell us the strength of the relationship. To put this into context, .2 is a small relationship, .5 is medium, and .8 is large.

11. For more examples of effect sizes in medicine and psychology, see Meyer, G. J. et al. (2001). Psychological testing and psychological assessment: A review of evidence and issues. *American Psychologist* 56, 128–65.

12. The correlation between smoking and lung cancer is .08, between alcohol use during pregnancy and premature birth is .09, and between taking aspirin to reduce the risk of death by a heart attack is .02. The effect of taking antihypertensive medication in reducing the risk of stroke is a cor-

relation of .03. In comparison, the correlation between heart rate and antisocial behavior is .22.

13. Raine, A., Venables, P. H. & Mednick, S. A. (1997). Low resting heart rate at age 3 years predisposes to aggression at age 11 years: Evidence from the Mauritius Child Health Project. *Journal of the American Academy of Child & Adolescent Psychiatry* 36, 1457–64.

14. Voors, A. W., Webber, L. S. & Berenson, B. S. (1982). Resting heart rate and pressure rate product of children in a total biracial community: The Bogalusa Heart study. *American Journal of Epidemiology* 116, 276–86.

15. Ibid. The effect size here is quite strong, at d = 0.36, p < .0001.

16. Shaw, D. S. & Winslow, E. B. (1997). Precursors and correlates of antisocial behavior from infancy to preschool. In D. M. Stoff, J. Breiling & J. D. Maser (eds.), *Handbook of Antisocial Behavior*, pp. 148–58. New York: Wiley.

17. Baker, L. A., Tuvblad, C., Reynolds, C., Zheng, M., Lozano, D. I., et al. (2009). Resting heart rate and the development of antisocial behavior from age 9 to 14: Genetic and environmental influences. *Development and Psychopathology*, 21, 939–60.

18. Farrington, D. P. (1987). Implications of biological findings for criminological research. In S. A. Mednick, T. E. Moffitt & S. A. Stack (eds.), *The Causes of Crime: New Biological Approaches*, pp. 42–64. New York: Cambridge University Press; Venables, P. H. (1987). Autonomic and central nervous system factors in criminal behavior. In Mednick et al., *The Causes of Crime*, pp. 110-36.

19. Farrington, D. P. (1997). The relationship between low resting heart rate and violence. In A. Raine, P. A. Brennan, D. P. Farrington & S. A. Mednick (eds.), *Biosocial Bases of Violence*, pp. 89–106. New York: Plenum.

20. The reason parental crime may be such a well-replicated risk factor for offspring crime is that it combines significant genetic and environmental risks. Criminal parents pass on the genetic risk for crime to their offspring, and they also give their children poor parenting, an unstable lifestyle, and abuse, important social risk factors for crime.

21. Farrington, The relationship between low resting heart rate and violence.

22. Raine, A., Venables, P. H. & Williams, M. (1995). High autonomic arousal and electrodermal orienting at age 15 years as protective factors against criminal behavior at age 29 years. *American Journal of Psychiatry* 152, 1595–1600.

23. Connor, D. F., Glatt, S. J., Lopez, I. D., Jackson, D. & Melloni, R. H. (2002). Psychopharmacology and aggression, vol. 1: A meta-analysis of stimulant effects on overt/covert aggression-related behaviors in ADHD. *Journal of the American Academy of Child and Adolescent Psychiatry* 41, 253–61.

24. Stadler, C., Grasmann, D., Fegert, J. M., Holtmann, M., Poustka, F., et al.

(2008). Heart rate and treatment effect in children with disruptive behavior disorders. *Child Psychiatry and Human Development* 39, 299–309.

25. Rogeness, G. A., Cepeda, C., Macedo, C. A., Fischer, C., et al. (1990). Differences in heart rate and blood pressure in children with conduct disorder, major depression, and separation anxiety. *Psychiatry Research* 33, 199–206.

26. Moffitt, T. E., Arseneault, L., Jaffee, S. R., Kim-Cohen, J., Koenen, K. C., et al. (2008). Research Review: DSM-V conduct disorder: Research needs for an evidence base. *Journal of Child Psychology and Psychiatry* 49, 3–33.

27. Raine, A. (1993). *The Psychopathology of Crime: Criminal Behavior as a Clinical Disorder.* San Diego: Academic Press.

28. Raine, A., Reynolds, C., Venables, P. H. & Mednick, S. A. (1997). Resting heart rate, skin conductance orienting, and physique. In Raine et al., *Biosocial Bases of Violence,* pp. 107–26.

29. Cox, D., Hallam, R., O'Connor, K. & Rachman, S. (1983). An experimental study of fearlessness and courage. *British Journal of Psychology* 74, 107–17; O'Connor, K., Hallam, R., and Rachman, S. (1985). Fearlessness and courage: A replication experiment. *British Journal of Psychology* 76, 187–97.

30. Scarpa, A., Raine, A., Venables, P. H. & Mednick, S. A. (1997). Heart rate and skin conductance in behaviorally inhibited Mauritian children. *Journal of Abnormal Psychology* 106, 182–90; Kagan, J. (1994). *Galen's Prophecy: Temperament in Human Nature.* New York: Basic Books.

31. Raine, A., Reynolds, C., Venables, P. H., Mednick, S. A. & Farrington, D. P. (1998). Fearlessness, stimulation-seeking, and large body size at age 3 years as early predispositions to childhood aggression at age 11 years. *Archives of General Psychiatry* 55, 745–51.

32. Oldehinkel, A. J., Verhulst, F. C. & Ormel, J. (2008). Low heart rate: A marker of stress resilience. The TRAILS Study. *Biological Psychiatry* 63, 1141–46.

33. Zahn-Waxler, C., Cole, P., Welsh, J. D. & Fox, N. A. (1995). Psychophysiological correlates of empathy and prosocial behaviors in preschool children with behavior problems. *Development and Psychopathology* 7, 27–48.

34. Lovett, B. J. & Sheffield, R. A. (2007). Affective empathy deficits in aggressive children and adolescents: A critical review. *Clinical Psychology Review* 27, 1–13.

35. Eysenck, H. J. (1997). Personality and the biosocial model of antisocial and criminal behavior. In Raine et al., *Biosocial Bases of Violence,* pp. 21–38.

36. Raine, A., Reynolds, C., Venables, P. H. & Mednick, S. A. (1997). Resting heart rate, skin conductance orienting, and physique.

37. El-Sheikh, M., Ballard, M. & Cummings, E. M. (1994). Individual differences in preschoolers' physiological and verbal responses to videotaped angry interactions. *Journal of Abnormal Child Psychology* 22, 303–20.

38. Raine et al., Fearlessness, stimulation-seeking, and large body size at age 3 years.

39. Zuckerman, M. (1994). *Behavioral Expressions and Biosocial Bases of Sensation Seeking.* Cambridge: Cambridge University Press.

40. Moffitt, T. E. (1993). Adolescence-limited and life-course persistent antisocial behavior: A developmental taxonomy. *Psychological Review* 100, 674–701.

41. Raine, A., Liu, J., Venables, P. H., Mednick, S. A. & Dalais, C. (2010). Cohort profile: The Mauritius Child Health Project. *International Journal of Epidemiology* 39, 1441–51.

42. WHO Scientific Group (1968). Neurophysiological and behavioural research in psychiatry. *WHO Technical Report No. 381.* Geneva: World Health Organization.

43. Raine, et al., Fearlessness, stimulation-seeking, and large body size at age 3 years.

44. Achenbach, T. M. (1991). *Manual for the Child Behavior Checklist/4-18.* Burlington, Vt.: Department of Psychiatry, University of Vermont.

45. *Over Aggressie* (2001). KRO network Amsterdam, Netherlands, http://www.kro.nl/.

46. Ibid.

47. Ibid.

48. Ibid.

49. Ibid.

50. Ibid.

51. Kenrick, D. T. & Sheets, V. (1993). Homicidal Fantasies. *Ethology and Sociobiology* 14, 231–46.

52. Crabb, P. B. (2000). The material culture of homicidal fantasies. *Aggressive Behavior* 26, 225–34.

53. Ibid.

54. Galvanic skin response (GSR) is an older term for skin conductance (SC), while electrodermal activity (EDA) is a more generic term encompassing both skin conductance and skin potential.

55. Dawson, M. E., Schell, A. M. & Filion, D. L. (2007). The electrodermal system. In J. T. Cacioppo, L. G. Tassinary & G. G. Berntson (eds.), *Handbook of Psychophysiology,* pp. 159-81. New York: Oxford University Press.

56. Williams, L. M., Felmingham, K., Kemp, A. H., Rennie, C., Brown, K. J., et al. (2007). Mapping frontal-limbic correlates of orienting to change detection. *Neuroreport* 18, 197–202.

57. Critchley, H. D. (2002). Electrodermal responses: What happens in the brain. *Neuroscientist* 8, 132–42.

58. Dawson, M. E. & Schell, A. M. (1987). Human autonomic and skeletal classical conditioning: The role of conscious cognitive factors. In G. Davey

(ed.), *Cognitive Processes and Pavlovian Conditioning in Humans*, pp. 27–55. New York: Wiley & Sons.

59. Raine, A. (1997). Crime, conditioning, and arousal. In H. Nyborg (ed.), *The Scientific Study of Human Nature: Tribute to Hans J. Eysenck*, pp. 122–41. Oxford: Elsevier.

60. For a detailed account of a conditioning theory of crime, see Eysenck, H. J. (1977). *Crime and Personality*. St. Albans, England: Paladin. Eysenck is debatably England's most influential and simultaneously controversial psychologist. His biosocial theory of crime did not sit well with many criminologists in the 1970s and still does not today.

61. Hare, R. D., Frazelle, J. & Cox, D. N. (1978). Psychopathy and physiological responses to threat of an aversive stimulus. *Psychophysiology* 15, 165–72; Lorber, M. F. (2004). Psychophysiology of aggression, psychopathy, and conduct problems: A meta-analysis. *Psychological Bulletin* 130, 531–52; Raine, A. (1993). *The Psychopathology of Crime: Criminal Behavior as a Clinical Disorder*. San Diego: Academic Press.

62. Gao, Y., Raine, A., Venables, P. H., Dawson, M. E. & Mednick, S. A. (2010). Association of poor childhood fear conditioning and adult crime. *American Journal of Psychiatry* 167, 56–60.

63. Ibid.

64. Hare, R. D. (1993). *Without Conscience: The Disturbing World of Psychopaths Amongst Us*. New York: Guilford Press.

65. Raine, A., Lencz, T., Bihrle, S., LaCasse, L. & Colletti, P. (2000). Reduced prefrontal gray matter volume and reduced autonomic activity in antisocial personality disorder. *Archives of General Psychiatry* 57, 119–27.

66. Meeting diagnostic criteria for DSM antisocial personality disorder requires that the individual also meet criteria for conduct disorder in childhood or adolescence.

67. The temporary-employment-agency workers who met the adult criteria for antisocial personality disorder lack the child criteria. That is, they are antisocial in adulthood, but they did not meet criteria for conduct disorder as children. We focused our research on those who met full criteria for antisocial personality disorder.

68. None had been convicted of either homicide, attempted homicide, or rape.

69. For the entire unselected sample, males reported an average of 16.1 criminal offenses while females reported 8.6 offenses. Rates of at least one seriously violent act were 55.7 percent in males and 42.9 percent in females. For males, 24.4 percent of the sample admitted to rape or sexual assault, while 34.8 percent admitted to assault on a stranger causing bodily injury, 13.3 percent had fired a gun at someone, and 8.9 percent had either attempted homicide or completed homicide. For females, 14.3 percent

admitted to assault on a stranger causing bodily injury, 7.1 percent had fired a gun at someone, and 7.1 percent had either attempted homicide or completed homicide.

70. Hare, R. D. (2003). *The Hare Psychopathy Checklist—Revised (PCL-R)*, 2nd ed. Toronto, Canada: Multi-Health Systems.

71. Ibid.

72. Rates of psychopathy for females were 8.3 percent (a score of 30 or more) and 16.7 percent (a score of 25 or more).

73. Widom, C. S. (1978). A methodology for studying non-institutionalized psychopaths. In R. D. Hare & D. Schalling (eds.), *Psychopathic Behavior: Approaches to Research*, p. 72. Chichester, England: Wiley.

74. Ibid., p. 83.

75. Widom, C. S. & Newman, J. P. (1985). Characteristics of non-institutionalized psychopaths. In D. P. Farrington and J. Gunn (eds.), *Aggression and Dangerousness*, pp. 57–80. London: Wiley.

76. This quasi-conditioning is very much like fear conditioning. Numbers are flashed on a screen counting down from 12 to 0. At the count of 0 the subject is blasted with a loud noise or given an electric shock. Between 12 and 0 (the anticipatory phase), most of us will give skin conductance "anticipatory" responses as we are somewhat anxious about the noise blast. Psychopaths give significantly fewer of these responses. The task differs from conditioning in that participants are told what will happen—there is cognitive awareness. In the classical conditioning paradigm, they are not told the association—that the CS+ tone predicts the aversive noise—and instead they must learn this association for themselves.

77. Ishikawa, S. S., Raine, A., Lencz, T., Bihrle, S. & LaCasse, L. (2001). Autonomic stress reactivity and executive functions in successful and unsuccessful criminal psychopaths from the community. *Journal of Abnormal Psychology* 110, 423–32.

78. Ibid.

79. Damasio, A. R. (1994). *Descartes' Error: Emotion, Reason, and the Human Brain*. New York: Grosset/Putnam.

80. We did not run a classical conditioning paradigm on the psychopaths because I felt at the time it was such a well-replicated finding that it did not need repeating, and that a paradigm with a social context that manipulated secondary emotions would be more novel. We have predicted that the successful psychopath would show better autonomic fear conditioning, and we have brought fear conditioning back into our research protocols.

81. Despite a dearth of systematic research studies, there has nevertheless been a great deal of speculation about what makes a serial killer; see, for example, Holmes, R. M. & Holmes, S. T. (1998). *Serial Murder*, 2nd ed.

Thousand Oaks, Calif.: Sage Publications; also Fox, J. A. & Levin, J. (2005). *Extreme Killing*. Thousand Oaks, Calif.: Sage Publications.

82. The executive-functioning task we gave our participants is the Wisconsin card-sorting task, a classic measure of executive functioning.

83. Strangulation as depicted in movies and TV does not take too long, but it is much harder in reality. It took Ross eight minutes to strangle one of his victims, as his fingers would cramp up. He had to stop and massage them before proceeding.

84. Berry-Dee, C. (2003). *Talking with Serial Killers*, p. 150. London: John Blake.

85. Scripps argued that he became annoyed with his victim in the hotel room when he suspected that Lowe was a homosexual and was making advances to him.

86. Berry-Dee, *Talking with Serial Killers*, p. 94. Scripps used a six-inch boning knife to systematically dismember his victims; he gives a systematic description of how he went about doing it. His skills are unusual but stem from the fact that Scripps worked in a butchery while serving a prior prison sentence.

87. Ibid.

88. Pontius, A. A. (1993). Neuropsychiatric update of the crime "profile" and "signature" in single or serial homicides: Rule out limbic psychotic trigger reaction. *Psychological Reports* 73, 875–92.

89. Carver, H. W. (2007). Reasonable doubt. *Scientific American* 297, 20–21.

90. Johnson, S. (1998). *Psychological Evaluation of Theodore Kaczynski*. Federal Correctional Institution, Butner, North Carolina. January 11–16, http://www.paulcooijmans.com/psychology/unabombreport2.html.

91. Ishikawa, S. S., Raine, A., Lencz, T., Bihrle, S. & LaCasse, L. (2001). Autonomic stress reactivity and executive functions in successful and unsuccessful criminal psychopaths from the community. *Journal of Abnormal Psychology* 110, 423–32.

92. Dan Rather had other risk factors for an antisocial behavior outcome, including bad spelling and coming from a working-class neighborhood. Interestingly it was a heart inflammation he had as a ten-year-old, confining him for weeks to bed, where he could only listen to World War II newscasts—that caused him to become fascinated by broadcasting.

93. Raine, A. (2006). *Crime and Schizophrenia: Causes and Cures*. New York: Nova Science Publishers.

94. Johnson, *Psychological Evaluation of Theodore Kaczynski*.

95. Raine, A., Brennan, P. & Mednick, S. A. (1994). Birth complications combined with early maternal rejection at age 1 year predispose to violent crime at age 18 years. *Archives of General Psychiatry* 51, 984–88.

96. If you see *The Hurt Locker*, note Sergeant James's thirst for vengeance when he believes that Beckham, a young boy he forms a fleeting rela-

tionship with, has suffered terribly at the hands of terrorists. Note also how he breaks down in the shower, haunted by guilt after his need for an adrenaline rush results in a comrade's leg being shattered. Despite the devil-may-care, stimulation-seeking cowboy persona that he presents, James has a conscience—he is neither a psychopath nor a "red-neck piece of trailer trash," as one of his disconcerted comrades calls him.

97. It should be recognized that there appears to be no unitary arousal system—measures of resting-state ANS correlate at a surprisingly low level, around .10. Arousal is clearly a complex and multifaceted construct, and low-arousal theory is perhaps too simplistic. Still, it is conceivable that an extreme (antisocial) group within this general population does have low arousal on multiple arousal measures. Evidence does exist for under-arousal on at least two separate measures of arousal in antisocial child and adolescent samples. Even with simple biological measures like heart rate, unfolding the "mechanism of action"—how low heart rate goes about producing individuals with antisocial and aggressive behavior—is likely highly complex, involving many different processes.

5. BROKEN BRAINS

1. Rojas-Burke, J. (1993). PET scan advance as tool in insanity defense: Debate erupts over capability of brain scanning technology. *Journal of Nuclear Medicine* 34, 13N–26N.

2. Rosen, J. (2007). The brain on the stand. *New York Times*. Sunday, March 11.

3. Rojas-Burke, PET scan advance as tool in insanity defense.

4. Ibid.

5. Raine, A., Buchsbaum, S., Stanley, J., et al. (1994). Selective reductions in prefrontal glucose metabolism in murderers. *Biological Psychiatry* 6, 365–73.

6. Raine, A., Lencz, T., Bihrle, S., Lacasse, L. & Colletti, P. (2000). Reduced prefrontal gray matter volume and reduced autonomic activity in antiso-cial personality disorder. *Archives of General Psychiatry* 57, 119–27.

7. Goodwin, R. D. & Hamilton, S. P. (2003). Lifetime comorbidity of anti-social personality disorder and anxiety disorders among adults in the community. *Psychiatry Research* 117, 159–66; Raine, A. (2005). *Crime and Schizophrenia*. New York: Nova Science Publishers.

8. Raine, A. et al., Reduced prefrontal gray matter volume and reduced autonomic activity in antisocial personality disorder.

9. Yang, Y. & Raine, A. (2009). Prefrontal structural and functional brain imaging findings in antisocial, violent, and psychopathic individuals: A meta-analysis. *Psychiatry Research: Neuroimaging* 174, 81–88.

10. Gansler, D. A., McLaughlin, N.C.R., Iguchia, L., et al. (2009). A multi-

variate approach to aggression and the orbital frontal cortex in psychiatric patients. *Psychiatry Research: Neuroimaging* 171, 145–54.

11. Damasio, A. (1994). *Descartes' Error: Emotion, Reason, and the Human Brain.* New York: GP Putnam's Sons.

12. Bechara, A., Damasio, H., Tranel, D. & Damasio, A. R. (1997). Deciding advantageously before knowing the advantageous strategy. *Science* 275, 1293–94.

13. McMillan, M. B. (1986). A wonderful journey through skull and brains: The travels of Mr. Gage's tamping iron. *Brain and Cognition* 5, 67–107.

14. Harlow, J. M. (1868). Recovery from the passage of an iron bar through the head. *Publications of the Massachusetts Medical Society,* 2, 327–47.

15. Such claims of drunkenness and sexual promiscuity in Gage have been questioned—see Malcolm Macmillan, The Phineas Gage information Page, http://www.deakin.edu.au/hmnbs/psychology/gagepage/Pgstory .php.

16. Glenn, A. L. & Raine, A. (2009). Neural circuits underlying morality and antisocial behavior. In J. Verplaetse and J. Braeckman (eds.), *The Moral Brain,* pp. 45–68. New York: Springer.

17. Butler, K., Rourke, B. P., Fuerst, D. R. & Fisk, J. L. (1997). A typology of psychosocial functioning in pediatric closed-head injury. *Child Neuropsychology* 3, 98–133.

18. Max, J. E., Koele, S. L., Smith, W. L., Sato, Y., Lindgren, S. D., et al. (1998). Psychiatric disorders in children and adolescents after severe traumatic brain injury: A controlled study. *Journal of the American Academy of Child & Adolescent Psychiatry* 37, 832–40.

19. Ibid.

20. Raine, A. (2002): Annotation: The role of prefrontal deficits, low autonomic arousal, and early health factors in the development of antisocial and aggressive behavior. *Journal of Child Psychology and Psychiatry* 43, 417–34.

21. Anderson, S. W., Behara, A., Damasio, H., Tranel, D. & Damasio, A. R. (1999). Impairment of social and moral behavior related to early damage in human prefrontal cortex. *Nature Neuroscience* 2, 1032–37. The female patient in Anderson's study had bilateral polar and ventromedial damage, while the male had damage localized to the right polar/medial-dorsal region.

22. Pennington, B. F. & Bennetto, L. (1993). Main effects or transactions in the neuropsychology of conduct disorder? Commentary on "The neuropsychology of conduct disorder." *Development and Psychopathology* 5, 153–64.

23. Damasio, A. R. (2000). A neural basis for sociopathy. *Archives of General Psychiatry* 57, 128–29.

24. Raine, et al. Reduced prefrontal gray matter volume and reduced autonomic activity in antisocial personality disorder, 119–27.

25. Damasio, A neural basis for sociopathy.

26. Damasio, H., Grabowski, T. J., Frank, R., Galaburda, A. M. & Damasio, A. (1994). The return of Phineas Gage—Clues about the brain from the skull of a famous patient. *Science* 264, 1102–5.

27. This ventromedial area is also known as gyrus rectus.

28. Knight, D. C., Cheng, D. T., Smith, C. N., Stein, E. A. & Helmstetter, F. J. (2004). Neural substrates mediating human delay and trace fear conditioning. *Journal of Neuroscience* 24, 218–28.

29. McNab, F., Leroux, G., Strand, F., Thorell, L., Bergman, S. & Klingberg, T. (2008). Common and unique components of inhibition and working memory: An fMRI, within-subjects investigation. *Neuropsychologia* 46, 2668–82.

30. Patrick C. J. (2008). Psychophysiological correlates of aggression and violence: An integrative review. *Philosophical Transactions of the Royal Society B-Biological Sciences* 363, 2543–55.

31. Raine, A. & Yang, Y. (2006). Neural foundations to moral reasoning and antisocial behavior. *Social, Cognitive, and Affective Neuroscience* 1, 203–13.

32. Blair, R.J.R. (2007). The amygdala and ventromedial prefrontal cortex in morality and psychopathy. *Trends in Cognitive Sciences* 11, 387–92.

33. McClure, S. M., Laibson, D. I., Loewenstein, G. & Cohen, J. D. (2004). Separate neural systems value immediate and delayed monetary rewards. *Science* 306, 503–7.

34. Dolan, M. & Fullam, R. (2004). Behavioural and psychometric measures of impulsivity in a personality disordered population. *Journal of Forensic Psychiatry & Psychology* 15, 426–50.

35. Miller, J. D. & Lynam, D. R. (2003). Psychopathy and the five-factor model of personality: A replication and extension. *Journal of Personality Assessment* 81, 168–78.

36. Gu, X. S. & Han, S. H. (2007). Attention and reality constraints on the neural processes of empathy for pain. *NeuroImage* 36, 256–67.

37. Sterzer, P., Stadler, C., Poustka, F. & Kleinschmidt, A. (2007). A structural neural deficit in adolescents with conduct disorder and its association with lack of empathy. *NeuroImage* 37, 335–42.

38. Ramnani, N. & Owen, A. M. (2004). Anterior prefrontal cortex: Insights into function from anatomy and neuroimaging. *Nature Reviews Neuroscience* 5, 184–94.

39. Happe, F. & Frith, U. (1996). Theory of mind and social impairment in children with conduct disorder. *British Journal of Developmental Psychology* 14, 385–98.

40. Rolls, E. T. (2000). The orbitofrontal cortex and reward. *Cerebral Cortex* 10, 284–94.

41. Ragozzino, M. E. (2007). The contribution of the medial prefrontal cortex, orbitofrontal cortex, and dorsomedial striatum to behavioral flexibility. *Annals of the New York Academy of Sciences* 1121, 355–75.

42. Seguin, J. R., Arseneault, L., Boulerice, B., Harden, P. W. & Tremblay, R. E. (2002). Response perseveration in adolescent boys with stable and unstable histories of physical aggression: The role of underlying processes. *Journal of Child Psychology and Psychiatry* 43, 481–94.

43. Fairchild, G., van Goozen, S. H., Stollery, S. J. & Goodyer, I. M. (2008). Fear conditioning and affective modulation of the startle reflex in male adolescents with early-onset or adolescence-onset conduct disorder and healthy control subjects. *Biological Psychiatry* 63, 279–85.

44. Toro, R., Leonard, G., Lerner, J. V., Lerner, R. M., Perron, M., et al. (2008). Prenatal exposure to maternal cigarette smoking and the adolescent cerebral cortex. *Neuropsychopharmacology* 33, 1019–27.

45. Schirmer, A., Escoffier, N., Zysset, S., Koester, D., Striano, T. & Friederici, A. D. (2008). When vocal processing gets emotional: On the role of social orientation in relevance detection by the human amygdala. *NeuroImage* 40, 1402–10.

46. Frick, P. J., Cornell, A. H., Bodin, S. D., Dane, H. E., Barry, C. T. & Loney, B. R. (2003). Callous-unemotional traits and developmental pathways to severe conduct problems. *Developmental Psychology* 39, 246–60.

47. Happe & Frith, U. Theory of mind and social impairment in children with conduct disorder.

48. Aron, A. R., Robbins, T. W. & Poldrack, R. A. (2004). Inhibition and the right inferior frontal cortex. *Trends in Cognitive Sciences* 8, 170–77.

49. Whittle, S., Yap, M.B.H., Yucel, M., Fornito, A., Simmons, J. G., et al. (2008). Prefrontal and amygdala volumes are related to adolescents' affective behaviors during parent-adolescent interactions. *Proceedings of the National Academy of Sciences, U.S.A.* 105, 3652–57.

50. Meyer-Lindenberg, A., Buckholtz, J. W., Kolachana, B., Hariri, A. R., Pezawas, L., et al. (2006). Neural mechanisms of genetic risk for impulsivity and violence in humans. *Proceedings of the National Academy of Sciences, U.S.A.* 103, 6269–74.

51. Davidson, R. J., Putnam, K. M., & Larson, C. L. (2000). Dysfunction in the neural circuitry of emotion regulation—a possible prelude to violence. *Science* 289, 591–94.

52. Raine, A., Yang, Y., Narr, K. & Toga, A. (2011). Sex differences in orbitofrontal gray as a partial explanation for sex differences in antisocial personality. *Molecular Psychiatry* 16, 227–236.

53. Ibid.

54. Goldstein, J. M., Seidman, L. J., Horton, N. J., Makris, N., Kennedy, D. N., et al. (2001). Normal sexual dimorphism of the adult human brain assessed by in vivo magnetic resonance imaging. *Cerebral Cortex* 11, 490–97.

55. Gur, R. C., Gunning-Dixon, F., Bilker, W. B. & Gur, R. E. (2002). Sex differences in temporo-limbic and frontal brain volumes of healthy adults. *Cerebral Cortex* 12, 998–1003; Garcia-Falgueras, A., Junque, C., Gimenez, M., Caldu, X., Segovia, S. & Guillamon, A. (2006). Sex differences in the human olfactory system. *Brain Research* 1116, 103–11.

56. Good, C. D., Johnsrude, I., Ashburner, J., Henson, R.N.A., Friston, K. J. & Frackowiak, R.S.J. (2001). Cerebral asymmetry and the effects of sex and handedness on brain structure: A voxel-based morphometric analysis of 465 normal adult human brains. *NeuroImage* 14, 685–700.

57. Schlosser, R., Hutchinson, M., Joseffer, S., Rusinek, H., Saarimaki, A., et al. (1998). Functional magnetic resonance imaging of human brain activity in a verbal fluency task. *Journal of Neurology, Neurosurgery, & Psychiatry* 64, 492–98.

58. Goldstein, J. M., Jerram, M., Poldrack, R., Ahern, T., Kennedy, D. N., et al. (2005). Hormonal cycle modulates arousal circuitry in women using functional magnetic resonance imaging. *Journal of Neuroscience* 25, 9309–16.

59. McClure, E. B., Monk, C. S., Nelson, E. E., Zarahn, E., Leibenluft, E., et al. (2004). A developmental examination of gender differences in brain engagement during evaluation of threat. *Biological Psychiatry* 55, 1047–55.

60. Koch, K., Pauly, K., Kellermann, T., Seiferth, N. Y., Reske, M., et al. (2007). Gender differences in the cognitive control of emotion: An fMRI study. *Neuropsychologia* 45, 2744–54.

61. Yang, Y. & Raine, A. (2009). Prefrontal structural and functional brain imaging findings in antisocial, violent, and psychopathic individuals: A meta-analysis. *Psychiatry Research: Neuroimaging* 174, 81–88.

62. Mataro, M., Jurado, M. A., García-Sanchez, C., Barraquer, L., Costa-Jussa, F. R. & Junque, C. (2001). Long-term effects of bilateral frontal brain lesion: 60 years after injury with an iron bar. *Archives of Neurology* 58, 1139–42.

63. Ibid.

64. Bigler, E. D. (2001). Frontal lobe pathology and antisocial personality disorder. *Archives of General Psychiatry* 58, 609–11.

65. Ellenbogen, J. M., Hurford, M. O., Liebeskind, D. S., Neimark, G. B. & Weiss, D. (2005). Ventromedial frontal lobe trauma. *Neurology* 64, 757.

66. Mataro et al., Long-term effects of bilateral frontal brain lesion.

67. Sarwar, M. (1989). The septum pellucidum—normal and abnormal. *American Journal of Neuroradiology* 10, 989–1005.

68. Raine, A., Lee, L., Yang, Y. & Colletti, P. (2010). Presence of a neurodevelopmental marker for limbic maldevelopment in antisocial personality disorder and psychopathy. *British Journal of Psychiatry* 197, 186–92.

69. Gao, Y., Glenn, A. L., Schug, R. A., Yang, Y. L. & Raine, A. (2009). The neurobiology of psychopathy: A neurodevelopmental perspective. *Canadian Journal of Psychiatry* 54, 813–23.

70. Swayze, V. W., Johnson, V. P., Hanson, J. W., Piven, J., Sato, Y., et al. (2006). Magnetic resonance imaging of brain anomalies in fetal alcohol syndrome. *Pediatrics* 99, 232–40.

71. Bodensteiner, J. & Schaefer, G. (1997). Dementia pugilistica and cavum septi pellucidi: Born to box. *Sports Medicine* 24, 361–65.

72. Yang, Y., Raine, A., Karr, K. L., Colletti, P. & Toga, A. (2009). Localization of deformations within the amygdala in individuals with psychopathy. *Archives of General Psychiatry* 66, 986–94.

73. Knapska, E., Radwanska, K., Werka, T. & Kaczmarek, L. (2007). Functional internal complexity of amygdala: Focus on gene activity mapping after behavioral training and drugs of abuse. *Physiological Reviews* 87, 1113–73.

74. Ibid.

75. Raine, A., Ishikawa, S. S., Arce, E., Lencz, T., Knuth, K. H., et al. (2004). Hippocampal structural asymmetry in unsuccessful psychopaths. *Biological Psychiatry* 55, 185–91. It should be noted that this structural abnormality was specific to unsuccessful or caught psychopaths—it was not observed for successful psychopaths, who seem to lack the classical brain abnormalities found in their unsuccessful counterparts.

76. Raine, A., Buchsbaum, M. & LaCasse, L. (1997). Brain abnormalities in murderers indicated by positron emission tomography. *Biological Psychiatry* 42, 495–508.

77. Verstynen, T., Tierney, R., Urbanski, T. & Tang, A. (2001). Neonatal novelty exposure modulates hippocampal volumetric asymmetry in the rat. *NeuroReport: For Rapid Communication of Neuroscience Research* 12, 3019–22.

78. Riikonen, R., Salonen, I., Partanen, K. & Verho, S. (1999). Brain perfusion SPECT and MRI in foetal alcohol syndrome. *Developmental Medicine & Child Neurology* 41, 652–59.

79. Laakso, M. P., Vaurio, O., Koivisto, E., Savolainen, L., Eronen, M., et al. (2001). Psychopathy and the posterior hippocampus. *Behavioural Brain Research* 118, 187–93.

80. Boccardi, M., Ganzola, R., Rossi, R., Sabattoli, F., Laakso, M. P., et al. (2010). Abnormal hippocampal shape in offenders with psychopathy. *Human Brain Mapping* 31, 438–47.

81. Yang, Y. L., Raine, A., Han, C. B., Schug, R. A., Toga, A. W. & Narr, K. L. (2010). Reduced hippocampal and parahippocampal volumes in mur-

derers with schizophrenia. *Psychiatry Research: Neuroimaging* 182, 9–13. It should be noted that these volume reductions in Chinese murderers were specific to those who also presented with schizophrenia.

82. LeDoux, J. (1996). *The Emotional Brain*. New York: Simon and Schuster.

83. Swanson, L. W. (1999). Limbic system. In G. Adelman & B. H. Smith (eds.), *Encyclopedia of Neuroscience*, pp. 1053–55. Amsterdam: Elsevier.

84. Lukas, T. R. & Siegel, A. (2001). Brain structures and neurotransmitters regulating aggression in cats: Implications for human aggression. *Progress in Neuro-Psychopharmacology & Biological Psychiatry* 25, 91–140.

85. Becker, A., Grecksch, G., Bernstein, H. G., Hollt, V. & Bogerts, B. (1999). Social behaviour in rats lesioned with ibotenic acid in the hippocampus: Quantitative and qualitative analysis. *Psychopharmacology* 144, 333–38.

86. Gorenstein, E. E. & Newman, J. P. (1980). Disinhibitory psychopathy—A new perspective and a model for research. *Psychological Review* 87, 301–15.

87. The dichotic listening task is a neuropsychological measure that presents consonant-vowel stimuli ("da," "ba") simultaneously to both left and right ears. Subjects who are more left-hemisphere dominant for language report more words from the right ear. Those less lateralized for language, who have language more equally represented in both hemispheres, show a reduction in this right-ear advantage.

88. Hare, R. D. & McPherson, L. M. (1984). Psychopathy and perceptual asymmetry during verbal dichotic listening. *Journal of Abnormal Psychology* 93, 141–49.

89. Raine, A., O'Brien, M., Smiley, N., Scerbo, A. & Chen, C. J. (1990). Reduced lateralization in verbal dichotic listening in adolescent psychopaths. *Journal of Abnormal Psychology* 99, 272–77.

90. Scerbo, A., Raine, A., O'Brien, M., Chan, C. J., Rhee, C. & Smiley, N. (1990). Reward dominance and passive avoidance learning in adolescent psychopaths. *Journal of Abnormal Child Psychology* 18, 451–63.

91. Quay, H. C. (1988). The behavioral reward and inhibition system in childhood behavior disorders. In L. M. Bloomingdale (ed.), *Attention Deficit Disorder*, vol. 3, pp. 176–86. Oxford: Pergamon Press.

92. Scerbo et al. Reward dominance and passive avoidance learning in adolescent psychopaths.

93. Glenn, A. L., Raine, A., Yaralian, P. S. & Yang, Y. (2010). Increased volume of the striatum in psychopathic individuals. *Biological Psychiatry* 67, 52–58.

94. Cohen, M. X., Schoene-Bake, J. C., Elger, C. E. & Weber, B. (2009). Connectivity-based segregation of the human striatum predicts personality characteristics. *Nature Neuroscience* 12, 32–34.

95. O'Doherty, J. (2004). Reward representations and reward-related learning in the human brain: Insights from neuroimaging. *Current Opinions in Neurobiology* 14, 769–76.

96. Barkataki, I., Kumari, V., Das, M., Taylor, P. & Sharma, T. (2006): Volumetric structural brain abnormalities in men with schizophrenia or antisocial personality disorder. *Behavioral Brain Research* 15, 239–47.

97. Tiihonen, J., Kuikka, J., Bergstrom, K., Hakola, P., Karhu, J., et al. (1995). Altered striatal dopamine re-uptake site densities in habitually violent and non-violent alcoholics. *Nature Medicine* 1, 654–57.

98. Amen, D. G., Stubblefield, M., Carmichael, B. & Thisted, R. (1996). Brain SPECT findings and aggressiveness. *Annals of Clinical Psychiatry* 8, 129–37.

99. Buckholtz, J. W., Treadway, M. T., Cowan, R. L., et al. (2010). Mesolimbic dopamine reward system hypersensitivity in individuals with psychopathic traits. *Nature Neuroscience.*

100. Williamson, S., Hare, R. D. & Wong, S. (1987). Violence: Criminal psychopaths and their victims. *Canadian Journal of Behavioral Sciences* 19, 454–62.

101. Glenn, A. L., Iyer, R., Graham, J., Koleva, S. & Haidt, J. (2010). Are all types of morality compromised in psychopathy? *Journal of Personality Disorders* 23, 384–98.

102. Decety, J., Michalska, K. J., Akitsuki, Y. & Lahey, B. B. (2009): Atypical empathic responses in adolescents with aggressive conduct disorder: A functional MRI investigation. *Biological Psychology* 80, 203–11.

103. Ekman, P. & O'Sullivan, M. (1991). Who can catch a liar? *American Psychologist* 46, 913–20.

104. Porter, S., Woodworth, M. & Birt, A. R. (2000). Truth, lies, and videotape: An investigation of the ability of federal parole officers to detect deception. *Law and Human Behavior* 24, 643–58.

105. DePaulo, B. M., Stone, J. L. & Lassiter, G. D. (1985). Deceiving and Detecting Deceit. In B. R. Schenkler (ed.), *The Self and Social Life*, pp. 323–70. New York: McGraw-Hill.

106. Leach, A. M., Talwar, V., Lee, K., Bala, N. & Lindsay, R.C.L. (2004). "Intuitive" lie detection of children's deception by law enforcement officials and university students. *Law and Human Behavior* 28, 661–85.

107. Ibid.

108. Yang, Y. L ., Raine, A., Lencz, T., Bihrle, S., Lacasse, L., et al. (2005). Prefrontal structural abnormalities in liars. *British Journal of Psychiatry* 187, 320–25.

109. Yang, Y., Raine, A., Narr, K., Lencz, T., Lacasse, L., Colletti, P. & Toga, A. W. (2007). Localization of increased prefrontal white matter in pathological liars. *British Journal of Psychiatry* 190, 174–75.

110. Spence, S. A. (2005). Prefrontal white matter—the tissue of lies? Invited commentary on . . . Prefrontal white matter in pathological liars. *British Journal of Psychiatry* 187, 326–27.

111. Lee, T.M.C., Liu, H. L., Tan, L. H., Chan, C.C.H., Mahankali, S., Feng, C.-M., Hou, J., Fox, P. T. & Gao, J. H. (2002). Lie detection by functional magnetic resonance imaging. *Human Brain Mapping* 15, 157–64.

112. Paus, T., Collins, D. L., Evans, A. C., Leonard, G., Pike, B. & Zijden-bos, A. (2001). Maturation of white matter in the human brain: A review of magnetic resonance studies. *Brain Research Bulletin* 54, 255–66.

113. McCann, J. T. (1998). *Malingering and Deception in Adolescents: Assessing Credibility in Clinical and Forensic Settings,* 1st ed. Washington, D.C.: American Psychological Press.

114. Yang, Y., Raine, A., Narr, K., Lencz, T., Lacasse, L., et al. (2007). Localization of increased prefrontal white matter in pathological liars. *British Journal of Psychiatry* 190, 174–75.

115. Bengtsson, S. I., Nagy, Z., Skare, S., et al. (2005). Extensive piano practice has regionally-specific effects on white matter development. *Nature Neuroscience* 8, 1148–50.

116. Maguire, E. A., Gadian, D. G., Johnsrude, I. S., Good, C. D., Ashburner, J., et al. (2000). Navigation-related structural change in the hippocampi of taxi drivers. *Proceedings of the National Academy of Sciences U.S.A.* 97, 4398–4403.

117. Maguire, E. A., Woollett, K. & Spiers, H. J. (2006). London taxi drivers and bus drivers: A structural MRI and neuropsychological analysis. *Hippocampus* 16.

118. Lombroso, C. (1968). *Crime: Its Causes and Remedies.* Translated by H. Horton. Montclair, N.J.: Patterson Smith (originally published 1911).

119. Langton, L. & Leeper-Piquero, N. L. (2007). Can general strain theory explain white-collar crime? A preliminary investigation of the relationship between strain and select white-collar offenses. *Journal of Criminal Justice* 35, 1–15.

120. Paternoster, R. & Simpson, S. (1993). A rational choice theory of corporate crime. In R.V.G. Clarke and M. Felson (eds.), *Routine Activities and Rational Choice Theory,* pp. 37–51. New Brunswick, N.J.: Transaction.

121. Sutherland, E. H. (1949). *White Collar Crime.* New York: Rinehart and Winston.

122. Wheeler, S., Weisburd, D. & Bode, N. (1982). Sentencing the white collar offender: Rhetoric and reality, *American Sociological Review* 47, 641–59.

123. Weisburd, D., Waring, E. & Chayet, E. J. (2001). *White Collar Crime and Criminal Careers.* New York: Cambridge University Press.

124. Raine, A., Laufer, W. S., Yang, Y., Narr, K. L. & Toga, A. W. (2012). Increased executive functioning, attention, and cortical thickness in white-collar criminals. *Human Brain Mapping,* 33, 2932–40.

125. Kongs, S. K., Thompson, L. L., Iverson, G. L., et al. (2000). *Wisconsin Card Sorting Test: 64 Card Version; Professional Manual.* Odessa, Fla.: Psychological Assessment Resources.

126. Williams, L. M., Brammer, M. J., Skerrett, D., Lagopolous, J., Rennie, C., et al. (2000). The neural correlates of orienting: An integration of fMRI and skin conductance orienting. *NeuroReport* 11, 3011–15.

127. Raine & Yang, Neural foundations to moral reasoning and antisocial behavior.

128. Tsujii, T., Okada, M. & Watanabe, S. (2010). Effects of aging on hemispheric asymmetry in inferior frontal cortex activity during belief-bias syllogistic reasoning: A near-infrared spectroscopy study. *Behavioral Brain Research* 210, 178–83; Hampshire, A., Chamberlain, S. R., Monti, M. M., Duncan, J. & Owen, A. M. (2010). The role of the right inferior frontal gyrus: Inhibition and attentional control. *NeuroImage* 50, 1313–19; Brass, M., Derrfuss, J., Forstmann, B. & von Cramon, D. Y. (2005). The role of the inferior frontal junction area in cognitive control. *Trends in Cognitive Sciences* 9, 314–16.

129. Shamay-Tsoory, S. G., Tomer, R., Berger, B. D., Goldsher, D. & Aharon-Peretz, J. (2005). Impaired "affective theory of mind" is associated with right ventromedial prefrontal damage. *Cognitive and Behavioral Neurology* 18, 55–67.; Goghari, V. M. & MacDonald, A. W. (2009). The neural basis of cognitive control: Response selection and inhibition. *Brain and Cognition* 71, 72–83.; Chikazoe, J. (2010). Localizing performance of go/no-go tasks to prefrontal cortical subregions. *Current Opinion in Psychiatry* 23, 267–72.

130. Bechara et al. Deciding advantageously; Bechara, A., Damasio, H. & Damasio, A. R. (2000). Emotion, decision making and the orbitofrontal cortex. *Cerebral Cortex* 10, 295–307.

131. Kringelbach, M. L. & Rolls, E. T. (2004). The functional neuroanatomy of the human orbitofrontal cortex: Evidence from neuroimaging and neuropsychology. *Progress in Neurobiology* 72, 341–72.

132. Ibid.

133. Kringelbach, M. L. (2005). The human orbitofrontal cortex: Linking reward to hedonic experience. *Nature Reviews Neuroscience* 6, 691–702.

134. Buch, E. R., Mars, R. B., Boorman, E. D. & Rushworth, M.F.S. (2010). A network centered on ventral premotor cortex exerts both facilitatory and inhibitory control over primary motor cortex during action reprogramming. *Journal of Neuroscience* 30, 1395–1401; Pardo-Vazquez, J. L., Leboran, V. & Acuna, C. (2009). A role for the ventral premotor cortex beyond performance monitoring. *Proceedings of the National Academy of Sciences, U.S.A.* 106, 18,815–19.

135. Iacoboni, M., Molnar-Szakacs, I., Gallese, V., Buccino, G., Mazziotta, J. C. & Rizzolatti, G. (2005). Grasping the intentions of others with one's own mirror neuron system. *PLOS Biology* 3, 529–35.

136. Lawrence, E. J., Shaw, P., Giampietro, V., Surguladze, S., Brammer, M. J., et al. (2006). The role of "shared representations" in social perception and empathy: An fMRI study. *NeuroImage* 29, 1173–84.

137. Damasio, *Descartes' Error*.

138. Bechara, A. & Damasio, A. R. (2005). The somatic marker hypothesis:

A neural theory of economic decision. *Games and Economic Behavior* 52, 336–72.

139. Decety, J. & Lamm, C. (2007). The role of the right temporo-parietal junction in social interaction: How low-level computational processes contribute to meta-cognition. *The Neuroscientist* 13, 580–93.

140. Hedden, T. & Gabrieli, J.D.E. (2010). Shared and selective neural correlates of inhibition, facilitation, and shifting processes during executive control. *NeuroImage* 51, 421–31.

141. Decety & Lamm, The role of the right temporo-parietal junction in social interaction.

6. NATURAL-BORN KILLERS

1. Jonnes, B. (1992). *Voices from an Evil God*, pp. 38–39. London: Blake.
2. It should be noted that while Sutcliffe believed his victims were prostitutes, not all of them were, including one of his first attacks.
3. This is not to say that we don't sorely need more good studies on the basic scientific question of what the genetic and biological correlates of violence are. There are many more questions to be answered on the neurobiology of violence. Nevertheless, we need to move away from the unproductive debates over whether there is a biological basis to violence. We need to take what knowledge we have and begin to understand the early factors in infancy, childhood, and adolescence that give rise to these biological risk factors.
4. The Centers for Disease Control and Prevention in the United States is a government agency that focuses on health promotion and disease prevention. It is one of the main components of the Department of Health and Human Services in the United States: http://www.cdc.gov/Violence Prevention/index.html.
5. Dahlberg, L. L. & Krug, E. G. (2002). Violence, a global public health problem. In E. G. Krug, L. L. Dahlberg, J. A. Mercy, A. B. Zwi & R. Lozano (eds.), *World Report on Violence and Health*, pp. 3–21. Geneva: World Health Organization.
6. Centers for Disease Control and Prevention. The cost of violence in the United States. http://www.cdc.gov/ncipc/factsheets/CostOfViolence.htm. See also Corso, P. S., Mercy, J. A., Simon, T. R., Finkelstein, E. A. & Miller, T. R. (2007). Medical costs and productivity losses due to interpersonal and self-directed violence in the United States. *American Journal of Preventive Medicine* 32, 474–82.
7. Corso et al., Medical costs and productivity losses.
8. Miller, T. R. & Cohen, M. A. (1997). Costs of gunshot and cut/stab wounds in the United States, with some Canadian comparisons. *Accident Analysis and Prevention* 29, 329–41.

9. World Health Organization (2004). Seventh World Conference on Injury Prevention and Safety Promotion, June 6–9, Vienna, Austria. See http://www.medicalnewstoday.com/articles/9312.php.

10. John Shepherd's achievements are truly significant in the field of crime prevention—they earned him the Stockholm Prize for Criminology in 2008.

11. Raine, A., Brennan, P. & Mednick, S. A. (1994). Birth complications combined with early maternal rejection at age 1 year predispose to violent crime at age 18 years. *Archives of General Psychiatry* 51, 984–88. Sarnoff Mednick, at the University of Southern California, should be credited with originally setting up this innovative study in 1969—it became one of many collaborative research works we had together.

12. Preeclampsia is hypertension that leads to hypoxia—a relative lack of oxygen, which damages the brain, especially the hippocampus, a control area for aggression.

13. While it may seem surprising, arrests are actually better assessments of who is a violent offender than convictions. About 90 percent of arrests never end up with a criminal conviction. Plea-bargaining results in many offenders' never coming to court. If we relied on conviction data, many truly violent offenders would be misclassified as "nonviolent" and placed in the control group. Even with arrests, we are really getting at the tip of the iceberg, as many violent offenders are never detected. Yet at least with the "softer" criterion of arrest, we can capture in our analyses more of the truly violent offenders than conviction data yields.

14. The fact that the group with both birth complications and maternal rejection accounts for 18 percent of all crimes committed by the entire population highlights the influence of these risk factors in predisposing individuals to crime, but also cautions that we cannot attribute all violence to these processes. Clearly, many other factors are responsible for the remaining 82 percent of the variance in violence.

15. Raine, A., Brennan, P. & Mednick, S. A. (1997). Interaction between birth complications and early maternal rejection in predisposing individuals to adult violence: Specificity to serious, early-onset violence. *American Journal of Psychiatry* 154, 1265–71.

16. The reason not wanting the pregnancy did not interact with birth complications in predisposing individuals to adult violence may be that some mothers who initially do not want the pregnancy end up changing their minds, and go on to become affectionate, caring mothers.

17. Piquero, A. & Tibbetts, S. G. (1999). The impact of pre/perinatal disturbances and disadvantaged familial environment in predicting criminal offending. *Studies on Crime & Crime Prevention* 8, 52–70.

18. Technically speaking, regression analyses are used to uncover the interac-

tion effects found in studies of birth complications and negative home environments. Breaking down the sample into four groups is used to help illustrate the nature and direction of the interaction effects.

19. Hodgins, S., Kratzer, L. & McNeil, T. F. (2001). Obstetric complications, parenting, and risk of criminal behavior. *Archives of General Psychiatry* 58, 746–52.

20. Arsenault, L., Tremblay, R. E., Boulerice, B. & Saucier, J. F. (2002). Obstetrical complications and violent delinquency: Testing two developmental pathways. *Child Development* 73, 496–508.

21. Unlike the other studies, in which more direct measures of family adversity were employed, being an only child is not obviously an indicator of psychosocial adversity, and the meaning of this interaction requires further elucidation.

22. Kemppainen, L., Jokelainen, J., Jaervelin, M. R., Isohanni, M. & Raesaenen, P. (2001). The one-child family and violent criminality: A 31-year follow-up study of the Northern Finland 1966 birth cohort. *American Journal of Psychiatry* 158, 960–62.

23. Werner, E. E., Bierman, J. M. & French, F. E. (1971). *The Children of Kauai: A Longitudinal Study from the Prenatal Period to Age Ten.* Honolulu: University of Hawaii Press.

24. Beck, J. E. & Shaw, D. S. (2005). The influence of perinatal complications and environmental adversity on boys' antisocial behavior. *Journal of Child Psychology and Psychiatry* 46, 35–46.

25. Although the birth complication–adverse home environment interaction effects have been replicated in several countries, a study from Germany found that perinatal insult did not interact with family adversity. This may be because the sample size was small (N=322), limiting the power to detect the interaction. Alternatively, whereas in other studies the outcome was adult offending, in this study the outcome was restricted to antisocial behavior at age eight. Neurological deficits stemming from birth complications may particularly influence the more severe outcome of life-course-persistent antisocial behavior rather than the more common outcome of child antisocial behavior.

26. Raine, A., Moffitt, T. E., Caspi, A., Loeber, R., Stouthamer-Loeber, M., et al. (2005). Neurocognitive impairments in boys on the life-course persistent antisocial path. *Journal of Abnormal Psychology* 114, 38–49.

27. Beaver, K. M. & Wright, J. P. (2005). Evaluating the effects of birth complications on low self-control in a sample of twins. *International Journal of Offender Therapy and Comparative Criminology* 49, 450–71.

28. Raine, A., Buchsbaum, M. & LaCasse, L. (1997). Brain abnormalities in murderers indicated by positron emission tomography. *Biological Psychi-*

atry 42, 495–508; Laakso, M. P., Vaurio, O., Koivisto, E., Savolainen, L., Eronen, M., et al. (2001). Psychopathy and the posterior hippocampus. *Behavioural Brain Research* 118, 187–93.

29. Liu, J. H., Raine, A., Venables, P. H., Dalais, C. & Mednick, S. A. (2004). Malnutrition at age 3 years and externalizing behavior problems at ages 8, 11 and 17 years. *American Journal of Psychiatry* 161, 2005–13.

30. Liu, J. H., Raine, A., Venables, P. H., Dalais, C. & Mednick, S. A. (2003). Malnutrition at age 3 years and lower cognitive ability at age 11 years— Independence from psychosocial adversity. *Archives of Pediatrics & Adolescent Medicine* 157, 593–600.

31. For reviews see Raine, A. (1993). *The Psychopathology of Crime: Criminal Behavior as a Clinical Disorder*. San Diego: Academic Press. And also Marsman, R., Rosmalen, J.G.M., Oldehinkel, A. J., Ormel, J. & Buitelaar, J. K. (2009). Does HPA-axis activity mediate the relationship between obstetric complications and externalizing behavior problems? The TRAILS study. *European Child Adolescent Psychiatry* 18, 565–73.

32. Batstra, L., Hadders-Algra, M., Ormel, J. & Neeleman, J. (2004). Obstetric optimality and emotional problems and substance use in young adulthood. *Early Human Development* 80, 91–101; Marsman et al., Does HPA-axis activity mediate the relationship between obstetric complications and externalizing behavior problems?

33. Wagner, A. I., Schmidt, N. L., Lemery-Chalfant, K., Leavitt, L. A. & Goldsmith, H. H. (2009). The limited effects of obstetrical and neonatal complications on conduct and attention-deficit hyperactivity disorder symptoms in Middle Childhood. *Journal of Developmental and Behavioral Pediatrics* 30, 217–25.

34. Schwartz, J. (1999). *Cassandra's Daughter: A History of Psychoanalysis*, p. 225. New York: Viking/Allen Lane.

35. Bowlby, J. (1946). *Forty-four Juvenile Thieves: Their Characters and Home-life*. London: Tindall and Cox.

36. Rutter, M. (1982). *Maternal Deprivation Reassessed*, 2nd ed. Harmondsworth: Penguin.

37. Stanford, M. S., Houston, R. J. & Baldridge, R. M. (2008). Comparison of impulsive and premeditated perpetrators of intimate partner violence. *Behavioral Sciences and the Law* 26, 709–22.

38. Genesis 4:10–12.

39. Abel, E. L. (1983). *Fetal Alcohol Syndrome*. New York: Plenum.

40. Ibid.

41. Waldrop, M. F., Bell, R. Q., McLaughlin, B. & Halverson, C. F. (1978). Newborn minor physical anomalies predict attention span, peer aggression, and impulsivity at age 3. *Science* 199, 563–65.

42. Paulus, D. L. & Martin, C. L. (1986). Predicting adult temperament from minor physical anomalies. *Journal of Personality and Social Psychology* 50, 1235–39.

43. Halverson, C. F. & Victor, J. B. (1976). Minor physical anomalies and problem behavior in elementary schoolchildren. *Child Development* 47, 281–85.

44. Arseneault, L., Tremblay, R. E., Boulerice, B., Seguin, J. R. & Saucier, J. F. (2000). Minor physical anomalies and family adversity as risk factors for violent delinquency in adolescence. *American Journal of Psychiatry* 157, 917–23.

45. Pine, D. S., Shaffer, D., Schonfeld, I. S. & Davies, M. (1997). Minor physical anomalies: Modifiers of environmental risks for psychiatric impairment? *Journal of the American Academy of Child & Adolescent Psychiatry* 36, 395–403.

46. Mednick, S. A. & Kandel, E. S. (1988). Congenital determinants of violence. *Bulletin of the American Academy of Psychiatry & the Law* 16, 101–9.

47. Although both hands show the dimorphism, it is stronger on the right hand than the left hand, and in general the correlation between psychological traits and finger-digit ratios are stronger for the right hand than the left hand.

48. The specific genes in question are HoxA and HoxD.

49. Kondo, T., Zakany, J., Innis, J. W. & Duboule, D. (1997). Of fingers, toes, and penises. *Nature* 390, 29.

50. Low estrogen exposure due to diminished placental production could also be a factor in the development of shorter finger-length ratios.

51. For ease of understanding, I will use the term "longer ring finger" to describe findings that go in the male direction. Bear in mind, however, that we are talking about the ring finger *relative* to the index finger, not the absolute length of the ring finger. In the research literature, scientists discuss ratios: they divide the length of the index finger by the length of the ring finger. Because men have a bigger denominator (ring finger) in this calculation, men are reported as having "smaller second-to-fourth digit ratios" compared with women—meaning a bigger ring-finger length compared with the index finger.

52. Congenital adrenal hyperplasia is caused by a deficiency in 21-hydroxylase, which converts progesterone into corticoids; the excess of progesterone results in high concentrations of adrenal androgens.

53. Brown, W. M., Hines, M., Fane, B. A. & Breedlove, S. M. (2002). Masculinized finger length patterns in human males and females with congenital adrenal hyperplasia. *Hormones and Behavior* 42, 380–86.

54. Manning, J. T., Trivers, R. L., Singh, D. & Thornhill, R. (1999). The mystery of female beauty. *Nature* 399, 214–15.

55. It is known that prenatal androgens in particular influence digit ratios

because this digit ratio is relatively stable after birth, and is not influenced by pubertal testosterone exposure.

56. Pokrywka, L., Rachon, D., Suchecka-Rachon, K. & Bitel, L. (2005). The second to fourth digit ratio in elite and non-elite female athletes. *American Journal of Human Biology* 17, 796–800.

57. The musicians who were high ranking also had lower digit ratios than low-ranking musicians—see Sluming, V. A. & Manning, J. T. (2000). Second to fourth digit ratio in elite musicians: Evidence for musical ability as an honest signal of male fitness. *Evolution and Human Behavior* 21, 1–9.

58. Manning, J. T., Taylor, R. P. (2001). Second to fourth digit ratio and male ability in sport: Implications for sexual selection in humans. *Evolution and Human Behavior* 22, 61–69.

59. Fink, B., Manning, J. T., Williams, J.H.G. & Podmore-Nappin, C. (2007). The 2nd to 4th digit ratio and developmental psychopathology in school-aged children. *Personality and Individual Differences* 42, 369–79; Austin, E. J., Manning, J. T., McInroy, K. & Mathews, E. (2002). A preliminary investigation of the associations between personality, cognitive ability and digit ratio. *Personality and Individual Differences* 33, 1115–24.

60. Hampson, E., Ellis, C. L. & Tenk, C. M. (2008). On the relation between 2D:4D and sex-dimorphic personality traits. *Archives of Sexual Behavior* 37, 133–44.

61. Bogaert, A. F., Fawcett, C. C. & Jamieson, L. K. (2009). Attractiveness, body size, masculine sex roles and 2D:4D ratios in men. *Personality and Individual Differences* 47, 273–78.

62. Martel, M. M., Klump, K., Nigg, J. T., Breedlove, S. M. & Sisk, C. L. (2009). Potential hormonal mechanisms of attention-deficit/hyperactivity disorder and major depressive disorder: A new perspective. *Hormones and Behavior* 55, 465–79.

63. McFadden, D. & Schubel, E. (2002). Relative length of fingers and toes in human males and females. *Hormones and Behavior* 42, 492–500.

64. Bailey, A. A. & Hurd, P. L. (2005). Finger length ratio (2D:4D) correlates with physical aggression in men but not in women. *Biological Psychology* 68, 215–22. It should be noted that while effects were found in males, females showed a nonsignificant trend in the predicted direction. Furthermore, effects in males were specific to physical aggression, with no effects found for verbal aggression.

65. Burton, L. A. (2009). Aggression, gender-typical childhood play, and a prenatal hormone index. *Social Behavior and Personality* 37, 105–16. Again, males show significant effects, with females showing trends in the predicted direction.

66. Liu, J., Portnoy, J. & Raine, A. (2010). Association between a marker for

prenatal testosterone exposure and externalizing behavior problems in children. *Development and Psychopathology* 24, 771–82.

67. Cousins, A. J., Fugère, M. A. & Franklin, M. (2009). Digit ratio (2D:4D), mate guarding, and physical aggression in dating couples. *Personality and Individual Differences* 46, 709–13.

68. Ibid.

69. Coyne, S. M., Manning, J. T., Ringer, L. & Bailey, L. (2007). Directional asymmetry (right–left differences) in digit ratio (2D:4D) predict indirect aggression in women. *Personality and Individual Differences* 43, 865–72.

70. Benderlioglu, Z. & Nelson, R. J. (2004). Digit length ratios predict reactive aggression in women, but not in men. *Hormones and Behavior* 46, 558–64.

71. McIntyre, M. H., Barrett, E. A., McDermott, R., Johnson, D.D.P., Cowden, J., et al. (2007). Finger length ratio (2D:4D) and sex differences in aggression during a simulated war game. *Personality and Individual Differences* 42, 755–64.

72. Potegal, M. & Archer, J. (2004). Sex differences in childhood anger and aggression. *Psychiatric Clinics of North America* 13, 513.

73. McIntyre et al., Finger length ratio (2D:4D) and sex differences in aggression during a simulated war game.

74. Smith, L. M., Cloak, C. C., Poland, R. E., Torday, J., Ross, M. G. (2003). Prenatal nicotine increases testosterone levels in the fetus and female offspring. *Nicotine & Tobacco Research* 5, 369–74.

75. Rizwan, S., Manning, J. T., Brabin, B. J. (2007). Maternal smoking during pregnancy and possible effects of in utero testosterone: Evidence from the 2D:4D finger length ratio. *Early Human Development* 83, 87–90.

76. Malas, M. A., Dogan, S., Evcil, E. H., Desdicioglu, K. (2006). Fetal development of the hand, digits and digit ratio (2D:4D). *Early Human Development* 82, 469–75.

77. Brennan, P., Grekin, E. & Mednick, S. (1999). Maternal smoking during pregnancy and adult male criminal outcomes. *Archives of General Psychiatry* 56, 215–19.

78. Rantakallio, P., Laara, E., Isohanni, M. & Moilanen, I. (1992). Maternal smoking during pregnancy and delinquency of the offspring: An association without causation? *International Journal of Epidemiology* 21, 1106–13.

79. Räsänen, P., Hakko, H., Isohanni, M., Hodgins, S., Järvelin, M. R. & Tiihonen, J. (1999). Maternal smoking during pregnancy and risk of criminal behavior among adult male offspring in the Northern Finland 1966 Birth Cohort. *American Journal of Psychiatry* 156, 857–62.

80. Weissman, M., Warner, V., Wickramaratne, P. & Kandel, D. (1999). Maternal smoking during pregnancy and psychopathology in offspring followed to adulthood. *Journal of the American Academy of Child and Adolescent Psychiatry* 38, 892–99.

81. Wakschlag, L. & Hans, S. (2002). Maternal smoking during pregnancy and conduct problems in high-risk youth: A developmental framework. *Development and Psychopathology* 14, 351–69.

82. Wakschlag, L. S. & Keenan, K. (2001). Clinical Significance and Correlates of Disruptive Behavior in Environmentally At-Risk Preschoolers. *Journal of Clinical Child Psychology* 30, 262–75.

83. Wakschlag, L., Lahey, B., Loeber, R., Green, S., Gordon, R., et al. (1997). Maternal smoking during pregnancy and the risk of conduct disorder in boys. *Archives of General Psychiatry* 54, 670–76.

84. Day, N. L., Richardson, G. A., Goldschmidt, L. & Cornelius, M. D. (2000). Effects of prenatal tobacco exposure on preschoolers' behavior. *Journal of Developmental and Behavioral Pediatrics* 21, 180–88.

85. Fergusson, D., Woodward, L. & Horwood, L. (1998). Maternal smoking during pregnancy and psychiatric adjustment in late adolescence. *Archives of General Psychiatry* 55, 721–27.

86. Button, T.M.M., Tharpar, A. & McGuffin, P. (2005). Relationship between antisocial behaviour, attention-deficit hyperactivity disorder and maternal prenatal smoking. *British Journal of Psychiatry* 187, 155–60.

87. Although many studies have controlled for multiple confounds, including maternal and paternal antisocial behavior, it is still possible that genes could play a role. Antisocial mothers who smoke could pass their antisocial genes on to their children. One study using a twin design concluded that while there is certainly a smoking-antisocial relationship in children, it is almost entirely genetically mediated. Even in this study, authors caution that findings do not preclude an independent causal role of cigarette smoking in the genesis of child antisocial behavior; also, that findings are limited to young children, aged five to seven. They may not apply to adult offending and violence. See Maughan, B., Taylor, A., Caspi, A. & Moffitt, T. E. (2004). Prenatal smoking and early childhood conduct problems. *Archives of General Psychiatry* 6, 836–84.

88. Gatzke-Kopp, L. M. & Beauchaine, T. P. (2007). Direct and Passive Prenatal Nicotine Exposure and the Development of Externalizing Psychopathology. *Child Psychiatry and Human Development* 38, 255–69.

89. Olds, D. (1997). Tobacco exposure and impaired development: A review of the evidence. *Mental Retardation and Developmental Disabilities Research Reviews* 3, 257–69.

90. Jaddoe, V.W.V., Verburg, B. O., de Ridder, M.A.J., et al. (2007). Maternal smoking and fetal growth characteristics in different periods of pregnancy: The Generation R Study. *American Journal of Epidemiology* 165, 1207–15.

91. Toro, R., Leonard, G., Lerner, J., et al. (2008). Prenatal exposure to maternal cigarette smoking and the adolescent cerebral cortex. *Neuropsychopharmacology* 33, 1019–27.

92. Cornelius, M. D. & Day, N. L. (2009). Developmental consequences of prenatal tobacco exposure. *Current Opinion in Neurology* 22, 121–25.

93. Batstra, L., Hadders-Algra, M. & Neeleman, J. (2003). Effect of antenatal exposure to maternal smoking on behavioural problems and academic achievement in childhood; prospective evidence from a Dutch birth cohort. *Early Human Development* 75, 21–33.

94. Levin, E. D., Wilkerson, A., Jones, J. P., Christopher, N. C. & Briggs, S. J. (1996). Prenatal nicotine effects on memory in rats: Pharmacological and behavioral challenges. *Developmental Brain Research* 97, 207–15.

95. Slotkin, T. A., Epps, T. A., Stenger, M. L., Sawyer, K. J. & Seidler, F. J. (1999). Cholinergic receptors in heart and brainstem of rats exposed to nicotine during development: Implications for hypoxia tolerance and perinatal mortality. *Brain Research* 113, 1–12.

96. Huizink, A. C. & Mulder, E.J.H. (2006). Maternal smoking, drinking or cannabis use during pregnancy and neurobehavioral and cognitive functioning in human offspring. *Neuroscience and Biobehavioral Reviews* 30, 24–41.

97. Wikipedia, http://en.wikipedia.org/wiki/Robert_Alton_Harris.

98. California Department of Corrections and Rehabilitation, http://www.cdcr.ca.gov/Reports_Research/robertHarris.html.

99. Jones, K. L. & Smith, D. W. (1973). Recognition of the fetal alcohol syndrome in early infancy. *Lancet* 2, 999–1012.

100. Sampson, P. D., Streissguth, A. P., Bookstein, F. L., Little, R. E., Clarren, S. K., et al. (1997). Incidence of fetal alcohol syndrome and prevalence of alcohol-related neurodevelopmental disorder. *Teratology* 56, 317–26.

101. Streissguth, A. P., Bookstein, F. L., Barr, H. M., Sampson, P. D., O'Malley, K. & Young, J. K. (2004). Risk factors for adverse life outcomes in fetal alcohol syndrome and fetal alcohol effects. *Developmental and Behavioral Pediatrics* 25, 228–38.

102. Fast, D. K., Conry, J. & Loock, C. A. (1999). Identifying fetal alcohol syndrome among youth in the criminal justice system. *Journal of Developmental & Behavioral Pediatrics* 20, 370–72.

103. Sowell, E. R., Johnson, A., Kan, E., Lu, L. H., Van Horn, J. D., et al. (2008). Mapping white matter integrity and neurobehavioral correlates in children with fetal alcohol spectrum disorders. *Journal of Neuroscience* 28, 1313–19.

104. Connor, P. D., Sampson, P. D., Bookstein, F. L., Barr, H. M. & Streissguth, A. P. (2000). Direct and indirect effects of prenatal alcohol damage on executive function. *Developmental Neuropsychology* 18, 331–54.

105. Batstra, L., et al., Effect of antenatal exposure to maternal smoking on behavioural problems and academic achievement in childhood.

106. Riikonen, R., Salonen, I., Partanen, K. & Verho, S. (1999). Brain perfusion

SPECT and MRI in foetal alcohol syndrome. *Developmental Medicine & Child Neurology* 41, 652–59.

107. Sood, B., Delaney-Black, V., Covington, C., Nordstrom-Klee, B., Ager, J., et al. (2001). Prenatal alcohol exposure and childhood behavior at age 6 to 7 years, vol. I, Dose–response effect. *Pediatrics* 108. doi:10.1542/peds.108.

108. Qiang, M., Wang, M. W. & Elberger, A. J. (2002). Second trimester prenatal alcohol exposure alters development of rat corpus callosum. *Neurotoxicology and Teratology* 6, 719–32.

7. A RECIPE FOR VIOLENCE

1. Van der Zee, H. A. (1998). *The Hunger Winter: Occupied Holland 1944–1945.* Lincoln: University of Nebraska Press.

2. Stein, Z. (1975). *Famine and Human Development: The Dutch Hunger Winter of 1944–1945.* New York: Oxford University Press.

3. Dutch Famine of 1944: http://en.wikipedia.org/wiki/Dutch_famine_of_1944.

4. The examining physicians diagnosed antisocial personality disorder using the sixth edition of the *International Classification of Diseases,* and these diagnoses would be very similar to those used today in the *Diagnostic and Statistical Manual for Mental Disorders.*

5. Neugebauer, R., Hoek, H. W. & Susser, E. (1999). Prenatal exposure to wartime famine and development of antisocial personality disorder in early adulthood. *Journal of the American Medical Association* 4, 479–81.

6. Wong, D. L. & Hess, C. S. (2000). *Clinical Manual of Pediatric Nursing.* St. Louis: Mosby.

7. Subotzky, E. F., Heese, H. D., Sive, A. A., Dempster, W. S., Sacks, R., et al. (1992). Plasma zinc, copper, selenium, ferritin and whole blood manganese concentrations in children with kwashiorkor in the acute stage and during refeeding. *Annals of Tropical Paediatrics* 12, 13–22.

8. Friedman, M. & Orraca-Tetteh, R. (1978). Hair as an index of protein malnutrition. *Advances in Experimental Medicine and Biology* 105, 131–54.

9. Spencer, L. V. & Callen, J. P. (1987). Hair loss in systemic disease. *Dermatologic Clinics* 5, 565–70.

10. Liu, J. H., Raine, A., Venables, P. H. & Mednick, S. A. (2004). Malnutrition at age 3 years and externalizing behavior problems at ages 8, 11 and 17 years. *American Journal of Psychiatry* 161, 2005–13.

11. Shankar, N., Tandon, O. P., Bandhu, R., Madan, N. & Gomber, S. (2000). Brainstem auditory evoked potential responses in iron-deficient anemic children. *Indian Journal of Physiology and Pharmacology* 44, 297–303.

12. Los Monteros, A. E., Korsak, R. A., Tran, T., Vu, D., de Vellis, J., et al.

(2000). Dietary iron and the integrity of the developing rat brain: A study with the artificially-reared rat pup. *Cellular and Molecular Biology* 46, 501–15.

13. Bruner, A. B., Joffe, A., Duggan, A. K., Casella, J. F. & Brandt, J. (1996). Randomised study of cognitive effects of iron supplementation in non-anaemic iron-deficient adolescent girls. *Lancet* 348, 992–96; van Stuijven-berg, M. E., Kvalsvig, J. D., Faber, M., Kruger, M., Kenoyer, D. G., et al. (1999). Effect of iron-, iodine-, and beta-carotene-fortified biscuits on the micronutrient status of primary school children: A randomized controlled trial. *American Journal of Clinical Nutrition* 69, 497–503.

14. Fishman, S. M., Christian, P. & West, K. P. (2000). The role of vitamins in the prevention and control of anaemia. *Public Health Nutrition* 3, 125–50.

15. Liu, J., Raine, A., Venables, P. H., Dalais, C. & Mednick, S. A. (2003). Malnutrition at age 3 years and lower cognitive ability at age 11: Independence from social adversity. *Archives of Pediatric and Adolescent Medicine* 157, 593–600.

16. LaFree, G. (1999). A summary and review of cross-national comparative studies of homicide. In M. D. Smith & M. A. Zahn (eds.), *Homicide: A Sourcebook of Social Research*, pp. 125–45. Thousand Oaks, Calif.: Sage Publications.

17. Hibbeln, J. R. (2001). Homicide mortality rates and seafood consumption: A cross-national analysis. *World Review of Nutrition and Dietetics* 88, 41–46. Due to space limitations the figure in the text provides data from twenty-one of the twenty-six countries that Hibbeln reported on, but it retains outliers to appropriately represent the relationship that Hibbeln documented.

18. Hibbeln, J. R., Davis, J. M., Steer, C., Emmett, P., Rogers, I., et al. (2007). Maternal seafood consumption in pregnancy and neurodevelopmental outcomes in childhood (ALSPAC study): An observational cohort study. *Lancet* 369, 578–85.

19. Iribarren, C., Markovitz, J. H., Jacobs, D. R., Schreiner, P. J., Daviglus, M., et al. (2004). Dietary intake of n-3, n-6 fatty acids and fish: Relationship with hostility in young adults—the CARDIA study. *European Journal of Clinical Nutrition* 58, 24–31.

20. Stevens, L. J., Zentall, S. S., Abate, M. L., Kuczek, T. & Burgess, J. R. (1996). Omega-3 fatty acids in boys with behavior, learning, and health problems. *Physiology and Behavior* 59, 915–20.

21. Buydens-Branchey, L., Branchey, M., McMakin, D. L. & Hibbeln, J. R. (2003). Polyunsaturated fatty acid status and aggression in cocaine addicts. *Drug and Alcohol Dependence* 71, 319–23.

22. Re, S., Zanoletti, M. & Emanuele, E. (2007). Aggressive dogs are characterized by low omega-3 polyunsaturated fatty acid status. *Veterinary Research Communications* 32, 225–30.

23. McNamara, R. K. & Carlson, S. E. (2006). Role of omega-3 fatty acids in brain development and function: Potential implications for the pathogenesis and prevention of psychopathology. *Prostaglandins, Leukotrienes and Essential Fatty Acids* 75, 329–49.

24. Kitajka, K., Sinclair, A. J., Weisinger, R. S., Weisinger, H. S., Mathai, M., et al. (2004). Effects of dietary omega-3 polyunsaturated fatty acids on brain gene expression. *Proceedings of the National Academy of Sciences of the United States of America* 101 (10) 931–36.

25. Das, U. N. (2003). Long-chain polyunsaturated fatty acids in the growth and development of the brain and memory. *Nutrition* 19, 62–65.

26. Stevens, L., Zhang, W., Peck, L., Kuczek, T., Grevstad, N., et al. (2003). EFA supplementation in children with inattention, hyperactivity, and other disruptive behaviors. *Lipids* 38, 1007–21.

27. Health Statistics: Obesity. http://www.nationmaster.com/graph/hea _obe-health-obesity.

28. Protein deficiency is more of a problem in developing countries, but even in developed countries protein deficiency can be an issue in poor areas. Protein provides essential amino acids for the rapid growth of fetal tissue and plays an important role in the antioxidant system.

29. World Health Organization (2001). Iron deficiency anaemia: Assessment, prevention and control. *A Guide for Program Managers*. Geneva: World Health Organization (WHO).

30. Takeda, A., Tamano, H., Kan, F., Hanajima, T., Yamada, K., et al. (2008). Enhancement of social isolation-induced aggressive behavior of young mice by zinc deficiency. *Life Sciences* 82, 909–14.

31. Halas, E. S., Reynolds, G. M. & Sandstead, H. H. (1977). Intra-uterine nutrition and its effects on aggression. *Physiology & Behavior* 19, 653–61.

32. Walsh, W. J., Isaacson, R., Rehman, F. & Hall, A. (1997). Elevated blood copper/zinc ratios in assaultive young males. *Physiology & Behavior* 62, 327–29. In this study zinc levels were low and copper levels were high. Copper is elevated because when zinc is low, there is more bioavailability for copper.

33. Tokdemir, M., Plota, S. A., Acik, Y., Gursu, F. & Cikim, G. (2003). Blood zinc and copper concentration in criminal and noncriminal schizophrenic men. *Archives of Andrology* 49, 365–68.

34. Werbach, M. R. (1992). Nutritional influences on aggressive behavior. *Journal of Orthomolecular Medicine* 7, 45–51.

35. Rosen, G. M., Deinard, A. S., Schwartz, S., Smith, C., Stephenson, B., et al. (1985): Iron deficiency among incarcerated juvenile delinquents. *Journal of Adolescent Health Care* 6, 419–23.

36. Lozoff, B., Clark, K. M., Jing, Y., Armony-Sivan, R. & Jacobsen, S. W. (2008). Dose-response relationships between iron deficiency with or with-

out anemia and infant social-emotional behavior. *Journal of Pediatrics* 152, 696–702.

37. McBurnett, K., Raine, A., Stouthamer-Loeber, M., Loeber, R., Kumar, A. M., et al. (2005). Mood and hormone responses to psychological challenge in adolescent males with conduct problems. *Biological Psychiatry* 57, 1109–16.

38. Bennis-Taleb, N., Remacle, C., Hoet, J. J. & Reusens, B. (1999). A low-protein isocaloric diet during gestation affects brain development and alters permanently cerebral cortex blood vessels in rat offspring. *Journal of Nutrition* 129, 1613–19.

39. Takeda, A. (2000). Movement of zinc and its functional significance in the brain. *Brain Research Reviews* 34, 137–48.

40. Newman, J. P. & Kosson, D. S. (1986). Passive avoidance learning in psychopathic and non-psychopathic offenders. *Journal of Abnormal Psychology* 95, 252–56.

41. Pfeiffer, C. C. & Braverman, E. R. (1982). Zinc, the brain and behavior. *Biological Psychiatry* 17, 513–32.

42. Arnold, L. E., Pinkham, S. M. & Votolato, N. (2000). Does zinc moderate essential fatty acid and amphetamine treatment of attention-deficit/hyperactivity disorder? *Journal of Child and Adolescent Psychopharmacology* 10, 111–17.

43. King, J. C. (2000). Determinants of maternal zinc status during pregnancy. *American Journal of Clinical Nutrition* 71, 1334–43.

44. Shea-Moore, M. M., Thomas, O. P. & Mench, J. A. (1996). Decreases in aggression in tryptophan-supplemented broiler breeder males are not due to increases in blood niacin levels. *Poultry Science* 75, 370–74.

45. In many laboratories the 100-gram drink that depletes tryptophan contains a mix of fifteen amino acids, none of which are tryptophan. This increases protein synthesis in the liver, which reduces tryptophan in the plasma. In addition, these amino acids compete with tryptophan for transportation across the blood-brain barrier. Essentially, what tryptophan the participants have available to them is swamped out by the other amino acids. The placebo drink is exactly the same except that the drink is balanced with the appropriate amount of tryptophan.

46. Bond, A. J., Wingrove, J. & Critchlow, D. G. (2001). Tryptophan depletion increases aggression in women during the premenstrual phase. *Psychopharmacology* 156, 477–80; Bjork, J. M., Dougherty, D. M., Moeller, F. G., Cherek, D. R. & Swann, A. C. (1999). The effects of tryptophan depletion and loading on laboratory aggression in men: Time course and a food-restricted control. *Psychopharmacology* 142, 24–30.

47. Cherek, D. R., Lane, S. D., Pietras, C. J. & Steinberg, J. L. (2002). Effects of chronic paroxetine administration on measures of aggressive and impul-

sive responses of adult males with a history of conduct disorder. *Psychopharmacologia* 159, 266–74.

48. Rubia, K., Lee, F., Cleare, A. J., Tunstall, N., Fu, C.H.Y., et al. (2005). Tryptophan depletion reduces right inferior prefrontal activation during response inhibition in fast, event-related fMRI. *Psychopharmacology* 179, 791–803.

49. Ledbetter, L. (1979). San Francisco Tense as Violence Follows Murder Trial. *New York Times*, May 23, A1, A18.

50. White Night Riots: http://en.wikipedia.org/wiki/White_Night_Riots.

51. Turner, W. (1979). Ex-official guilty of manslaughter in slayings on coast; 3,000 protest. *New York Times*, May 22, A1, D17.

52. White Night Riots: http://en.wikipedia.org/wiki/White_Night_Riots.

53. Schoenthaler, S. J. (1982). The effect of sugar on the treatment and control of anti-social behavior: A double-blind study of an incarcerated juvenile population. *International Journal of Biosocial Research* 3, 1–9.

54. Venables, P. H. & Raine, A. (1987). Biological theory. In B. McGurk, D. Thornton & M. Williams (eds.), *Applying Psychology to Imprisonment: Theory and Practice*, pp. 3–28. London: HMSO.

55. Pelto, P. (1967). Psychological anthropology. In A. Beals & B. Stegel (eds.), *Biennial Review of Anthropology*, pp. 151–55. Stanford, Calif.: Stanford University Press.

56. Bolton, R. (1973). Aggression and hypoglycemia among the Quolla: A study in psycho-biological anthropology. *Ethology* 12, 227–57.

57. Bolton, R. (1979). Hostility in fantasy: A further test of the hypoglycaemia-aggression hypothesis. *Aggressive Behavior* 2, 257–74.

58. For a review of these studies, see Venables & Raine, Biological theory.

59. Virkkunen, M., Rissanen, A., Naukkarinen, H., Franssila-Kallunki, A., Linnoila, M., et al. (2007). Energy substrate metabolism among habitually violent alcoholic offenders having antisocial personality disorder. *Psychiatry Research* 150, 287–95.

60. Virkkunen, M., Rissanen, A., Franssila-Kallunki, A. & Tiihonen, J. (2009). Low non-oxidative glucose metabolism and violent offending: An 8-year prospective follow-up study. *Psychiatry Research* 168, 26–31.

61. McCrimmon, R. J., Ewing, F.M.E., Frier, B. M. & Deary, I. J. (1999). Anger state during acute insulin-induced hypoglycaemia. *Physiology and Behavior* 67, 35–39.

62. Moore, S. C., Carter, L. M. & van Goozen, S.H.M. (2009). Confectionery consumption in childhood and adult violence. *British Journal of Psychiatry* 195, 366–67.

63. Stewart, W. F., Schwartz, B. S., Davatzikos, C., et al. (2006). Past adult lead exposure is linked to neurodegeneration measured by brain MRI. *Neurology* 66, 1476–84.

64. CDC safety level values for bone and lead levels are somewhat different. In this case we are dealing with bone lead, and CDC safety levels for bone are defined as <15. Consequently, the average person in this study was at the very top of that safety level. Put another way, about half of the sample exceeded CDC-defined safe bone-lead levels.

65. Other affected structures included the cingulate and insula. Within the frontal lobe, the middle frontal gyrus was the area most reduced in volume.

66. Cecil, K. M., Brubaker, C. J., Adler, C. M., Dietrich, K. N., Altaye, M., et al. (2008). Decreased brain volume in adults with childhood lead exposure. *PLOS Medicine* 5, 741–50.

67. One caveat is that this sample was 90 percent African-American and these prospective brain-imaging findings could be usefully replicated on a Caucasian sample. One would expect the same findings in other ethnic groups, although it is conceivable that poorer neighborhood conditions could result in greater exposure to lead in this community sample, and possibly stronger brain-lead relationships. The ethnicity of the sample of lead workers was not reported in Cecil et al. (2008).

68. For a detailed review see Needleman, H. L., Riess, J. A., Tobin, M. J., Biesecker, G. E. & Greenhouse, J. B. (1996). Bone lead levels and delinquent behavior. *Journal of the American Medical Association* 275, 363–69.

69. Delville, Y. (1999). Exposure to lead during development alters aggressive behavior in golden hamsters. *Neurotoxicology and Teratology* 21, 445–49.

70. Wright, J. P., Dietrich, K. N., Ris, M. D., Hornung, R. W., Wessel, S. D., et al. (2008). Association of prenatal and childhood blood lead concentrations with criminal arrests in early adulthood. *PLOS Medicine* 5, 732–40.

71. These findings on early lead exposure and adult crime applied to women as well as men, with careful control for potential confounds such as maternal smoking, alcohol use, and drug use in addition to the usual social suspects such as low income.

72. Wright, J. P. et al. (2008). Association of prenatal and childhood blood lead concentrations with criminal arrests in early adulthood.

73. Wasserman, G., Staghezza-Jaramillo, B., Shrout, P., Popovac, D. & Graziano, J. (1998). The effect of lead exposure on behavior problems in preschool children. *American Journal of Public Health* 88, 481–86.

74. Chen, A., Cai, B., Dietrich, K. N., Radcliffe, J. & Rogan, W. J. (2007). Lead exposure, IQ, and behavior in urban 5- to 7-year-olds: Does lead affect behavior only by lowering IQ? *Pediatrics* 119, 650–58.

75. Nevin, R. (2000). How lead exposure relates to temporal changes in IQ, violent crime, and unwed pregnancy. *Environmental Research*, 83, 1–22.

76. Nevin, R. (2007). Understanding international crime trends: The legacy of preschool lead exposure. *Environmental Research*, 104, 315–36.

77. Reyes, J. W. (2007). Environmental policy as social policy? The impact of childhood lead exposure on crime. *BE Journal of Economic Analysis & Policy*, 7, Issue 1, Article 51, 1–41.

78. Mielke, H. W. & Zahran, S. (2012). The urban rise and fall of air lead (Pb) and the latent surge and retreat of societal violence. *Environment International*, 43, 48–55.

79. Drum, K. (2013). America's real criminal element: Lead. *Mother Jones*. January/February issue. http://www.motherjones.com/environment/2013/01/lead-crime-link-gasoline.

80. San Ysidro McDonald's Massacre: http://en.wikipedia.org/wiki/San_Ysidro_McDonald's_massacre.

81. Wilson, J. (1998). Science: The chemistry of violence. *Popular Mechanics*, April, 42–43.

82. Ibid.

83. Gottschalk, L. A., Rebello, T., Buchsbaum, M. S. & Tucker, H. G. (1991). Abnormalities in hair trace elements as indicators of aberrant behavior. *Comprehensive Psychiatry* 32, 229–37.

84. Masters, R. D., Hone, B. & Doshi, A. (1998). Environmental pollution, neurotoxicity, and criminal violence. In J. Rose (ed.), *Environmental Toxicology: Current Developments*, pp. 13–48. New York: Gordon and Breach.

85. Masters, R. D. & Coplan, M. (1999). A dynamic, multifactorial model of alcohol, drug abuse, and crime: Linking neuroscience and behavior to toxicology. *Social Science Information* 38, 591–624.

86. Pihl, R. O. & Ervin, F. (1990). Lead and cadmium in violent criminals. *Psychological Reports* 66, 839–44.

87. Marlowe, M., Cossairt, A., Moon, C., Errera, J., MacNeel, A., et al. (1985). Main and interaction effects of metallic toxins on classroom behavior. *Journal of Abnormal Child Psychology* 13, 185–98.

88. Bao, Q. S., Lu, C. Y., Song, H., Wang, M., Ling, W., et al. (2009). Behavioural development of school-aged children who live around a multimetal sulphide mine in Guangdong province, China: A cross-sectional study. *BMC Public Health* 9, 1–8.

89. Absorption of cadmium from the lungs is much more efficient—five times better—than absorption by the gut, which also helps explain why smokers have such high cadmium levels compared with nonsmokers, even when both groups have the same food intake.

90. Jarup, L. (2003). Hazards of heavy metal contamination. *British Medical Bulletin* 68, 167–82.

91. Hubbs-Tait, L., Nation, J. R., Krebs, N. F. & Bellinger, D. C. (2005). Neurotoxicants, micronutrients, and social environments: Individual and combined effects on children's development. *Psychological Science in the Public Interest* 6, 57–121.

92. Van Assche, F. J. (1998). *A Stepwise Model to Quantify the Relative Contribution of Different Environmental Sources to Human Cadmium Exposure*. Paper presented at NiCad '98, Prague, Czech Republic, September 21–22.

93. Flanagan, P. R., McLellan, J. S., Haist, J., Cherian, M. G., Chamberlain, M. J., et al. (1978). Increased dietary cadmium absorption in mice and human subjects with iron deficiency. *Gastroenterology* 74, 841–46.

94. Blum, D. (1995). Manganese an evil player in criminal urges, experts say. *The Sacramento Bee*, November 27, p. 1.

95. Gottschalk, et al., Abnormalities in hair trace elements as indicators of aberrant behavior.

96. Masters, R., Way, B., Hone, B., Grelotti, D., Gonzalez, D., et al. (1998). Neurotoxicity and violence. *Vermont Law Review* 22, 358–82.

97. Finlay, J. W. (2007). Does environmental exposure to manganese pose a health risk to healthy adults? *Nutrition Reviews* 62, 148–53.

98. Ericson, J., Crinella, F., Clarke-Stewart, K. A., Allhusen, V., Chan, T., et al. (2007). Prenatal manganese levels linked to childhood behavioral disinhibition. *Neurotoxicology and Teratology* 29, 181–87.

99. Finley, J. W. (1999). Manganese absorption and retention by young women is associated with serum ferritin concentration. *American Journal of Clinical Nutrition* 70, 37–43.

100. Zhang, G., Liu, D. & He, P. (1995). Effects of manganese on learning abilities in school children. *Zhonghua Yufang Yixue Zazhi* 29, 156–58.

101. Bowler, R. M., Mergler, D., Sassine, M. P., Laribbe, F. & Kudnell, K. (1999). Neuropsychiatric effects of manganese on mood. *Neurotoxicology* 20, 367–78.

102. Ibid.

103. Hubbs-Tait et al., Neurotoxicants, micronutrients, and social environments.

104. Grandjean, P., Weihe, P., White, R. F., Debes, F., Araki, S., et al. (1997). Cognitive deficit in 7-year-old children with prenatal exposure to methylmercury. *Neurotoxicology and Teratology* 19, 417–28.

105. Myers, G. J., Davidson, P. W., Cox, C., Shamlaye, C. F., Palumbo, D., et al. (2003). Prenatal methylmercury exposure from ocean fish consumption in the Seychelles child development study. *Lancet* 361, 1686–92.

106. Justin, H. G. & Williams, L. R. (2007). Consequences of prenatal toxin exposure for mental health in children and adolescents: A systematic review. *European Child and Adolescent Psychiatry* 16, 243–53.

107. Laing, R. D. & Esterson, A. (1970). *Sanity, Madness, and the Family: Families of Schizophrenics*. Oxford: Pelican.

108. Reiss, A. J. & Roth, J. A. (eds.). *Understanding and Preventing Violence*. Washington, D.C.: National Academy Press.

109. Raine, A. (2002). Annotation: The role of prefrontal deficits, low auto-

nomic arousal, and early health factors in the development of antisocial and aggressive behavior. *Journal of Child Psychology and Psychiatry* 43, 417–34.

110. Brennan, P. A. & Alden, A. (2005). Schizophrenia and violence: The overlap. In A. Raine (ed.), *Crime and Schizophrenia: Causes and Cures*, pp. 15–28. New York: Nova Science Publishers.

111. Torrey, E. F. (2011). Stigma and violence: Isn't it time to connect the dots? *Schizophrenia Bulletin* 37, 892–96.

112. Fazel, S., Gulati, G., Linsell, L., Geddes, J. R. & Grann, M. (2009). Schizophrenia and violence: Systematic review and meta-analysis. *PLOS Medicine* 6, 1–15.

113. Cannon, T. D. & Raine, A. (2006). Neuroanatomical and genetic influences on schizophrenia and crime: The schizophrenia-crime association. In Raine, *Crime and Schizophrenia*, pp. 219–46.

114. Raine, A. (2006). Schizotypal personality: Neurodevelopmental and psychosocial trajectories. *Annual Review of Clinical Psychology* 2, 291–326.

115. Raine, A. (1991). The Schizotypal Personality Questionnaire (SPQ): A measure of schizotypal personality based on DSM-III-R criteria. *Schizophrenia Bulletin* 17, 555–64.

116. Ibid.

117. Siever, L. J. & Davis, K. L. (2004). The pathophysiology of schizophrenia disorders: Perspectives from the spectrum. *American Journal of Psychiatry* 161, 398–413.

118. Wahlund, K. & Kristiansson, M. (2009). Aggression, psychopathy and brain imaging: Review and future recommendations. *International Journal of Law and Psychiatry* 32, 266–71.

119. Cannon & Raine, Neuroanatomical and genetic influences on schizophrenia and crime, pp. 219–46.

120. Raine, A., Fung, A. L. & Lam, B.Y.H. (2011). Peer victimization partially mediates the schizotypy—aggression relationship in children and adolescents. *Schizophrenia Bulletin*, 37, 937–45.

121. Norris, J. (1988). *Serial Killers*. New York: Anchor Books.

122. Leonard Lake: http://en.wikipedia.org/wiki/Leonard_Lake.

123. The name of the location is ominous—Calaveras is the Spanish word for skulls, and over forty-five pounds of bones were eventually excavated at Lake's hideout, the remains of many of his victims.

124. Henry, J. D., Bailey, P. E., Rendell, P. G. (2008). Empathy, social functioning and schizotypy. *Psychiatry Research* 160, 15–22.

125. Norris, *Serial Killers*, p. 152.

126. Suhr, J. A., Spitznagel, M. B. & Gunstad, J. (2006). An obsessive-compulsive subtype of schizotypy: Evidence from a nonclinical sample. *Journal of Nervous and Mental Disease* 194, 884–86.

127. Fenton, W. S., McGlashan, T. H., Victor, B. J., et al. (1997). Symptoms, subtype, and suicidality in patients with schizophrenia spectrum disorders. *American Journal of Psychiatry* 154, 199–204.

128. Raine, *Crime and Schizophrenia*.

129. Torrey, Stigma and violence.

8. THE BIOSOCIAL JIGSAW PUZZLE

1. Norris, J. (1988). *Serial Killers*. New York: Anchor Books.

2. Jones, R. G. (1992). *Lambs to the Slaughter*. London: BCA.

3. Norris, *Serial Killers*.

4. Ibid.

5. Jones, *Lambs to the Slaughter*.

6. Berry-Dee, C. (2003). *Talking with Serial Killers*. London: John Blake.

7. Hare, R. D. (1965). Acquisition and generalization of a conditioned-fear response in psychopathic and non-psychopathic criminals. *Journal of Psychology* 59, 367–70; Hare, R. D. (1970). *Psychopathy: Theory and Practice*. New York: Wiley.

8. Raine, A. & Venables, P. H. (1981). Classical conditioning and socialization—a biosocial interaction. *Personality & Individual Differences* 2, 273–83; Raine, A. & Venables, P. H. (1984). Tonic heart rate level, social class and antisocial behaviour in adolescents. *Biological Psychology* 18, 123–32.

9. Rafter, N. H. (2006). H. J. Eysenck in Fagin's kitchen: The return to biological theory in 20th-century criminology. *History of the Human Sciences* 19, 37–56.

10. Raine, A., Brennan, P. & Mednick, S. A. (1994). Birth complications combined with early maternal rejection at age 1 year predispose to violent crime at age 18 years. *Archives of General Psychiatry* 51, 984–88.

11. Raine, A. (2002). Biosocial studies of antisocial and violent behavior in children and adults: A review. *Journal of Abnormal Child Psychology* 30, 311–26.

12. Mednick, S. A. & Kandel, E. (1988). Genetic and perinatal factors in violence. In S. A. Mednick & T. Moffitt (eds.), *Biological Contributions to Crime Causation*, pp. 121–34. Dordrecht, Holland: Martinus Nijhoff.

13. Pine, D. S., Shaffer, D., Schonfeld, I. S. & Davies, M. (1997). Minor physical anomalies: Modifiers of environmental risks for psychiatric impairment? *Journal of the American Academy of Child & Adolescent Psychiatry* 36, 395–403.

14. Norris, *Serial Killers*.

15. Rose, D. (2000). *The Big Eddy Club: The Stocking Stranglings and Southern Justice*. New York: The New Press.

16. Jordan, B. L. (2000). *Murder in the Peach State*. Atlanta: Midtown Publishing Corp.

17. Norris, *Serial Killers*.

18. Bowlby, J. (1946). *Forty-four Juvenile Thieves: Their Characters and Home-Life*. London: Tindall and Cox.

19. Norris, *Serial Killers*, p. 131.

20. Raine, A., Brennan, P., Mednick, B. & Mednick, S. A. (1996). High rates of violence, crime, academic problems, and behavioral problems in males with both early neuromotor deficits and unstable family environments. *Archives of General Psychiatry* 53, 544–49.

21. While cluster analysis does not exactly carve nature at its joints, it statistically seeks out naturally occurring homogenous subgroups within a population on the basis of social and neurological risk factors in order to identify naturally occurring discrete groups. The emergence of a biosocial group with a combination of social and biological risk confirms the presence within the general population of a biosocial "at risk" group.

22. The biosocial group was specifically characterized by neurological problems, parental crime, family instability, marital conflict, and maternal rejection of the child.

23. Raine, A., Brennan, P., Mednick, B. & Mednick, S. A. (1996). High rates of violence, crime, academic problems, and behavioral problems in males with both early neuromotor deficits and unstable family environments. *Archives of General Psychiatry* 53, 544–49.

24. Brennan, P. A., Hall, J., Bor, W., Najman, J. M. & Williams, G. (2003). Integrating biological and social processes in relation to early-onset persistent aggression in boys and girls. *Developmental Psychology* 39, 309–23.

25. Räsänen, P., Hakko, H., Isohanni, M., Hodgins, S., Järvelin, M. R., et al. (1999). Maternal smoking during pregnancy and risk of criminal behavior among adult male offspring in the northern Finland 1966 birth cohort. *American Journal of Psychiatry* 156, 857–62.

26. Brennan, P. A., Grekin, E. R. & Mednick, S. A. (1999). Maternal smoking during pregnancy and adult male criminal outcomes. *Archives of General Psychiatry* 56, 215–19.

27. Gibson, C. L. & Tibbetts, S. G. (2000). A biosocial interaction in predicting early onset of offending. *Psychological Reports* 86, 509–18.

28. Caspi, A., McClay, J., Moffitt, T., Mill, J., Martin, J., et al. (2002). Role of genotype in the cycle of violence in maltreated children. *Science* 297, 851–54.

29. Farrington, D. P. (1997). The relationship between low resting heart rate and violence. In A. Raine, P. A. Brennan, D. Farrington & S. A. Mednick (eds.), *Biosocial Bases of Violence*, pp. 89–105. New York: Plenum.

30. Raine, A., Park, S., Lencz, T., Bihrle, S., LaCasse, L., et al. (2001). Reduced right hemisphere activation in severely abused violent offenders during a working memory task: An fMRI study. *Aggressive Behavior* 27, 111–29.

31. Rowe, R., Maughan, B., Worthman, C. M., Costello, E. J. & Angold, A. (2004). Testosterone, antisocial behavior, and social dominance in boys: Pubertal development and biosocial interaction. *Biological Psychiatry* 55, 546–52.

32. Feinberg, M. E., Button, T.M.M., Neiderhiser, J. M., Reiss, D. & Hetherington, E. M. (2007). Parenting and adolescent antisocial behavior and depression: Evidence of genotype x parenting environment interaction. *Archives of General Psychiatry* 64, 457–65.

33. Eysenck, H. J. (1977). *Crime and Personality*, 3rd ed. St. Albans, England: Paladin.

34. Ibid.

35. Here "good home" is a relative term, and was defined in my early fear-conditioning study as children from high social classes.

36. Rafter, H. J. Eysenck in Fagin's kitchen.

37. Other international scholars in addition to Hans Eysenck who should be acknowledged as having shaped a biosocial perspective on crime in the 1970s include Sarnoff Mednick (United States), Karl Christiansen (Denmark), Michael Wadsworth (England), and David Farrington (England). As discussed in the chapter on genetic influences, Avshalom Caspi and Terrie Moffitt took this perspective much further in their far-reaching work on the interaction between severe child abuse and the genotype conferring low levels of MAOA in predisposing to offending. Eysenck himself was half a century ahead of his time in suggesting a biosocial approach to crime, for it is only now that this approach is beginning to be embraced by a wider scientific community.

38. Raine, Biosocial studies of antisocial and violent behavior in children and adults.

39. Raine, A., Buchsbaum, M. & LaCasse, L. (1997). Brain abnormalities in murderers indicated by positron emission tomography. *Biological Psychiatry* 42, 495–508.

40. Raine, A., Stoddard, J., Bihrle, S. & Buchsbaum, M. (1998). Prefrontal glucose deficits in murderers lacking psychosocial deprivation. *Neuropsychiatry, Neuropsychology & Behavioral Neurology* 11, 1–7.

41. Damasio, A. R., Tranel, D. & Damasio, H. (1990). Individuals with sociopathic behavior caused by frontal damage fail to respond autonomically to social stimuli. *Behavioural Brain Research* 41, 81–94.

42. Miller, M. (2010). Inside the mind of a serial killer: Interview with Michael Stone. July 27. http://bigthink.com/ideas/21782.

43. Raine, Biosocial studies of antisocial and violent behavior in children and adults.

44. Mednick, S. A. (1977). A bio-social theory of the learning of law-abiding

behavior. In S. A. Mednick & K. O. Christiansen (eds.), *Biosocial Bases of Criminal Behavior.* New York: Gardner Press.

45. Raine, A. & Venables, P. H. (1981). Classical conditioning and socialization—a biosocial interaction. *Personality & Individual Differences* 2, 273–83.

46. Raine, Biosocial studies of antisocial and violent behavior in children and adults.

47. Raine, A. & Venables, P. H. (1984). Electrodermal nonresponding, antisocial behavior, and schizoid tendencies in adolescents. *Psychophysiology* 21, 424–33.

48. Maliphant, R., Hume, F. & Furnham, A. (1990). Autonomic nervous system (ANS) activity, personality characteristics, and disruptive behaviour in girls. *Journal of Child Psychology & Psychiatry & Allied Disciplines* 31, 619–28.

49. Wadsworth, M.E.J. (1976). Delinquency, pulse rate and early emotional deprivation. *British Journal of Criminology* 16, 245–56.

50. Hemming, J. H. (1981). Electrodermal indices in a selected prison sample and students. *Personality & Individual Differences* 2, 37–46.

51. Buikhuisen, W., Bontekoe, E. H., Plas-Korenhoff, C. & Van Buuren, S. (1984). Characteristics of criminals: The privileged offender. *International Journal of Law & Psychiatry* 7, 301–13.

52. Raine, A., Reynolds, C., Venables, P. H. & Mednick, S. A. (1997). Biosocial bases of aggressive behavior in childhood: Resting heart rate, skin conductance orienting and physique. In A. Raine, P. A. Brennan, D. Farrington & S. Mednick (eds.), *Biosocial Bases of Violence*, pp. 107–260. New York: Plenum.

53. Raine, A. (1987). Effect of early environment on electrodermal and cognitive correlates of schizotypy and psychopathy in criminals. *International Journal of Psychophysiology* 4, 277–87.

54. Tuvblad, C., Grann, M. & Lichtenstein, P. (2006). Heritability for adolescent antisocial behavior differs with socioeconomic status: Gene-environment interaction. *Journal of Child Psychology and Psychiatry* 47, 734–43. In this study, the negative home environment was defined on the basis of socioeconomic status. Furthermore, this moderating effect was particularly found for boys.

55. The genotype in question consisted of those with the A1 allelic form of the DRD2 gene.

56. Bechara, A., Damasio, H., Tranel, D. & Damasio, A. R. (1997). Deciding advantageously before knowing the advantageous strategy. *Science* 275, 1293–94.

57. Baker, L. A., Barton, M. & Raine, A. (2002). The Southern California Twin Register at the University of Southern California. *Twin Research* 5, 456–59.

58. Gao, Y., Baker, L. A., Raine, A., Wu, H. & Bezdjian, S. (2009). Brief Report: Interaction between social class and risky decision-making in children with psychopathic tendencies. *Journal of Adolescence* 32, 409–14.

59. Delamater, A. R. (2007). The role of the orbitofrontal cortex in sensory-specific encoding of associations in Pavlovian and instrumental conditioning. *Annals of the New York Academy of Sciences* 1121, 152–73.

60. In outlining how different brain structures may give rise to the emotional, cognitive, and behavioral risk factors for violence, I have been relatively simplistic. Limbic abnormalities may, for example, in part give rise to the more affective, emotional components of violence, but clearly there will be *interactions* between multiple brain circuits—including the orbitofrontal cortex—giving rise to any one risk factor for violence.

61. Meyer-Lindenberg, A., Buckholtz, J. W., Kolachana, B., Hariri, A. R., Pezawas, L., et al. (2006). Neural mechanisms of genetic risk for impulsivity and violence in humans. *Proceedings of the National Academy of Sciences of the United States of America* 103, 6269–74.

62. Huang, E. J. & Reichardt, L. F. (2001). Neurotrophins: Roles in neural development and function. *Annual Review of Neuroscience* 24, 677–736.

63. Gorski, J. A., Zeiler, S. R., Tamowski, S. & Jones, K. R. (2003). Brain-derived neurotrophic factor is required for the maintenance of cortical dendrites. *Journal of Neuroscience* 23, 6856–65.

64. Bueller, J. A., Aftab, M., Sen, S., Gomez-Hassan, D., Burmeister, M., et al. (2006). BDNF val(66)met allele is associated with reduced hippocampal volume in healthy subjects. *Biological Psychiatry* 59, 812–15.

65. Goldberg, T. E. & Weinberger, D. R. (2004). Genes and the parsing of cognitive processes. *Trends in Cognitive Sciences* 8, 325–35.

66. Soliman, F., Glatt, C. E., Bath, K. G., Levita, L., Jones, R. M., et al. (2010). A genetic variant BDNF polymorphism alters extinction learning in both mouse and human. *Science* 327, 863–66.

67. Oades, R. D., Lasky-Su, J., Christiansen, H., Faraone, S. V., Sonuga-Barke, E. J., et al. (2008). The influence of serotonin and other genes on impulsive behavioral aggression and cognitive impulsivity in children with attention-deficit/hyperactivity disorder (ADHD): Findings from a family-based association test (FBAT) analysis. *Behavioral and Brain Functions* 4, 48.

68. Einat, H., Manji, H. K., Gould, T. D., Du, J. & Chen, G. (2003). Possible involvement of the ERK signaling cascade in bipolar disorder: Behavioral leads from the study of mutant mice. *Drug News & Perspectives* 16, 453–63.

69. Earls, F. J., Brooks-Gunn, J., Raudenbush, S. W. & Sampson, R. J. (2002). *Project on Human Development in Chicago Neighborhoods (PHDCN): Longitudinal Cohort Study, Waves 1–3, 1994–2002.* Computer file. Ann Arbor, Mich.: Inter-University Consortium for Political and Social Research (distributor).

70. Sharkey, P. (2010). The acute effect of local homicides on children's cognitive performance. *Proceedings of the National Academy of Sciences of the United States of America* 107 (11) 733–38.

71. Meyer, G. J. et al. (2001). Psychological testing and psychological assessment: A review of evidence and issues. *American Psychologist* 56, 128–65.

72. Ibid.

73. Sharkey, The acute effect of local homicides on children's cognitive performance.

74. Oitzl, M. S., Champagne, D. L., van der Veen, R. & de Kloet, E. R. (2010). Brain development under stress: Hypotheses of glucocorticoid actions revisited. *Neuroscience and Biobehavioral Reviews* 34, 853–66.

75. Sharkey, The acute effect of local homicides on children's cognitive performance.

76. McNulty, T. L. & Bellair, P. E. (2003). Explaining racial and ethnic differences in adolescent violence: Structural disadvantage, family well-being, and social capital. *Justice Quarterly* 20, 1–31.

77. Moffitt, T. E. & Silva, P. A. (1987). WISC-R verbal and performance IQ discrepancy in an unselected cohort: Clinical significance and longitudinal stability. *Journal of Consulting & Clinical Psychology* 55, 768–74.

78. Rowe, D. C. (2002). IQ, birth weight, and number of sexual partners in white, African American, and mixed race adolescents. *Population and Environment* 23, 513–24.

79. Centers for Disease Control (2007). Youth Violence: National Statistics. http://www.cdc.gov/ViolencePrevention/youthviolence/stats_at-a_glance/hr_age-race.html.

80. Sampson, R. J., Sharkey, P. & Raudenbush, S. W. (2008). Durable effects of concentrated disadvantage on verbal ability among African-American children. *Proceedings of the National Academy of Sciences of the United States of America* 105, 845–52.

81. Winship, C. & Korenman, S. (1997). In B. Devlin, S. E. Fienberg, D. P. Resnick & K. Roeder (eds.), *Intelligence, Genes, and Success: Scientists Respond to the Bell Curve*, pp. 215–34. New York: Springer.

82. Pat Sharkey and Rob Sampson's work on ethnicity, violence, neighborhoods, and verbal ability has been confined to African-Americans and Hispanics because exposure to homicide was rare among Caucasians in their sample. Nevertheless, it is conceivable that the same negative effects on cognitive ability—and, by proxy, brain functioning—could equally apply to Caucasians.

83. Kellerman, J. (1977). Behavioral treatment of a boy with 47, XYY Karyotype. *Journal of Nervous and Mental Disease* 165, 67–71.

84. Each cell in the body has about five feet of DNA, but once it is spooled around the histone proteins it is tiny, less than 100 micrometers. Redon,

C., Pilch, D., Rogakou, E., Sedelnikova, O., Newrock, K., et al. (2002). Histone H2A variants H2AX and H2AZ. *Current Opinion in Genetics & Development* 12, 162–69.

85. Liu, D., Diorio, J., Tannenbaum, B., Caldji, C., Francis, D., et al. (1997). Maternal care, hippocampal glucocorticoid receptors, and hypothalamic-pituitary-adrenal responses to stress. *Science* 277, 1659–62.

86. Weaver, I.C.G., Meaney, M. J. & Szyf, M. (2006). Maternal care effects on the hippocampal transcriptome and anxiety-mediated behaviors in the offspring that are reversible in adulthood. *Proceedings of the National Academy of Sciences of the United States of America* 103, 3480–85.

87. Murgatroyd, C., Patchev, A. V., Wu, Y., Micale, V., Bockmühl, Y., et al. (2009). Dynamic DNA methylation programs persistent adverse effects of early-life stress. *Nature Neuroscience* 12, 1559–68.

88. Mill, J. & Petronis, A. (2008). Pre- and peri-natal environmental risks for attention-deficit hyperactivity disorder (ADHD): The potential role of epigenetic processes in mediating susceptibility. *Journal of Child Psychology and Psychiatry* 49, 1020–30.

89. Tremblay, R. E. (2010). Developmental origins of disruptive behaviour problems: The "original sin" hypothesis, epigenetics and their consequences for prevention. *Journal of Child Psychology and Psychiatry* 51, 341–67.

90. Champagne, F. A. (2010). Epigenetic influence of social experiences across the lifespan. *Developmental Psychobiology* 52, 299–311.

91. Zambrano, E., Martinez-Samayoa, P. M., Bautista, C. J., Deas, M., Guillen, L., et al. (2005). Sex differences in transgenerational alterations of growth and metabolism in progeny (F2) of female offspring (F1) of rats fed a low protein diet during pregnancy and lactation. *Journal of Physiology* 566, 225–36.

92. Chugani, H. T., Behen, M. E., Muzik, O., Juhasz, C., Nagy, F., et al. (2001). Local brain functional activity following early deprivation: A study of postinstitutionalized Romanian orphans. *NeuroImage* 14, 1290–1301.

93. Eluvathingal, T. J., Chugani, H. T., Behen, M. E., Juhasz, C., Muzik, O., et al. (2006). Abnormal brain connectivity in children after early severe socioemotional deprivation: A diffusion tensor imaging study. *Pediatrics* 117, 2093–2100.

94. Oitzl, M. S., Champagne, D. L., van der Veen, R. & de Kloet, E. R. (2010). Brain development under stress: Hypotheses of glucocorticoid actions revisited. *Neuroscience and Biobehavioral Reviews* 34, 853–66.

95. Andersen, S. L., Tomada, A., Vincow, E. S., Valente, E., Polcari, A., et al. (2008). Preliminary evidence for sensitive periods in the effect of childhood sexual abuse on regional brain development. *Journal of Neuropsychiatry and Clinical Neurosciences* 20, 292–301.

96. Kaldy, Z. & Sigala, N. (2004). The neural mechanisms of object working memory: What is where in the infant brain? *Neuroscience and Biobehavioral Reviews* 28, 113–21.

97. Alexander, G. E. & Goldman, P. S. (1978). Functional development of the dorsolateral prefrontal cortex: An analysis utilizing reversible cryogenic depression. *Brain Research* 14, 233–49.

98. Ganzel, B. L. et al. (2008). Resilience after 9/11: Multimodal neuroimaging evidence for stress-related change in the healthy adult brain. *NeuroImage* 40, 788–95.

99. Blair, R.J.R. (2007). The amygdala and ventromedial prefrontal cortex in morality and psychopathy. *Trends in Cognitive Sciences* 11, 387–92.

100. Raine, A. & Yang, Y. (2006). The neuroanatomical bases of psychopathy: A review of brain imaging findings. In C. J. Patrick (ed.), *Handbook of Psychopathy*, pp. 278–95. New York: Guilford; Raine, A. & Yang, Y. (2006). Neural foundations to moral reasoning and antisocial behavior. *Social, Cognitive, and Affective Neuroscience* 1, 203–13.

101. Blair, R.J.R. (2008). The amygdala and ventromedial prefrontal cortex: Functional contributions and dysfunction in psychopathy. *Philosophical Transactions of the Royal Society B: Biological Sciences* 363, 2557–65; Davidson, R. J., Putnam, K. M. & Larson, C. L. (2000). Dysfunction in the neural circuitry of emotion regulation—a possible prelude to violence. *Science* 289, 591–94; Kiehl, K. A. (2006). A cognitive neuroscience perspective on psychopathy: Evidence for paralimbic system dysfunction. *Psychiatry Research* 142, 107–28; Raine & Yang, The neuroanatomical bases of psychopathy; Raine & Yang, Neural foundations to moral reasoning and antisocial behavior.

102. The angular gyrus fundamentally has a cognitive role, but as with other brain regions it is multifunctional. The functional neuroanatomical model of violence is heuristic, illustrating the complexity of brain-behavior relationships.

103. Freedman, M. et al. (1998). Orbitofrontal function, object alternation and perseveration. *Cerebral Cortex* 8, 18–27.

104. Ochsner, K. N. et al. (2002). Rethinking feelings: An fMRI study of the cognitive regulation of emotion. *Journal of Cognitive Neuroscience* 14, 1215–29.

105. Bechara, A. (2004). The role of emotion in decision-making: Evidence from neurological patients with orbitofrontal damage. *Brain and Cognition* 55, 30–40.

106. Gusnard, D. A. et al. (2001). Medial prefrontal cortex and self-referential mental activity: Relation to a default mode of brain function. *Proceedings of the National Academy of Sciences of the United States of America* 98, 4259–64.

107. Rolls, E. T. (2000). The orbitofrontal cortex and reward. *Cerebral Cortex* 10, 284–94.

108. Dolan, M. & Park, I. (2002). The neuropsychology of antisocial personality disorder. *Psychological Medicine* 32, 417–27.

109. Larden, M. et al. (2006). Moral judgment, cognitive distortions and empathy in incarcerated delinquent and community control adolescents. *Psychology, Crime & Law* 12, 453–62.

110. Dinn, W. M. & Harris, C. L. (2000). Neurocognitive function in antisocial personality disorder. *Psychiatry Research* 97, 173–90.

111. Blair, K. S. et al. (2006). Impaired decision-making on the basis of both reward and punishment information in individuals with psychopathy. *Personality & Individual Differences* 41, 155–65.

112. Davidson, R. J. et al. (2000). Dysfunction in the neural circuitry of emotion regulation—a possible prelude to violence. *Science* 289, 591–94.

113. Dodge, K. A. & Frame, C. L. (1982). Social cognitive biases and deficits in aggressive boys. *Child Development* 53, 620–35.

114. Moll, J. et al. (2005). The moral affiliations of disgust: A functional MRI study. *Cognitive Behavioral Neurology* 18, 68–78.

115. Jarrard, L. E. (1993). On the role of the hippocampus in learning and memory in the rat. *Behavioral Neural Biology* 60, 9–26.

116. Greene, J. D. et al. (2001). An fMRI investigation of emotional engagement in moral judgment. *Science* 293, 2105–8.

117. Takahashi, H. et al. (2004). Brain activation associated with evaluative processes of guilt and embarrassment: An fMRI study. *NeuroImage* 23, 967–74.

118. Decety, J. & Jackson, P. L. (2006). A social-neuroscience perspective on empathy. *Current Directions in Psychological Science* 15, 54–58.

119. LeDoux, J. E. (2000). Emotion circuits in the brain. *Annual Review of Neuroscience* 23, 155–84.

120. Ochsner, K. N. et al. (2005). The neural correlates of direct and reflected self-knowledge. *NeuroImage* 28, 797–814.

121. Moll, J. et al. (2002). The neural correlates of moral sensitivity: A functional magnetic resonance imaging investigation of basic and moral emotions. *The Journal of Neuroscience* 22, 2730–36.

122. Kosson, D. S. et al. (2002). Facial affect recognition in criminal psychopaths. *Emotion* 2, 398–411.

123. Jolliffe, D. & Farrington, D. P. (2004). Empathy and offending: A systematic review and meta-analysis. *Aggression and Violent Behavior* 9, 441–76.

124. Birbaumer, N. et al. (2005). Deficient fear conditioning in psychopathy: A functional magnetic resonance imaging study. *Archives of General Psychiatry* 62, 799–805.

125. Keltner, D. et al. (1995). Facial expressions of emotion and psychopathology in adolescent boys. *Journal of Abnormal Psychology* 104, 644–52.

126. Lombardi, W. J. et al. (1999). Wisconsin card sorting test performance following head injury: Dorsolateral fronto-striatal circuit activity predicts perseveration. *Journal of Clinical and Experimental Neuropsychology* 21, 2–16.

127. Tekin, S. & Cummings, J. L. (2002). Frontal-subcortical neuronal circuits and clinical neuropsychiatry: An update. *Journal of Psychosomatic Research* 532, 647–54.

128. Antonucci, A. S. et al. (2006). Orbitofrontal correlates of aggression and impulsivity in psychiatric patients. *Psychiatry Research* 147, 213–20.

129. Dias, R. et al. (1996). Dissociation in prefrontal cortex of affective and attentional shifts. *Nature* 380, 69–72.

130. Makris, N., Biedferman, J., Velera, E. M., et al. (2007). Cortical thinning of the attention and executive function networks in adults with attention-deficit/hyperactivity disorder. *Cerebral Cortex* 17, 1364–75.

131. Dolan, M. & Park, I. (2002). The neuropsychology of antisocial personality disorder. *Psychological Medicine* 32, 417–27.

132. Seguin, J. R. et al. (2002). Response perseveration in adolescent boys with stable and unstable histories of physical aggression: The role of underlying processes. *Journal of Child Psychology & Psychiatry* 43, 481–94.

133. Völlm, B. et al. (2004). Neurobiological substrates of antisocial and borderline personality disorders: Preliminary result of a functional MRI study. *Criminal Behavior and Mental Health* 14, 39–54.

134. Simonoff, E. et al. (2004). Predictors of antisocial personality: Continuities from childhood to adult life. *British Journal of Psychiatry* 184, 118–27.

135. Raine, A., Lee, L., Yang, Y. & Colletti, P. (2010). Neurodevelopmental marker for limbic maldevelopment in antisocial personality disorder and psychopathy. *British Journal of Psychiatry* 197, 186–92.

136. George, D. T., Rawlings, R. R., Williams, W. A., Phillips, M. J., Fong, G., et al. (2004). A select group of perpetrators of domestic violence: Evidence of decreased metabolism in the right hypothalamus and reduced relationships between cortical/subcortical brain structures in position emission tomography. *Psychiatry Research: Neuroimaging* 130, 11–25.

137. Glenn, A. L., Raine, A., Yaralian, P. S. & Yang, Y. L. (2010). Increased volume of the striatum in psychopathic individuals. *Biological Psychiatry* 67, 52–58.

138. Widom, C. S. (1989). Child-abuse, neglect, and adult behavior: Research design and findings on criminality, violence, and child-abuse. *American Journal of Orthopsychiatry* 59, 355–67.

139. Fox, A. L. & Levine, B. (2005). *Extreme Killing: Understanding Serial and Mass Murder*. Thousand Oaks, Calif.: Sage Publications.

140. Norris, *Serial Killers*.
141. Ibid.
142. Lucas was released after serving only ten years of his twenty-year sentence, apparently due to prison overcrowding.
143. Norris, *Serial Killers*.
144. Hare, R. D. (1999). *Without Conscience*. New York: Guilford Press.
145. Norris, *Serial Killers*, p. 109.
146. ABC News. (2001). Henry Lee Lucas Dies in Prison. http://abcnews.go .com/US/story?id=93864&page=1#.T2aIWBGmi8A.

9. CURING CRIME

1. Moir, A. (1996). *A Mind to Crime: The Dangerous Few*. TV documentary.
2. Ibid.
3. Raine, A., Venables, P. H. & Williams, M. (1990). Relationships between central and autonomic measures of arousal at age 15 years and criminality at age 24 years. *Archives of General Psychiatry* 47, 1003–7.
4. Bouchard, T. J. (2004). Genetic influence on human psychological traits: A survey. *Current Directions in Psychological Science* 13, 148–51.
5. Räsänen, P., Hakko, H., Isohanni, M., Hodgins, S., Järvelin, M. R., et al. (1999). Maternal smoking during pregnancy and risk of criminal behavior among adult male offspring in the northern Finland 1966 birth cohort. *American Journal of Psychiatry* 156, 857–62.
6. Liu, J., Raine, A., Wuerker, A., Venables, P. H. & Mednick, S. (2009). The association of birth complications and externalizing behavior in early adolescents: Direct and mediating effects. *Journal of Research on Adolescence* 19, 93–111.
7. Neugebauer, R., Hoek, H. W. & Susser, E. (1999). Prenatal exposure to wartime famine and development of antisocial personality disorder in early adulthood. *Journal of the American Medical Association* 282, 455–62.
8. Weaver, I.C.G., Meaney, M. J. & Szyf, M. (2006). Maternal care effects on the hippocampal transcriptome and anxiety-mediated behaviors in the offspring that are reversible in adulthood. *Proceedings of the National Academy of Sciences of the United States of America* 103, 3480–85.
9. Streissguth, A. P., Bookstein, F. L., Barr, H. M., Sampson, P. D., O'Malley, K., et al. (2004). Risk factors for adverse life outcomes in fetal alcohol syndrome and fetal alcohol effects. *Journal of Developmental and Behavioral Pediatrics* 25, 228–38.
10. Olds, D., Henderson, C. R., Cole, R., et al. (1998). Long-term effects of nurse home visitation on children's criminal and antisocial behavior: 15-year follow-up of a randomized controlled trial. *Journal of the American Medical Association* 280, 1238–44.
11. Olds, D. L., Kitzman, H., Cole, R., Robinson, J., Sidora, K., et al. (2004).

Effects of nurse home-visiting on maternal life course and child development: Age 6 follow-up results of a randomized trial. *Pediatrics* 114, 1550–59.

12. Olds, D. L., Kitzman, H. J., Cole, R. E., Hanks, C. A., Arcoleo, K. J., et al. (2010). Enduring effects of prenatal and infancy home visiting by nurses on maternal life course and government spending: Follow-up of a randomized trial among children at age 12 years. *Archives of Pediatrics & Adolescent Medicine* 164, 419–24.

13. Ibid.

14. Venables, P. H. (1978). Psychophysiology and psychometrics. *Psychophysiology* 15, 302–15.

15. Raine, A., Venables, P. H., Dalais, C., Mellingen, K., Reynolds, C., et al. (2001). Early educational and health enrichment at age 3–5 years is associated with increased autonomic and central nervous system arousal and orienting at age 11 years: Evidence from the Mauritius Child Health Project. *Psychophysiology* 38, 254–66.

16. Ibid.

17. Ibid.

18. Hugdahl, K. (1995). *Psychophysiology: The Mind-Body Perspective.* Cambridge: Harvard University Press.

19. Raine, A., Mellingen, K., Liu, J. H., Venables, P. & Mednick, S. A. (2003). Effects of environmental enrichment at ages 3–5 years on schizotypal personality and antisocial behavior at ages 17 and 23 years. *American Journal of Psychiatry* 160, 1627–35.

20. Matousek, M. & Petersen, P. (1973). Frequency analysis of the EEG in normal children and adolescents. In P. Kellaway & I. Petersen (eds.), *Automation of Clinical Encephalography,* pp. 75–101. New York: Raven Press.

21. Raine et al., Effects of environmental enrichment at ages 3–5 years on schizotypal personality.

22. Elliott, D. S., Ageton, S., Huizinga, D., Knowles, B. & Canter, R. (1983). *The Prevalence and Incidence of Delinquent Behavior: 1976–1980. National Youth Survey, Report No. 26.* Boulder, Colo.: Behavior Research Institute.

23. Raine et al., Effects of environmental enrichment at ages 3–5 years on schizotypal personality.

24. The p value for the difference in court convictions between the children in the enriched group compared with controls was .07, and employed a two-tailed test. We elected to be conservative, but given the a priori prediction that crime would be reduced (and not increased) by the intervention we could have argued for the use of a one-tailed test of significance, and thus the results would have been significant at p < .035.

25. Raine et al., Effects of environmental enrichment at ages 3–5 years on schizotypal personality.

26. Gomez-Pinilla, F., Dao, L. & So, V. (1997). Physical exercise induces FGF-2 and its mRNA in the hippocampus. *Brain Research* 764, 1–8.

27. Van Praag, H., Christie, B. R., Sejnowski, T. J. & Gage, F. H. (1999). Running enhances neurogenesis, learning, and long-term potentiation in mice. *Proceedings of the National Academy of Sciences of the United States of America* 96 (13) 427–31.

28. Raine et al., Effects of environmental enrichment at ages 3–5 years on schizotypal personality.

29. Murphy, J. M., Wehler, C. A., Pagano, M. E., Little, M., Kleinman, R. E. & Jellinek, M. S. (1998). Relationship between hunger and psychosocial functioning in low-income American children. *Journal of the American Academy of Child & Adolescent Psychiatry* 37, 163–70; Smith, J., Lensing, S., Horton, J. A., Lovejoy, J., Zaghloul, S., et al. (1999). Prevalence of self-reported nutrition-related health problems in the Lower Mississippi Delta. *American Journal of Public Health* 89, 1418–21.

30. UNESCO (2007). *EFA Global Monitoring Report 2007: Strong Foundations: Early Childhood Care and Education.* Paris: UNESCO Publishing.

31. Carroll, L. (1865). *Alice's Adventures in Wonderland.* London: MacMillan.

32. Reuters (2012). Germany urged to halt castration of sex offenders. February 22. http://www.reuters.com/article/2012/02/22/us-germany-castration -idUS TRE81L18G20120222.

33. Bilefsky, D. (2009). Europeans debate castration of sex offenders. *New York Times*, March 10. http://www.nytimes.com/2009/03/11/world/europe /11castrate.html?_r=2&pagewanted=1&hp.

34. Wille, R. & Beier, K. M. (1989). Castration in Germany. *Annals of Sex Research* 2, 103–34.

35. Bradford, J. (1990). The antiandrogen and hormonal treatment of sex offenders. In W. Marshall, D. Laws & H. Barbaree (eds.), *Handbook of Sexual Assault: Issues, Theories, and Treatment of the Offender*, pp. 297–310. New York: Plenum.

36. Weinberger, L. E., Sreenivasan, S., Garrick, T. & Osran, H. (2005). The impact of surgical castration on sexual recidivism risk among sexually violent predatory offenders. *Journal of the American Academy of Psychiatry and the Law* 33, 16–36. The quote can be found on page 34.

37. Berlin, F. S. (2005). Commentary: The impact of surgical castration on sexual recidivism risk among civilly committed sexual offenders. *Journal of the American Academy of Psychiatry and the Law* 33, 37–41.

38. Lösel, F. & Schmucker, M. (2005). The effectiveness of treatment for sexual offenders: A comprehensive meta-analysis. *Journal of Experimental Criminology* 1, 117–46.

39. Reuters (2009). Poland okays forcible castration for pedophiles. Septem-

ber 25. http://www.reuters.com/article/2009/09/25/us-castration-idUSTRE 58O4LE20090925.

40. Poland to castrate sex offenders. (2008). *Belfast Telegraph*, September 26. http://www.belfasttelegraph.co.uk/news/world-news/poland-to-castrate -sex-offenders-13985385.html.

41. RT (2011). Russia introduces chemical castration for pedophiles. October 4. http://rt.com/news/pedophilia-russia-chemical-castration-059/.

42. Norman-Eady, S. (2006). OLR research report: castration of sex offenders. http://www.cga.ct.gov/2006/rpt/2006-R-0183.htm.

43. A child sex offender in Wisconsin's Penal Code 302.11 is defined as someone having intercourse with a child under the age of thirteen.

44. Grubin, D. & Beech, A. (2010). Chemical castration for sex offenders. *British Medical Journal* 340, 433–34.

45. *The Adventures of Tintin.* http://en.wikipedia.org/wiki/The_Adventures_of _Tintin.

46. Lekhwani, M., Nair, C., Nikhinson, I. & Ambrosini, P. J. (2004). Psychotropic prescription practices in child psychiatric inpatients 9 years old and younger. *Journal of Child and Adolescent Psychopharmacology* 14, 95–103; Gilligan, J. & Lee, B. (2004). The psychopharmacological treatment of violent youth. *Annals of the New York Academy of Sciences* 1036, 356–81.

47. Jensen, P. S., Youngstrom, E. A., Steiner, H., Findling, R. L., Meyer, R. E., et al. (2007). Consensus report on impulsive aggression as a symptom across diagnostic categories in child psychiatry: Implications for medication studies. *Journal of the American Academy of Child and Adolescent Psychiatry* 46, 309–22.

48. Pappadopulos, E., Woolston, S., Chait, A., Perkins, M., Connor, D. F. & Jensen, P. S. (2006). Pharmacotherapy of aggression in children and adolescents: Efficacy and effect size. *Journal of the Canadian Academy of Child and Adolescent Psychiatry* 15, 27–39.

49. The effect size quoted here for pharmacological treatment effects are Cohen's *d*.

50. "Atypical antipsychotics" is the more formal term used to describe "newer generation" antipsychotics. While originally developed for the treatment of psychotic disorders such as schizophrenia and bipolar depression, over the past fifteen years they have been increasingly used to treat childhood aggression. Examples of atypicals would be risperidone and olanzapine. Their advantage, relative to their efficacy, is that atypical antipsychotics do not have the more severe side effects of more traditional antipsychotic medications such as tardive dyskinesia. Nevertheless they do have some side effects, including weight gain.

51. Effect sizes reported in this review of psychopharmacology of aggression are Cohen's *d*.

52. Connor, D. F., Glatt, S. J., Lopez, I. D., Jackson, D. & Melloni, R. H. (2002). Psychopharmacology and aggression, vol. 1, A meta-analysis of stimulant effects on overt/covert aggression-related behaviors in ADHD. *Journal of the American Academy of Child and Adolescent Psychiatry* 41, 253–61.

53. Connor, D. F., Carlson, G. A., Chang, K. D., Daniolos, P. T., Ferziger, R., et al. (2006). Juvenile maladaptive aggression: A review of prevention, treatment, and service configuration and a proposed research agenda. *Journal of Clinical Psychiatry* 67, 808–20.

54. Lopez-Larson, M. & Frazier, J. A. (2006). Empirical evidence for the use of lithium and anticonvulsants in children with psychiatric disorders. *Harvard Review of Psychiatry* 14, 285–304.

55. Soller, M. V., Karnik, N. S. & Steiner, H. (2006). Psychopharmacologic treatment in juvenile offenders. *Child and Adolescent Psychiatric Clinics of North America* 15, 477–99.

56. Connor, D. F., Boone, R. T., Steingard, R. J., Lopez, I. D. & Melloni, R. H. (2003). Psychopharmacology and aggression, vol. 2: A meta-analysis of nonstimulant medication effects on overt aggression-related behaviors in youth with SED. *Journal of Emotional and Behavioral Disorders* 11, 157–68.

57. Connor et al., Juvenile maladaptive aggression; Pappadopulos et al., Pharmacotherapy of aggression in children and adolescents; Jensen et al., Consensus report on impulsive aggression as a symptom across diagnostic categories in child psychiatry.

58. Maughan, D. R., Christiansen, E., Jenson, W. R. & Clark, E. (2005). Behavioral parent training as a treatment for externalizing behaviors and disruptive behavior disorders: A meta-analysis. *School Psychology Review* 34, 267–86.

59. Connor et al., Juvenile maladaptive aggression.

60. Connor et al., Psychopharmacology and aggression, vol. 2.; Connor et al., Juvenile maladaptive aggression; Pappadopulos et al., Pharmacotherapy of aggression in children and adolescents.

61. Connor et al., Psychopharmacology and aggression, vol. 1.

62. Staller, J. A. (2007). Psychopharmacologic treatment of aggressive preschoolers: A chart review. *Progress in Neuro-Psychopharmacology & Biological Psychiatry* 31, 131–35.

63. The anticonvulsants used in this community study consisted of phenytoin, carbamazepine, and valproate.

64. Stanford, M. S., Helfritz, L. E., Conklin, S. M., Villemarette-Pittman, N. R., Greve, K. W., et al. (2005). A comparison of anticonvulsants in the treatment of impulsive aggression. *Experimental and Clinical Psychopharmacology* 13, 72–77.

65. Barratt, E. S., Stanford, M. S., Felthous, A. R. & Kent, T. A. (1997). The effects of phenytoin on impulsive and premeditated aggression: A controlled study. *Journal of Clinical Psychopharmacology* 17, 341–49; Stanford, M. S., Houston, R. J., Mathias, C. W., Greve, K. W., Villemarette-Pittman, N. R., et al. (2001). A double-blind placebo-controlled crossover study of phenytoin in individuals with impulsive aggression. *Psychiatry Research* 103, 193–203.

66. Stoll, A. L. (2001). *The Omega-3 Connection*. New York: Simon and Schuster.

67. Ibid., p. 150.

68. Raine, A. & Mahoomed, T. (2012). *A Randomized, Double-blind, Placebo-controlled Trial of Omega-3 on Aggression and Delinquency*. Paper presentation, the Stockholm Symposium, Stockholm, Sweden, June 13.

69. Hibbeln, J. (2012). Personal communication. Philadelphia, April 12.

70. Food for court: Diet and crime. (2005). *Magistrate* 61, 5.

71. Gesch, C. B., Hammond, S. M., Hampson, S. E., Eves, A. & Crowder, M. J. (2002). Influence of supplementary vitamins, minerals and essential fatty acids on the antisocial behaviour of young adult prisoners: Randomised, placebo-controlled trial. *British Journal of Psychiatry* 181, 22–28.

72. Zaalberg, A., Nijman, H., Bulten, E., Stroosma, L. & van der Staak, C. (2010). Effects of nutritional supplements on aggression, rule-breaking, and psychopathology among young adult prisoners. *Aggressive Behavior* 36, 117–26.

73. Clayton, E. H., Hanstock, T. L., Hirneth, S. J., Kable, C. J., Garg, M. L., et al. (2009). Reduced mania and depression in juvenile bipolar disorder associated with long-chain omega-3 polyunsaturated fatty acid supplementation. *European Journal of Clinical Nutrition* 63, 1037–40.

74. Fontani, G., Corradeschi, F., Felici, A., Alfatti, F., Migliorini, S., et al. (2005). Cognitive and physiological effects of omega-3 polyunsaturated fatty acid supplementation in healthy subjects. *European Journal of Clinical Investigation* 35, 691–99.

75. Hamazaki, T., Sawazaki, S., Itomura, M., Asaoka, E., Nagao, Y., et al. (1996). The effect of docosahexaenoic acid on aggression in young adults: A placebo-controlled double-blind study. *Journal of Clinical Investigation* 97, 1129–33.

76. Gustafsson, P. A., Birberg-Thornberg, U., Duchen, K., Landgren, M., Malmberg, K., et al. (2010). EPA supplementation improves teacher-rated behaviour and oppositional symptoms in children with ADHD. *Acta Paediatrica* 99, 1540–49.

77. Hamazaki, T., Thienprasert, A., Kheovichai, K., Samuhaseneetoo, S., Nagasawa, T., et al.(2002). The effect of docosahexaenoic acid on aggression in elderly Thai subjects: a placebo-controlled double-blind study. *Nutritional Neuroscience* 5, 37–41. It should be noted that although DHA

reduced aggression in university workers, this effect was not observed in villagers.

78. Zanarini, M. C. & Frankenburg, F. R. (2003). Omega-3 fatty acid treatment of women with borderline personality disorder: A double-blind, placebo-controlled pilot study. *American Journal of Psychiatry* 160, 167.

79. Stevens, L., Zhang, W., Peck, L., Kuczek, T., Grevstad, N., et al. (2003). EFA supplementation in children with inattention, hyperactivity, and other disruptive behaviors. *Lipids* 38, 1007–21.

80. Shoham, S. & Youdim, M. B. (2002). The effects of iron deficiency and iron and zinc supplementation on rat hippocampus ferritin. *Journal of Neural Transmission* 109, 1241–56.

81. Smit, E. N., Muskiet, F. A. & Boersma, E. R. (2004). The possible role of essential fatty acids in the pathophysiology of malnutrition: A review. *Prostaglandins, Leukorienes and Essential Fatty Acids* 71, 241–50.

82. Not all studies have found that omega-3 supplementation reduces antisocial behavior; see, for example: Hirayama, S., Hamazaki, T. & Terasawa, K. (2004). Effect of docosahexaenoic acid-containing food administration on symptoms of attention-deficit/hyperactivity disorder: A placebo-controlled double-blind study. *European Journal of Clinical Nutrition* 58, 467–73. Another study failed to find effects in English schoolchildren: Kirby, A., Woodward, A., Jackson, S., Wang, Y. & Crawford, M. (2010). A double-blind, placebo-controlled study investigating the effects of omega-3 supplementation in children aged 8–10 years from a mainstream school population. *Research in Developmental Disabilities* 31, 718–30. Nevertheless, some studies with statistically nonsignificant findings still find a 29 percent reduction in aggression. See, for example: Hallahan, B., Hibbeln, J. R., Davis, J. M. & Garland, M. R. (2007). Omega-3 fatty acid supplementation in patients with recurrent self-harm: Single-centre double-blind randomised controlled trial. *British Journal of Psychiatry* 190, 118–22.

83. Amminger, G. P., Schafer, M. R., Papageorgiou, K., Klier, C. M., Cotton, S. M., et al. (2010). Long-chain omega-3 fatty acids for indicated prevention of psychotic disorders: A randomized, placebo-controlled trial. *Archives of General Psychiatry* 67, 146–54.

84. Raine et al., Effects of environmental enrichment at ages 3–5 years on schizotypal personality.

85. Surmeli, T. & Edem, A. (2009). QEEG guided neurofeedback therapy in personality disorders: 13 case studies. *Clinical EEG and Neuroscience* 40, 5–10.

86. Davidson, R. J., Kabat-Zinn, J., Schumacher, J., Rosenkranz, M., Muller, D., et al. (2003). Alterations in brain and immune function produced by mindfulness meditation. *Psychosomatic Medicine* 65, 564–70.

87. Holzel, B. K., Carmody, J., Vangel, M., Congleton, C., Yerramsetti, S. M.,

et al. (2011). Mindfulness practice leads to increases in regional brain gray matter density. *Psychiatry Research: Neuroimaging* 191, 36–43.

88. Davidson et al., Alterations in brain and immune function produced by mindfulness meditation.

89. Lutz, A., Brefczynski-Lewis, J., Johnstone, T. & Davidson, R. J. (2008). Regulation of the neural circuitry of emotion by compassion meditation: Effects of meditative expertise. *PLOS One* 3.

90. Brefczynski-Lewis, J. A., Lutz, A., Schaefer, H. S., Levinson, D. B. & Davidson, R. J. (2007). Neural correlates of attentional expertise in long-term meditation practitioners. *Proceedings of the National Academy of Sciences of the United States of America* 104 (11), 483–88.

91. Lutz, A., Greischar, L. L., Rawlings, N. B., Ricard, M. & Davidson, R. J. (2004). Long-term meditators self-induce high-amplitude gamma synchrony during mental practice. *Proceedings of the National Academy of Sciences of the United States of America* 101 (16), 369–73.

92. Holzel, B. K., Carmody, J., Vangel, M., Congleton, C., Yerramsetti, S. M., et al. (2011). Mindfulness practice leads to increases in regional brain gray matter density. *Psychiatry Research: Neuroimaging* 191, 36–43.

93. Gregg, T. R. & Siegel, A. (2001). Brain structures and neurotransmitters regulating aggression in cats: Implications for human aggression. *Progress in Neuro-Psychopharmacology & Biological Psychiatry* 25, 91–140.

94. Oitzl, M. S., Champagne, D. L., van der Veen, R. & de Kloet, E. R. (2010). Brain development under stress: Hypotheses of glucocorticoid actions revisited. *Neuroscience and Biobehavioral Reviews* 34, 853–66.

95. Kaldy, Z. & Sigala, N. (2004). The neural mechanisms of object working memory: What is where in the infant brain? *Neuroscience and Biobehavioral Reviews* 28, 113–21.

96. Lazar, S. W., Kerr, C. E., Wasserman, R. H., Gray, J. R., Greve, D. N., et al. (2005). Meditation experience is associated with increased cortical thickness. *NeuroReport* 16, 1893–97.

97. Abrams, A. I. & Siegel, L. M. (1978). The Transcendental Meditation program and rehabilitation at Folsom State Prison: A cross validation study. *Criminal Justice and Behavior* 5, 3–20.

98. Orme-Johnson, D. W. & Moore, R. M. (2003). First prison study using the Transcendental Meditation program: La Tuna Federal Penitentiary, 1971. *Journal of Offender Rehabilitation* 36, 89–95.

99. Samuelson, M., Carmody, J., Kabat-Zinn, J. & Bratt, M. A. (2007). Mindfulness-based stress reduction in Massachusetts correctional facilities. *The Prison Journal* 87, 254–68.

100. Chandiramani, K., Verma, S. K. & Dhar, P. L. (1995). *Psychological Effects of Vipassana on Tihar Jail Inmates: Research Report*. Igatpuri, Maharashtra, India: Vipassana Research Institute.

101. Himelstein, S. (2011). Meditation research: The state of the art in correctional settings. *International Journal of Offender Therapy and Comparative Criminology* 55, 646–61.
102. Wupperman, P., Marlatt, G. A., Cunningham, A., Bowen, S., Berking, M., et al. (2012). Mindfulness and modification therapy for behavioral dysregulation: Results from a pilot study targeting alcohol use and aggression in women. *Journal of Clinical Psychology* 68, 50–66.
103. Robins, C. J., Keng, S. L., Ekblad, A. G. & Brantley, J. G. (2012). Effects of mindfulness-based stress reduction on emotional experience and expression: A randomized controlled trial. *Journal of Clinical Psychology* 68, 117–31.
104. Warnecke, E., Quinn, S., Ogden, K., Towle, N. & Nelson, M. R. (2011). A randomised controlled trial of the effects of mindfulness practice on medical student stress levels. *Medical Education* 45, 381–88.
105. Witkiewitz, K. & Bowen, S. (2010). Depression, craving, and substance use following a randomized trial of mindfulness-based relapse prevention. *Journal of Consulting and Clinical Psychology* 78, 362–74.
106. Brewer, J. A., Mallik, S., Babuscio, T. A., Nich, C., Johnson, H. E., et al. (2011). Mindfulness training for smoking cessation: Results from a randomized controlled trial. *Drug and Alcohol Dependence* 119, 72–80.
107. Geschwind, N., Peeters, F., Drukker, M., van Os, J. & Wichers, M. (2011). Mindfulness training increases momentary positive emotions and reward experience in adults vulnerable to depression: A randomized controlled trial. *Journal of Consulting and Clinical Psychology* 79, 618–28.
108. Kabat-Zinn, J. (2005). *Coming to Our Senses: Healing Ourselves and the World Through Mindfulness*. New York: Hyperion.
109. Davidson, R. J. (1992). Emotion and affective style: Hemispheric substrates. *Psychological Science* 3, 39–43.
110. Kabat-Zinn, J., Massion, A. O., Kristeller, J., Peterson, L. G., Fletcher, K. E., et al. (1992). Effectiveness of a meditation-based stress reduction program in the treatment of anxiety disorders. *American Journal of Psychiatry* 149, 936–43.
111. Sherman, L. W., Gottfredson, D., MacKenzie, D., Reuter, P., Eck, J. & Bushway, S. (1997). *Preventing Crime: What Works, What Doesn't, What's Promising*. A Report to the U.S. Congress. Washington, D.C.: U.S. Department of Justice.

10. THE BRAIN ON TRIAL

1. Burns, J. M. & Swerdlow, R. H. (2003). Right orbitofrontal tumor with pedophilia symptom and constructional apraxia sign. *Archives of Neurology* 60, 437–40.
2. The pseudonym "Mr. Oft" was coined by my good friend and colleague

Dr. Stephen Morse, a professor of law at the University of Pennsylvania. Stephen first introduced me to the case. "Oft" is an acronym for "orbito-frontal tumor."

3. Burns & Swerdlow, Right orbitofrontal tumor with pedophilia symptom and constructional apraxia sign.

4. Crick, F. (1994). *The Astonishing Hypothesis: The Scientific Search for the Soul.* New York: Touchstone.

5. Documents obtained by the defense team from Children's National Medical Center in Washington, D.C., were at trial to prove how brutal his childhood beatings were.

6. Gusnard, D. A. et al. (2001). Medial prefrontal cortex and self-referential mental activity: Relation to a default mode of brain function. *Proceedings of the National Academy of Sciences of the United States of America* 98, 4259–64; Antonucci, A. S. et al. (2006). Orbitofrontal correlates of aggression and impulsivity in psychiatric patients. *Psychiatry Research* 147, 213–20.

7. Freedman, M. et al. (1998). Orbitofrontal function, object alternation and perseveration. *Cerebral Cortex* 8, 18–27; Shamay-Tsoory, S. G. et al. (2005). Impaired "affective theory of mind" is associated with right ventromedial prefrontal damage. *Cognitive Behavioral Neurology* 18, 55–67.

8. Bechara, A., Damasio, H., Tranel, D. & Damasio, A. R. (1997). Deciding advantageously before knowing the advantageous strategy. *Science* 275, 1293–94; Damasio, A. R., Tranel, D. & Damasio, H. (1990). Individuals with sociopathic behavior caused by frontal damage fail to respond autonomically to social stimuli. *Behavioural Brain Research* 41, 81–94.

9. Raine, A., Meloy, J. R., Bihrle, S., Stoddard, J., LaCasse, L., et al. (1998). Reduced prefrontal and increased subcortical brain functioning assessed using positron emission tomography in predatory and affective murderers. *Behavioral Sciences & the Law* 16, 319–32.

10. It also has to be recognized that Donta Page could have left the house after going outside to take money from the car. The fact that he came back into the house instead of leaving the scene of the burglary suggests some degree of premeditation. This admixture of lack of planning combined with some degree of regulatory control is not uncommon in murderers, and it is a difficult task to clearly divide murders into "impulsive" versus "planned."

11. Centers for Disease Control and Prevention. Sexually transmitted diseases (STDs). http://www.cdc.gov/std/pregnancy/STDFact-Pregnancy .htm.

12. Raine, A., Brennan, P. & Mednick, S. A. (1994). Birth complications combined with early maternal rejection at age 1 year predispose to violent crime at age 18 years. *Archives of General Psychiatry* 51, 984–88.

13. Farrington, D. P. (2005). Childhood origins of antisocial behavior. *Clini-*

cal Psychology & Psychotherapy 12, 177–90; Loeber, R. & Farrington, D. P. (2000). Young children who commit crime: Epidemiology, developmental origins, risk factors, early interventions, and policy implications. *Development & Psychopathology* 12, 737–62.

14. Jackson, S. (2001). Dead reckoning. *Denver Westward News.* June 28.
15. Ibid.
16. Federal Bureau of Investigation (2011). *Uniform Crime Reports.* http://www.fbi.gov/about-us/cjis/ucr/ucr#ucr_cius.
17. Jenkins, A. C. & Mitchell, J. P. (2011). Medial prefrontal cortex subserves diverse forms of self-reflection. *Social Neuroscience* 6, 211–18. While the medial prefrontal cortex is the area most robustly associated with self-reflection and self-referential thinking, other brain areas have been implicated as well, including the anterior cingulate and the posterior cingulate, areas also found to be dysfunctional in offenders. In particular, the medial prefrontal cortex and the anterior cingulate cortex appear to underlie thinking about aspirations and hopes, while the posterior cingulate is particularly activated when reflecting on one's duties and obligations. It has been hypothesized that the medial prefrontal cortex is more linked to an inward-directed focus, whereas the posterior cingulate is more associated with an outward-directed, social, or contextual focus. See also Johnson, M. K., Raye, C. L., Mitchell, K. J., et al. (2006). Dissociating medial frontal and posterior cingulate activity during self-reflection. *Social, Cognitive, and Affective Neuroscience* 1, 56–64.
18. You will recall that Mr. Oft did indeed tell the hospital authorities that he felt that if released he would rape his landlady. Bear in mind, however, that he was about to go to prison, and this could be construed as a simple con to remain in the more benign environment of a psychiatric hospital.
19. Gorman-Smith, D., Henry, D. B. & Tolan, P. H. (2004). Exposure to community violence and violence perpetration: The protective effects of family functioning. *Journal of Clinical and Adolescent Psychology* 33, 439–49.
20. Raine, A., Venables, P. H. & Williams, M. (1996). Better autonomic conditioning and faster electrodermal half-recovery time at age 15 years as possible protective factors against crime at age 29 years. *Developmental Psychology* 32, 624–30.
21. Raine, A., Venables, P. H. & Williams, M. (1995). High autonomic arousal and electrodermal orienting at age 15 years as protective factors against criminal behavior at age 29 years. *American Journal of Psychiatry* 152, 1595–1600.
22. It is ironic that the stimulus that led Page to slitting Peyton Tuthill's throat was the fact that he could not stand her screaming. It was his own screaming as a baby that caused his mother to shake him vigorously and repeatedly. This shaking is one cause of prefrontal dysfunction, the risk factor

we documented in Page and that we believe was instrumental in his killing of Peyton Tuthill.

23. Kershaw, I. (2008). *Hitler: A Biography*. New York: W. W. Norton & Company.

24. Ibid.

25. Fulda, B. (2009). *Press and Politics in the Weimar Republic*. Oxford University Press.

26. CNN (2007). Amish grandfather: "We must not think evil of this man." December 10.

27. Gottlieb, D. (2006). Not Always Divine. *Cross-Currents*. October 17. http://www.cross-currents.com/archives/2006/10/17/not-always-divine/.

28. Jacoby, J. (2006). Undeserved forgiveness. *Boston Globe*. October 8.

29. This seminar series was conducted by AAAS in collaboration with the Federal Judicial Center and the National Center for State Courts, with funding from the Dana Foundation. It has been ongoing since 2006, and there are concerted efforts to bring the judiciary up to speed on the latest advances in neuroscience, how new techniques and knowledge may inform legal decision making, and what their limitations are.

30. Pedophilia is indeed viewed by the medical profession as a clinical disorder, and is a condition outlined in detail in *DSM*-4.

31. Morse, S. J. (2011). Mental disorder and the criminal law. *Journal of Criminal Law and Criminology* 101, 885–968.

32. Such affirmative defenses are not restricted to mental illness. Another example is acting in self-defense. You may have the "mens rea," or guilty mind—you may know what you are doing—but you are acting in self-defense against a deadly aggressor. In this case, you are not a "responsible agent."

33. You would also not be held responsible if you were sufficiently coerced or compelled to commit a criminal act. For example, if somebody holds a gun to your head and threatens to kill you unless you perform a sex act on a third party.

34. *Mindshock: Sex on the Brain*. (2006). Channel Four. Tiger Aspect Productions.

35. Ibid.

36. Ibid.

37. Damasio, A. R. (2000). A neural basis for sociopathy. *Archives of General Psychiatry* 57, 128–29.

38. Morse, S. J. (2008). Psychopathy and criminal responsibility. *Neuroethics* 1, 205–12.

II. THE FUTURE

1. Kip's anxiety after shooting his father was documented by a friend, Tony, who by happenstance called Kip after his father had been killed but before

his mother arrived home. The call lasted an hour, and Tony sensed that Kip was on edge, pacing around his room, and repeatedly commenting that his mother was not home yet.

2. Although Kip Kinkel injured twenty-five students at school, he was charged with an additional assault on an arresting police officer who had to use pepper spray to subdue him. This assault has been construed as an attempt by Kip to goad the police officer into shooting him.

3. Konkol, R. J. (1999). *Expert Witness Testimony*. November. http://www .pbs.org/wgbh/pages/frontline/shows/kinkel/trial/konkol.html.

4. The scan in question was a SPECT scan (single-photon emission computed tomography), a nuclear-medicine technique that uses a gamma camera and gamma rays to create a three-dimensional image of the functioning of the brain.

5. *Frontline: The Killer at Thurston High*. (2000). WGBH Educational Foundation. http://www.pbs.org/wgbh/pages/frontline/shows/kinkel/.

6. Ibid.

7. Fitzgibbon, C. (2007). *Sunshine and Shadows: Reflections of a Macmillan Nurse*, pp. 31–32. Doncaster: Encircling Publications.

8. The rapidity of Roma's death may likely be the result of her having had acute myeloid leukemia, which is the most common acute form of leukemia that affects adults and is known to kill within weeks. See Vardiman, J. W., Harris, N. L. & Brunning, R. D. (2002). The World Health Organization (WHO) classification of the myeloid neoplasms. *Blood* 100, 292–302.

9. Sontag, S. (1978). *Illness as Metaphor*. New York: Picador.

10. Fitzgibbon, *Sunshine and Shadows*, p. 32.

11. Raine, A. (1993). *The Psychopathology of Crime: Criminal Behavior as a Clinical Disorder*. San Diego: Academic Press.

12. Ibid.

13. Spitzer, R. L. (1999). Harmful dysfunction and the DSM definition of mental disorder. *Journal of Abnormal Psychology* 108, 430–32.

14. At the time of writing, *DSM-4* is being revised for the 2013 publication of *DSM-5*.

15. American Psychiatric Association (2012). *DSM-5 Development: Definition of a Mental Disorder*. http://www.dsm5.org/ProposedRevisions/Pages /proposedrevision.aspx?rid=465.

16. The proposed revision to *DSM-5* further adds to its definition of mental disorder as follows: "A mental disorder is not merely an expectable or culturally sanctioned response to a specific event such as the death of a loved one. Neither culturally deviant behavior (e.g., political, religious, or sexual) nor a conflict that is primarily between the individual and society is a mental disorder unless the deviance or conflict results from a dysfunction in the individual, as described above." This in no way rules out violence

as a disorder, unless one is specifically referring to terrorism, which can be construed as a conflict between the individual and society. Nevertheless, if such terrorists were found to also present with dysfunction that is caused by a biological or psychological condition, then even they would be viewed as clinically disordered. The question that we do not have an answer to at present is whether terrorists do have the type of biological characteristics we see in violent offenders.

17. Manderscheid, R. W., Ryff, C. D., Freeman, E. J., McKnight-Eily, L. R., Dhingra, S., et al. (2010). Evolving definitions of mental illness and wellness. *Preventing Chronic Disease* 7. http://www.cdc.gov/pcd/issues/2010/jan/09_0124.htm.

18. Raine, *The Psychopathology of Crime.*

19. Herbert, E., Kennedy, M., Licht, J. & Mandra, J. (2008). Using genetics to treat leukemia: How Gleevec works. *Science in Society,* Northwestern University. scienceinsociety.northwestern.edu/sites/default/files/chisholmani1.swf.

20. Lichtenstein, P., Holm, N. V., Verkasalo, P. K., et al. (2000). Environmental and heritable factors in the causation of cancer: analyses of cohorts of twins from Sweden, Denmark, and Finland. *New England Journal of Medicine* 343, 78–85.

21. Lodish, H., Berk, A., Matsudaira, P., Kaiser, C. A., Krieger, M., et al. (2004). *Molecular Biology of the Cell,* 5th ed. New York: W. H. Freeman. Technically, the changes taking place consist of damage to the molecular DNA that produces errors in DNA synthesis, and this in turn results in mutations.

22. Landenberger, N. A. & Lipsey, M. W. (2005). The positive effects of cognitive–behavioral programs for offenders: A meta-analysis of factors associated with effective treatment. *Journal of Experimental Criminology* 1, 451–76.

23. Sampson, R. (2012). *Great American Cities: Chicago and the Enduring Neighborhood Effect.* Chicago: University of Chicago Press.

24. Raine, A., Brennan, P. A., Farrington, D. P. & Mednick, S. A., eds. (1997). *Biosocial Bases of Violence.* New York: Plenum.

25. Laub, J. & Sampson, R. J. (2003). *Shared Beginnings, Divergent Lives: Delinquent Boys to Age 70.* Cambridge, Mass.: Harvard University Press.

26. Federal Bureau of Investigation (2010). *Uniform crime reports: Offenses cleared.* http://www.fbi.gov/about-us/cjis/ucr/crime-in-the-u.s/2010/crime-in-the-u.s.-2010/clearances.

27. Blow, C. M. (2010). The high cost of crime. *New York Times.* October 8, editorial, p. A21. http://www.nytimes.com/2010/10/09/opinion/09blow.html?_r=1.

28. Anderson, D. A. (1999). The aggregate burden of crime, *Journal of Law and Economics* 42, 611–42.

29. Malvestuto, R. J. (2007). *Testimony to Committee on Public Safety*. Council of the City of Philadelphia, February 13.

30. Berk, R., Sherman, L., Barnes, G., Kurtz, E. & Ahlman, L. (2009). Forecasting murder within a population of probationers and parolees: A high stakes application of statistical learning. *Journal of the Royal Statistical Society: Series A (Statistics in Society)* 172, 191–211.

31. In the study by Berk, Sherman, and colleagues, the authors recognized their high false-positive rate. For every true-positive case identified, there were twelve false-positives. They point out, however, that there is nevertheless an eightfold increase in prediction accuracy using random-forest statistical learning procedures.

32. The futuristic scenario I have drawn out was inspired by a novel that Marty Seligman sent me after I visited the University of Pennsylvania in October 1994 to give a talk on brain imaging and homicide. The novel, by Philip Kerr, is set in London in 2013 and involves a cat-and-mouse game between a detective and a serial killer typed under the Lombroso program as a potentially dangerous killer. Kerr, P. (1993). *A Philosophical Investigation*. New York: Farrar, Straus & Giroux.

33. The Academy of Experimental Criminology. http://www.crim.upenn.edu/aec/index.html.

34. Kringelbach, M. L., Jenkinson, N., Owen, S.L.F. & Aziz, T. Z. (2007). Translational principles of deep brain stimulation. *Nature Reviews Neuroscience* 8, 623–35.

35. Ridding, M. C. & Rothwell, J. C. (2007). Perspectives: Opinion—Is there a future for therapeutic use of transcranial magnetic stimulation? *Nature Reviews Neuroscience* 8, 559–67.

36. Department of Justice (2012). Assistant Attorney General Laurie Robinson announces departure from office of justice programs. *Office of Public Affairs*. Tuesday, January 3. http://www.justice.gov/opa/pr/2012/January/12-ag-005.html.

37. Mitchell, O. (2005). A meta-analysis of race and sentencing research: Explaining the inconsistencies. *Journal of Quantitative Criminology* 21, 439–66.

38. Office of Justice Programs (2012). Homicide trends in the U.S.: Trends by race. *Bureau of Justice Statistics*. http://bjs.ojp.usdoj.gov/content/homicide/race.cfm.

39. Cohen, A. (2011). Licensing parents. *Bleeding Heart Libertarians*. December 27. http://bleedingheartlibertarians.com/2011/12/licensing-parents-2/.

40. State of California Department of Justice (2009). Megan's Law homepage. http://www.meganslaw.ca.gov/.

41. A conker is the seed from the horse-chestnut tree, about 1.5 inches in diameter. This traditional British game is played by putting a hole through

the conker, threading it onto a piece of string, and swinging it to break the opponent's conker, held vertically and stationary on a string. Health and safety concerns that have led to it being banned at school include the shards of the conker flying off into a child's eye, and nut allergies, although personally I never had any problem myself playing conkers at school.

42. Strickland, P. (2011). Sentences of imprisonment for public protection: Commons Library standard note. October 19. http://www.parliament.uk/briefing-papers/SN06086.

43. Although judges were originally mandated to impose a life sentence if the offender met IPP criteria, the Criminal Justice and Immigration Act of 2008 allowed judges more discretion.

44. Taylor, R., Wasik, M. & Leng, R. (2004). *The Criminal Justice Act 2003: Blackstone's Guide*. Oxford: Oxford University Press.

45. Jacobson, J. & Hough, M. (2010). *Unjust Deserts: Imprisonment for Public Protection*. London: Prison Reform Trust.

46. Ibid., p. 8.

47. Duggan, C. (2011). Dangerous and severe personality disorder. *British Journal of Psychiatry* 198, 431–33.

48. Buchanan, A. & Grounds, A. (2011). Forensic psychiatry and public protection. *British Journal of Psychiatry* 198, 420–23.

49. Ibid.

50. Verkaik, R. (2004). Blair has not been tough on the causes of crime, says Woolf. *The Independent* (London), April 23.

51. Mackintosh, N., Baddeley, A., Brownsworth, R., et al. (2011). *Brain Waves Module 4: Neuroscience and the Law*. London: The Royal Society.

52. Profiling school shooters (2000). *Frontline: The Killer at Thurston High*. web article. http://www.pbs.org/wgbh/pages/frontline/shows/kinkel/profile/.

53. Kellerman, J. (1999). *Savage Spawn: Reflections on Violent Children*. New York: Ballantine.

54. Developmental and social-emotional screening of young children (0-6 years of age) in Minnesota. http://www.health.state.mn.us/divs/fh/mch/devscrn/.

55. Krug, E. G., Dahlberg, L. L., Mercy, J. A., Zwi, A. B. & Lozano, R. (2002). *World Report on Violence and Health*. Geneva: World Health Organization.

56. Centers for Disease Control (2008). *The Public Health Approach to Violence Prevention*. http://www.cdc.gov/ViolencePrevention/overview/public health approach.html.

57. Social Finance (2012). *About Us*. http://www.socialfinanceus.org/about.

58. Social Finance (2012). *History*. http://www.socialfinanceus.org/work/history.

59. Commonwealth of Massachusetts (2012). *Massachusetts First State in the Nation to Pursue "Pay For Success" Social Innovation Contracts.* Press release, January 18. http://www.mass.gov/anf/press-releases/ma-first-to-pursue -pay-for-success-contracts.html.

60. Belkin, L. (2009). Should parenting require a license? *New York Times.* January 8. http://parenting.blogs.nytimes.com/2009/01/08/should-parenting -require-a-license/.

61. Tittle, P. (2004). *Should Parents Be Licensed?* Buffalo, N.Y.: Prometheus Books.

62. Leading articles (2012). Parental guidance suggested. *The Times* (London), p. 2, May 19.

63. Farah, M. J. (2012). Neuroethics: The ethical, legal, and societal impact of neuroscience. *Annual Review of Psychology* 63, 571–91.

64. Sterzer, P. (2010). Born to be criminal?: What to make of early biological risk factors for criminal behavior. *American Journal of Psychiatry* 167, 1, ajp .psychiatryonline.org.

65. Kellerman, *Savage Spawn*, pp. 109-11.

66. Farrington, D. P. & Welsh, B. C. (2007). *Saving Children from a Life of Crime: Early Risk Factors and Effective Interventions.* Oxford: Oxford University Press.

67. U.S. Department of Health and Human Services (2006). *Child Maltreatment.* Washington, D.C.

68. Kahneman, D. (2011). *Thinking, Fast and Slow.* New York: Farrar, Straus & Giroux.

69. LaFollette, H. (2010). Licensing parents revisited. *Journal of Applied Philosophy* 27, 327–43.

70. Ibid.

71. Couple who made boy, 11, live in a coal bunker jailed. (2012). *The Independent* (London), Courts section, May 29.

72. Shami Chakrabarti was previously a barrister working for the Home Office, and later the director of Liberty, a civil-liberties pressure group in England. She is currently the chancellor of Oxford Brookes University in England. She is widely recognized as one of the most influential and effective public-affairs lobbyists in the U.K.

73. *The "If" Debate: A Newsnight Special* (2004). BBC2. December 22. Prog ID 50/and/PS34L/77.

74. In considering whether to kill Adolph Hitler you could also consider Mao, Stalin, Pol Pot, or a host of other leaders who have been responsible for enormous loss of life in many countries throughout the world. Pol Pot is thought to have killed about 20 percent of the Cambodian population in the later 1970s. Stalin executed nearly a million Russians just before World War II.

75. Baynes, N. H. (1942). *The Speeches of Adolf Hitler*. Oxford: Oxford University Press.

76. Almost weekly vitriolic columns were written by Piet Grijs against Buikhuisen's biological perspective in the influential and highly valued *Vrij Nederland*, a socialist critical weekly magazine.

77. Raine, A., Brennan, P. & Mednick, S. A. (1994). Birth complications combined with early maternal rejection at age 1 year predispose to violent crime at age 18 years. *Archives of General Psychiatry* 51, 984–88.

78. Mann, C. (1994). War of words continues in violence research. *Science* 263, 1375.

79. Ibid.

80. Raine, A. & Venables, P. H. (1981). Classical conditioning and socialization—a biosocial interaction. *Personality & Individual Differences* 2, 273–83.

81. The reconciliation of both the public and officials in the Netherlands with Wouter Buikhuisen is striking. In recognition of the earlier unfair ostracization of Buikhuisen for his biological perspective, a symposium was held to "rehabilitate" him on April 16, 2010, organized by criminology students with the support of the dean of the law school of Leiden University. On April 17, 2009, he gave a lecture on the amygdala to a fully packed college hall at Leiden University, and in November 2009 Leiden University formally reconciled with him. Full details of what is known in the Netherlands as the "Buikhuisen Affair" may be found in Keijning, L. (2006). *Buikhuisen had wel wat uit te leggen. De affaire-Buikhuisen en de ontwikkeling van biosociaal onderzoek naar criminaliteit* (Buikhuisen did have something to explain. The Buikhuisen affair and the development of biosocial research of criminality). Master's thesis for Science and Technology Studies, Amsterdam University, 2006.

82. Mann, War of words continues.

83. Quotation from Winston Churchill in Bottomly, P., et al. (2011). Outdated approach to votes for prisoners. *The Guardian*. Letters, January 11.

84. Pinker, S. (2011). *The Better Angels of Our Nature: Why Violence Has Declined*. New York: Viking.

85. Raine, *The Psychopathology of Crime*.

86. Gutmann, A. & Thompson, D. (2012). *The Spirit of Compromise: Why Governing Demands It and Campaigning Undermines It*. Princeton, N.J.: Princeton University Press.

87. Carlyle, T. (1855). *Oliver Cromwell's Letters and Speeches*, p. 448. New York: Harper.

Index

Page numbers in *italics* refer to figures.